Springer Series on
ATOMIC, OPTICAL, AND PLASMA PHYSICS 37

Springer Series on
ATOMIC, OPTICAL, AND PLASMA PHYSICS

The Springer Series on Atomic, Optical, and Plasma Physics covers in a comprehensive manner theory and experiment in the entire field of atoms and molecules and their interaction with electromagnetic radiation. Books in the series provide a rich source of new ideas and techniques with wide applications in fields such as chemistry, materials science, astrophysics, surface science, plasma technology, advanced optics, aeronomy, and engineering. Laser physics is a particular connecting theme that has provided much of the continuing impetus for new developments in the field. The purpose of the series is to cover the gap between standard undergraduate textbooks and the research literature with emphasis on the fundamental ideas, methods, techniques, and results in the field.

27 **Quantum Squeezing**
   By P.D. Drumond and Z. Ficek

28 **Atom, Molecule, and Cluster Beams I**
   Basic Theory, Production and Detection of Thermal Energy Beams
   By H. Pauly

29 **Polarization, Alignment and Orientation in Atomic Collisions**
   By N. Andersen and K. Bartschat

30 **Physics of Solid-State Laser Physics**
   By R.C. Powell
   (Published in the former Series on Atomic, Molecular, and Optical Physics)

31 **Plasma Kinetics in Atmospheric Gases**
   By M. Capitelli, C.M. Ferreira, B.F. Gordiets, A.I. Osipov

32 **Atom, Molecule, and Cluster Beams II**
   Cluster Beams, Fast and Slow Beams, Accessory Equipment and Applications
   By H. Pauly

33 **Atom Optics**
   By P. Meystre

34 **Laser Physics at Relativistic Intensities**
   By A.V. Borovsky, A.L. Galkin, O.B. Shiryaev, T. Auguste

35 **Many-Particle Quantum Dynamics in Atomic and Molecular Fragmentation**
   Editors: J. Ullrich and V.P. Shevelko

36 **Atom Tunneling Phenomena in Physics, Chemistry and Biology**
   Editor: T. Miyazaki

37 **Charged Particle Traps**
   Physics and Techniques of Charged Particle Field Confinement
   By F.G. Major, V.N. Gheorghe, G. Werth

Vols. 1–26 of the former Springer Series on Atoms and Plasmas are listed at the end of the book

F.G. Major   V.N. Gheorghe   G. Werth

Charged Particle Traps

Physics and Techniques
of Charged Particle Field Confinement

With 187 Figures

Dr. Fouad G. Major
284 Michener Court E, Severna Park, MD, USA
E-mail: fmmajor@earthlink.net

Professor Dr. Viorica N. Gheorghe
Professor Dr. Günther Werth
Johannes Gutenberg Universität, Fachbereich Physik (18), Institut für Physik
Staudingerweg 7, 55099 Mainz, Deutschland
E-mail: gheorghe@mail.uni-mainz.de, werth@mail.uni-mainz.de

ISSN 1615-5653

ISBN 3-540-22043-7 Springer Berlin Heidelberg New York

Library of Congress Control Number: 2004107650

This work is subject to copyright. All rights are reserved, whether the whole or part of the material is concerned, specifically the rights of translation, reprinting, reuse of illustrations, recitation, broadcasting, reproduction on microfilm or in any other way, and storage in data banks. Duplication of this publication or parts thereof is permitted only under the provisions of the German Copyright Law of September 9, 1965, in its current version, and permission for use must always be obtained from Springer-Verlag. Violations are liable to prosecution under the German Copyright Law.

Springer is a part of Springer Science+Business Media.
springeronline.com

© Springer-Verlag Berlin Heidelberg 2005
Printed in Germany

The use of general descriptive names, registered names, trademarks, etc. in this publication does not imply, even in the absence of a specific statement, that such names are exempt from the relevant protective laws and regulations and therefore free for general use.

Typesetting: Camera-ready copies by the author
Final Processing: PTP-Berlin Protago-T$_E$X-Production GmbH, Germany
Cover concept by eStudio Calmar Steinen
Cover design: *design & production* GmbH, Heidelberg

Printed on acid-free paper   SPIN: 10876306   57/3141/YU - 5 4 3 2 1 0

Preface

Over the last quarter of this century, revolutionary advances have been made both in kind and in precision in the application of particle traps to the study of the *physics* of charged particles, leading to intensified interest in, and wide proliferation of, this topic. This book is intended as a timely addition to the literature, providing a systematic unified treatment of the subject, from the point of view of the application of these devices to fundamental atomic and particle physics.

The technique of using electromagnetic fields to confine and isolate atomic particles in vacuo, rather than by material walls of a container, was initially conceived by W. Paul in the form of a 3D version of the original rf quadrupole mass filter, for which he shared the 1989 Nobel Prize in physics [1], whereas H.G. Dehmelt who also shared the 1989 Nobel Prize [2] saw these devices (including the Penning trap) as a way of isolating electrons and ions, for the purposes of high resolution spectroscopy. These two broad areas of application have developed more or less independently, each attaining a remarkable degree of sophistication and generating widespread interest and experimental activity.

In the case of mass spectrometry, starting in the 1960s there was initially a rapid proliferation of the use of the 3D rf quadrupole in many fields, such as residual gas analysis, upper atmospheric research, environmental studies, gas chromatography. Since then the field has continued to grow and become refined along differentiated specialized directions, for example sequential ion mass spectrometry. The extant literature on the mass spectrometry uses of ion traps is comprehensive, both in the form of monographs and published proceedings of conferences, such as [3, 4].

On the other hand, it was not until tunable laser radiation sources became available that the application of particle traps to the study of atomic and particle physics saw an explosive expansion in interest and laboratory activity. By combining laser techniques with those of particle trapping it became possible to fully exploit the particle isolation property of the latter. Before that, with the notable exception of the exquisitely precise work on the free electron spectrum by Dehmelt's group, the early difficult experiments to exploit the long perturbation-free spectral observation time in ion traps were severely handicapped by small signal-to-noise ratios. These experiments were carried out by students of Dehmelt and Major on the magnetic resonance spectrum of He^+, and Jefferts on H_2^+. The first successful attempt to detect optical resonance fluorescence from trapped ions using a conventional light source was achieved at NASA Goddard by Major and Werth in measuring the hyperfine interval in Hg^+. It was first shown by Werth at Mainz that the scattering of laser light, even from a diffuse distribution of trapped ions, could be readily detected. The ultimate break-through in the laser detection of trapped ions came in the work of Toschek et al., in which single ions were visually observed. This, combined with the demonstration of laser cooling by Wineland et al. led to the incorporation of laser technology into ion trapping, from which evolved a technique in which not only is the signal-to-noise ratio problem eliminated, but also through laser cooling, the Doppler broadening of the particle spectrum is effectively annulled, ultimately leading to the formation of ion crystals as first observed by Walther and coworkers and transforming the technique into one of great power and elegance.

Unlike the application of ion traps to mass spectrometry, the literature on ion trap physics is diffuse, covering many aspects in the form of extensive review articles, including for example [5–8]. Also an overview on different aspects of ion trap physics can be found in the form of conference proceedings, such as [9–11]. A single monograph *Ion Traps* by P. Ghosh (Oxford) appeared in 1995.

Nevertheless in view of the accelerated advances in the technique in recent years, and the fundamental importance of the many applications, it is evident that a serious gap in the literature exists, which this volume is meant to fill. The treatment of the subject matter is designed, on the one hand, to develop an appreciation of the practical evolution of the technique, its current power and limitations, and, on the other hand, to provide the necessary theoretical underpinning, also through appendices and a comprehensive bibliography. It is left for a future volume to deal with the many important applications, such as ultrahigh resolution spectroscopy, atomic frequency/time standards, particle physics, and quantum computation. Having been associated as experimentalists with the development and application of ion trapping from the time of its inception, F.G. Major and G. Werth have a natural desire to attempt an integrated treatment of the subject, which it is hoped will prove authoritative and useful. With the cooperation of V.N. Gheorghe the treatment of the experimental areas is nicely complemented by supporting theory.

V.N. Gheorghe acknowledges support from the Johannes Gutenberg University, Mainz, Germany and the Alexander von Humboldt Foundation, enabling fruitful international cooperation while on leave from the National Institute for Laser, Plasma, and Radiation and the Physics Department at the University in Bucharest, Romania.

Mainz, *Fouad G. Major*
July 2004 *Viorica N. Gheorghe*
  Günter Werth

Contents

Part I Trap Operation Theory

1 Introduction 3
 1.1 Historical Background 3
 1.2 Principles of Particle Confinement 10

2 The Paul Trap 17
 2.1 Theory of the Ideal Paul Trap 17
 2.2 Motional Spectrum in Paul Trap 23
 2.3 Adiabatic Approximation 24
  2.3.1 Potential Depth 25
  2.3.2 Optimum Trapping Conditions 26
 2.4 Real Paul Traps 27
  2.4.1 Models for Ion Clouds 28
 2.5 Instabilities in an Imperfect Paul Trap 33
 2.6 The Role of Collisions in a Paul Trap 36
 2.7 Quantum Dynamics in Paul Traps 39
  2.7.1 Quantum Parametric Oscillator 39
  2.7.2 Quantum Dynamics in Ideal Paul Trap 43
  2.7.3 Effective Potentials 46

3 The Penning Trap 51
 3.1 Theory of the Ideal Penning Trap 51
 3.2 Motional Spectrum in Penning Trap 56
 3.3 Real Penning Traps 57
 3.4 Shift of the Eigenfrequencies 59
  3.4.1 Electric Field Imperfections 59
  3.4.2 Magnetic Field Inhomogeneities 62
  3.4.3 Distortions and Misalignments 64
  3.4.4 Space Charge Shift 67
  3.4.5 Image Charges 68
 3.5 Instabilities of the Ion Motion 68
 3.6 Tuning the Trap 70
 3.7 Quantum Dynamics in Ideal Penning Trap 72
  3.7.1 Spinless Particle Dynamics 72
  3.7.2 Spin Motion 78

3.8 Quantum Dynamics in Real Penning Traps 81
    3.8.1 Electric Field Perturbations 81
    3.8.2 Magnetic Field Perturbations 83
    3.8.3 The General Hamiltonian 84

4 Other Traps 87
4.1 Combined Traps.. 87
    4.1.1 Equations of Motion 87
    4.1.2 Magnetron-free Operation 90
    4.1.3 Quantum Dynamics in Combined Traps 91
4.2 Cylindrical Traps 95
    4.2.1 Electrostatic Field in a Cylindrical Trap............ 96
    4.2.2 Inherent Anharmonicity of the Field 98
    4.2.3 Control for Anharmonicity 99
    4.2.4 Dipole Field in a Cylindrical Trap 102
    4.2.5 Open-ended Cylindrical Traps 105
4.3 Nested Traps 107
4.4 Multipolar Traps 108
4.5 Linear Traps 109
    4.5.1 The Ideal Linear Trap 109
    4.5.2 Electrostatic Field in a Linear Quadrupole Trap...... 113
    4.5.3 Electric Field in a Linear Multipole Trap 115
4.6 Ring Traps 117
4.7 Planar Paul Traps 118
4.8 Electrostatic Traps 123
4.9 Kingdon Trap 124

Part II Trap Techniques

5 Loading of Traps 131
5.1 Ion Creation Inside Trap 131
5.2 Ion Injection from Outside the Trap 133
5.3 Positron Loading 135

6 Trapped Charged Particle Detection 139
6.1 Destructive Detection 139
    6.1.1 Nonresonant Ejection 139
    6.1.2 Resonant Ejection 140
6.2 Nondestructive Detection 141
    6.2.1 Electronic Detection 141
    6.2.2 Bolometric Detection 142
    6.2.3 Fourier Transform Detection 145
    6.2.4 Optical Detection 146

Part III Nonclassical States of Trapped Ions

7 Quantum States of Motion 153
  7.1 Fock States 153
  7.2 Oscillator Coherent States 154
    7.2.1 The Ideal Penning Trap 155
    7.2.2 The Harmonic Paul Trap 156
  7.3 Squeezed States 159

8 Coherent States for Dynamical Groups 161
  8.1 Trap Symmetries 161
  8.2 Quasienergy States for Combined Traps 162
    8.2.1 A Single Trapped Charged Particle 162
    8.2.2 Quantum Multiparticle States..................... 165

9 State Engineering and Reconstruction 169
  9.1 Trapped Ion-Laser Interaction........................... 169
    9.1.1 Atom-Field Hamiltonians.......................... 169
    9.1.2 Two-Level Approximation 170
  9.2 State Creation 173
    9.2.1 Number States 173
    9.2.2 Coherent States 174
    9.2.3 Squeezed States 175
    9.2.4 Arbitrary States................................... 176
    9.2.5 Thermal States 177
    9.2.6 Schrödinger-Cat States............................ 178
  9.3 State Reconstruction..................................... 183
    9.3.1 Wigner Functions 183
    9.3.2 Experimental State Reconstruction 185

Part IV Cooling of Trapped Charged Particles

10 Trapped Ion Temperature 193
  10.1 Measurement of Ion Temperature........................ 194

11 Radiative Cooling 197

12 Buffer Gas Cooling 203
  12.1 Paul Trap 204
  12.2 Penning Trap 206

13 Resistive Cooling 211
  13.1 Negative Feedback...................................... 215
  13.2 Stochastic Cooling..................................... 216

14 Laser Cooling ... 221
- 14.1 Physical Principles ... 221
- 14.2 Doppler Cooling: Semi-classical Theory ... 223
- 14.3 Resolved Sideband Cooling ... 226
- 14.4 EIT Cooling ... 233
- 14.5 Sisyphus Cooling ... 236
- 14.6 Stimulated Raman Cooling ... 246
- 14.7 Sympathetic Cooling ... 250

15 Adiabatic Cooling ... 257

Part V Trapped Ions as Nonneutral Plasma

---

16 Plasma Properties ... 263
- 16.1 Coulomb Correlation Parameter ... 263
- 16.2 Weakly Coupled Plasmas ... 263
  - 16.2.1 Penning Traps ... 263
  - 16.2.2 Paul Traps ... 266

17 Plasma Oscillations ... 269
- 17.1 Rotating Wall Technique ... 272

18 Plasma Crystallization ... 275
- 18.1 Phase Transitions ... 275
- 18.2 Chaos and Order ... 278
- 18.3 Crystalline Structures ... 280
  - 18.3.1 Crystals in Paul Traps ... 281
  - 18.3.2 Crystals in Penning Traps ... 290
  - 18.3.3 Crystals in Storage Rings ... 291

19 Sympathetic Crystallization ... 295

A Mathieu Equations ... 299
- A.1 Parametric Oscillators ... 299

B Orbits of Trapped Ions ... 303

C Nonlinear Oscillator ... 309
- C.1 Multipole Expansions ... 309
- C.2 Normal Forms ... 310
- C.3 Nonlinear Resonances ... 312

D	Generating Functions for Quantum States 315
	D.1 Uncertainty Relations 315
	D.2 Generating Functions 316
	D.2.1 Hermite functions 316
	D.2.2 Laguerre Polynomials 317
	D.3 Displacement Operators 319
	D.4 Time Dependent Oscillators 320
	D.4.1 Gaussian Packets 320
	D.4.2 Linear Invariants 322
	D.4.3 Quadratic Invariants 322
	D.5 Coherent States for Symplectic Groups 323
	D.5.1 $Sp(2, \mathbf{R})$ Coherent States 323
	D.5.2 Linear Dynamical Systems 324
E	Trap Design and Electronics 327
F	Charged Microparticle Trapping 331

References 335

Index 349

Part I

Trap Operation Theory

1 Introduction

The subject of this book is the study of charged particles suspended in vacuum, their motion constrained by combinations of electric and magnetic fields, rather than collisions with material particles or walls. The ability to confine individual atomic and subatomic particles in vacuum with relatively simple fields achieves three fundamental objectives: thermal isolation for reaching low temperatures, simple motional spectrum, storage of highly charged ions and antiparticles. It makes possible measurements on such uncommon particles at extremely low temperatures with unprecedented observation times and high precision. By avoiding containment by collisions with material walls or particles, it is free from extraneous and often undeterminable perturbations, which would limit the accuracy of measurements. Moreover, prior to the introduction of particle trapping techniques in the 1950s, only statistical average quantities could be derived from the study of enembles of particles, quantities from which the intrinsic properties of individual, isolated particles can be derived only with a degree of uncertainty.

The confinement of charged particles, in a broad sense, would include such endeavors as the stable confinement and hence thermal isolation of high temperature plasmas, using magneto–hydrodynamic forces to achieve a temperature and confinement time sufficient to sustain a thermonuclear reaction. Similarly, particle accelerators such as the synchrotron are based on the confinement of high energy particles in stable orbits, using specially designed magnetic fields. However these examples of particle confinement relate in a sense to the opposite energy extreme to the kind of particle trapping that is the subject of the present work, which deals with the suspension of particles near the absolute zero of temperature.

1.1 Historical Background

The origins of one class of particle trap, what has come to be called the *Penning trap*, are found in the invention of the "cold cathode" high vacuum ionization gauge, called the Penning (or Philips) ionization gauge [12]. In this, an axial magnetic field is used to slow the diffusion of electrons to the walls in an electrical discharge tube, having a high voltage anode ring set between two grounded cathode plates symmetrically mounted normal to the ring axis. Once the discharge is ignited, the magnetic field causes the electrons to describe tight cyclotron orbits around the magnetic field lines, radically slowing their diffusion rate across the magnetic field. This allows the build up of a high electron flux in the discharge, and the sustaining of the discharge at much lower gas pressures than would be possible without the magnetic field. Therefore a significant degree of ionization, and hence measurable ion current is observable well beyond the normal "black out" vacuum. The gauge typically operated with an anode voltage of around 2×10^3 V and a magnetic field of 0.07 T. In the original form they operated in the pressure range $10^{-2} - 10^{-4}$ Pa; at these gas pressures, the additional presence of spurious negative ions could not be ignored.

The explicit description of a harmonic electron trap using a field geometry consisting of a pure quadrupole electric field produced with hyperbolic electrodes and a superposed uniform, axial magnetic field was given by J.R. Pierce [13] in 1949. The design should rightly be called the *Penning–Pierce trap*.

About this time W. Paul et al. [14,15] were investigating the use of multipole electric and magnetic fields to focus beams of neutral particles having an electric or magnetic dipole moment, in the context of molecular beam spectroscopy. One product of that research is the well–known focusing hexapole magnet that was critical to the successful development of the hydrogen maser. Important spectroscopic work on molecular beams was also conducted using electric multipole fields. Out of this work arose the question of the motion of a beam of ions in such a multipole electric field. It was realized that in a *static* electric quadrupole field, for example, ions traveling along the z-axis could be focused in one transverse direction (say) along the x-axis, but not along the y-axis at the same time.

At this point the research took a direction guided by a principle that was being proposed concerning particle optics [16], in the context of high energy particle beams in accelerators: the so-called *strong focusing principle*. The principle concerns the focusing in the transverse plane of charged particles in a beam passing through a regular sequence of alternately converging and diverging electric or magnetic lens. It states that such a sequence will exhibit a net convergence, provided a certain condition on the focal length and spacing is fulfilled. Following Pierce [17], consider for example the focusing properties of the quadrupole electrostatic field given by:

$$V(x,y) = V_0 + V_1(x^2 - y^2) \ . \tag{1.1}$$

A particle carrying a charge Q initially moving parallel to the z-axis will experience a force whose components are given by:

$$F_x = -2QV_1 x \ , \quad F_y = +2QV_1 y \ , \tag{1.2}$$

which, for positive V_1, results in the particle motion converging toward the z-axis in the x-direction, but diverging away from the z-axis in the y-direction.

Assume for simplicity that the quadrupole field exists only over a short interval along the z-axis, so that it may be regarded as a "thin lens", and further assume a particle travels through a large number of such lenses, equally spaced along the z-axis with V_1 alternating in sign. Then if r_n is the radial displacement of a particle at the nth pair of converging and diverging lens, the following recursion formula applies:

$$r_{n+2} - \left[2 - (L/f)^2\right] r_{n+1} + r_n = 0 , \qquad (1.3)$$

where L is the spacing between the lens, and $\pm f$ is the focal length of each lens. The general solution of this ray equation can be written in the following form:

$$r_n = A\cos(n\theta) + B\sin(n\theta) , \qquad (1.4)$$

provided $L/f < \sqrt{2}$, and θ is given by $\cos\theta = 1 - (L/f)^2/2$. In this case the motion in the x–y-plane is bounded. If, on the other hand, $L/f > \sqrt{2}$ then the solution involves the hyperbolic functions and the motion will diverge without limit. The net focusing behavior can be understood as resulting from the particle in a converging lens being deflected by a stronger field toward the axis where the field is weaker as it enters the diverging lens, and the divergence consequently less.

A similar analysis applies to a sequence of magnetic quadrupoles having the field lines along the equipotential surfaces of the electrostatic case, again alternating in polarity, or equivalently, consecutive ones rotated through 90°.

In the frame of reference of a given particle moving with nearly constant longitudinal velocity along the axis, the spatially periodic focusing and defocusing regions produce a periodicity *in time*. Such an explicit oscillation in time is precisely the conclusion reached by Paul et al. and thus the Paul mass filter was born, which does away with the traditional magnetic mass separation of ions. It was realized from the beginning that the two-dimensional dynamic stabilization of the ion motion in a linear mass filter could be generalized to three-dimensions. Thus if the potential is assumed to have the general form:

$$\Phi = \alpha x^2 + \beta y^2 + \gamma z^2 , \qquad (1.5)$$

then to satisfy the Laplace equation we must have $\alpha + \beta + \gamma = 0$. The linear mass filter corresponds to the choice $\alpha = -\beta, \gamma = 0$ [18, 19], whereas the choice $\alpha = \beta, \gamma = -2\beta$ leads to the axisymmetric three-dimensional trap, which was originally called *Ionenkäfig* [20]. The initial context of the development of the Paul trap was in the area of mass spectrometry; its relative simplicity and capability of accumulating ions of a given species, and the extraordinary sensitivity of ion detection quickly ensured its exploitation in residual gas analysis and vacuum leak detection.

The first application of the Paul trap to spectroscopy of free ions was reported in 1962 [21], in which ^{4}He$^+$ ions, confined in what would now be considered a large rf-quadrupole trap, interacted with a spin polarized Cs atomic beam, causing the ions to become spin polarized through spin exchange collisions with the atoms. This enabled magnetic resonance to be observed on free ^{4}He$^+$ ions and their ground state g-factor to be determined. Contemporaneous with this work another experiment using the Paul trap was ongoing to study the rf-spectrum of the simplest molecule, H$_2^+$ [22]. The ^{4}He$^+$ apparatus was subsequently adapted to observe microwave resonances in ^{3}He$^+$ arising from hyperfine transitions in the ground state [23].

The application of the Penning trap to measure the g-factor of free electrons was first reported in 1968 [24], using an apparatus equipped with a Na atomic beam magnetic state selector, whose beam traversed the trap in order to polarize the trapped electrons by spin exchange with the polarized atoms. The detection of magnetic resonance transitions, induced by an applied rf-field, required special measures, since the effect on the polarization of the emerging atomic beam is not detectable in practice. The problem was solved by taking advantage of the spin dependence of *inelastic collisions* between the electrons and the polarized Na atoms, and the consequent dependence of the cooling rate of the electrons on the their spin polarization.

An early attempt at alternative field geometries is exemplified by the work published in 1969 [25] describing a circular "race track" form of trap, which is in effect a Paul linear mass filter curved to form a complete circle. This has an interesting property, which was to become of particular interest when efficient methods became known of cooling particles in the trap. That property is that the limiting space available to the particles as their energy is lowered is a circle, rather a single point at the center of the Paul trap. The trap was used to demonstrate the method of ion cooling by induced currents in an external *LC* circuit.

In 1972 was published the observation of the first high resolution microwave spectrum of ions stored in a Paul trap, using optical fluorescence pumping of the ground state hyperfine structure sublevels [26]. The work was undertaken to establish a frequency standard for a new generation spacecraft atomic clocks. The heavy ion, ^{199}Hg$^+$ was chosen because of the simplicity of its hyperfine structure in the ground state, and the convenience of being able to pump its hyperfine states using the 194 nm UV resonance radiation from a mercury lamp filled with the ^{202}Hg – enriched vapor – it must be remembered that this long preceded the availability of suitable laser sources.

In 1973 an important experimental development in the methods of cooling and detection of electrons in a Penning trap was reported, culminating in the observation of individual electrons [27]. In the same year, Dehmelt et al. proposed a method of measuring the magnetic moment of free electrons based on what they called *the continuous Stern–Gerlach effect*, which was successfully accomplished in 1976 [28].

The first attempt to use a *laser* in a resonance fluorescence experiment on trapped ions was reported in 1977 on Ba$^+$ ions in a Paul trap [29]. By scanning the position of the laser beam across the ion population, the den-

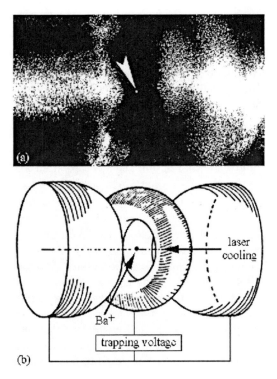

Fig. 1.1. The first observation of an atomic particle, as a single Ba$^+$ ion traped in a miniaturized Paul trap presented in Fig. E.2. (a) Photographic image of the Ba$^+$ ion (indicated by *arrow*) localized at the center of the trap; (b) the drawing of the trap electrodes in the same orientation as in the photograph [32]

sity distribution of the ions was determined. This marked the introduction of laser sources in experiments involving optical interactions with trapped particles, a development which was to transform the whole particle trapping technique, since it made possible an incomparably greater signal-to-noise ratio in the fluorescence than with conventional lamps. It was no longer a great accomplishment just to detect fluorescence after lengthy integrations, as was the case with the Hg$^+$ experiment.

In 1975 the possible application of lasers to the cooling of atomic particles was discussed by Hänsch and Schawlow [30], and specifically proposed for the cooling of ions oscillating in a trap by Wineland and Dehmelt [31]. It was in 1978 that optical sideband cooling and the optical observation of a single ion was demonstrated by Toschek et al. at Heidelberg [32], an advance that literally transformed ion trapping, making the whole field far more interesting and productive in the field of precision measurements on isolated particles.

The technique is based on the resonance fluorescence excitation of the ion by a laser tuned to a lower Doppler sideband $\omega_L = \omega_0 - \omega_V$ in its frequency modulated absorption spectrum, as it oscillates in the trap. It has proven to be an efficient method of cooling ions by laser induced fluorescence.

The theory of sidebands in the frequency modulated spectrum of ions oscillating in a trap, as well as their experimental detection in the microwave spectrum of ^{199}Hg$^+$ had been well established. Nevertheless the achievement of cooling ions in the trap to such a degree and rendering visible a single ion by intense laser resonance scattering was truly remarkable. Henceforth the evolution of the state of a single trapped ion could be followed rather than statistical averages over an ensemble of ions. For example, by appropriate laser excitation and detection of emitted photons from an individual ion, it is possible to repetitively place a particular ion in a given state and measure each time how long it remains unperturbed in that state before decaying to another state, in a scheme called *ion shelving* [33].

In the following year Dehmelt et al. reported on the successful trapping of a detectable number of positrons in a high B-field Penning trap [34]. The main experimental challenge, of course, is to decelerate the relatively high energy positrons emitted from natural radioactive β^+ emitters. The first slowing down was done by a known technique of scattering from a moderating foil, from which the positrons emerge nearly parallel to the plane surface of the foil. By mounting the plane of the foil normal to the intense B-field of the Penning trap, the positron longitudinal velocity component along the field lines is low enough that they can be injected into the interior of the trap and rely on energy loss by synchrotron radiation to permanently capture them in the trap.

The ability to accumulate, suspend and detect small numbers of ions, whose charge-to-mass ratio falls in a chosen range, has been exploited in a number of studies, including spectroscopic studies of highly charged ions, including g-factor determinations [35], and the precise measurement of the masses of unstable radioactive isotopes [36].

When clouds of trapped ions are subjected to the laser cooling process, by which their oscillation is dampened, it was reported in 1988 [37, 38] that a point can be reached when the ion kinetic energy falls below the mean Coulomb energy, and the random oscillatory secular motion of individual ions is replaced with an ordered *crystalline* structure. A similar phenomenon had been observed in 1959 [39] with macroscopic particles under certain operating conditions of the Paul trap, in the presence of a background gas providing sufficient damping of the motion.

A remarkable experimental development occurred in 1988 laying the foundation for the recently achieved production of antihydrogen: it was the first capture and storage of a measurable number of antiprotons. In the later development at CERN reported in 2001 [40] the trap used was a multiring Penning with 32 rings vertically stacked coaxially with a B-field of 6 T. Since these particles of antimatter are formed at high energy, their entrapment requires considerable precooling. The source of antiprotons available in the year 2000 was CERN's "Antiproton Decelerator", which could deliver pulses of 3×10^7 antiprotons each at a repetition rate of one pulse per 110 s. The 5 MeV antiprotons are slowed, accumulated, and sympathetically cooled by interaction first with cold electrons, with which they do not annihilate, then made to interract with positrons from a ^{22}Na source similarly accumulated and cooled by synchrotron radiation to a temperature ultimately of around 4.2 K.

In the intervening years significant progress was made in the field of laser cooling of atoms reaching the limit of photon recoil energy and beyond. For harmonically bound ions vibrating in a trap at frequency ω_V higher than the absorption linewidth, the quantum zero-point energy $\frac{1}{2}\hbar\omega_V$ is the limit to the particle energy. Laser cooling under these conditions, where the Doppler sidebands are resolved, termed *resolved sideband cooling*, has been applied first to a single ^{199}Hg$^+$ ion in a Paul trap, reaching an ultimate energy corresponding to the $n = 0$ zero point oscillation 95% of the time [41]. The technique is based on the resonance fluorescence excitation of the ion by a laser tuned to a lower sideband $\omega_L = \omega_0 - \omega_V$ in its frequency modulated absorption spectrum. In order that the Doppler sideband be resolved without requiring impractically high vibrational ion frequencies, an optical transition must be chosen to have a relatively narrow linewidth. In the case of the mercury ion the electric quadrupole transition $^2S_{1/2} -^2 D_{5/2}$ was chosen at $\lambda = 281$ nm. The cooling is done in two stages: first Doppler cooling is applied using the allowed dipole transition at $\lambda = 194$ nm until the Doppler limit ($T = \hbar\gamma/2k_B$) is reached, followed by the resolved sideband cooling on the quadrupole transition. Subsequent work on the $^{199}Hg^+$ using a cryogenic linear Paul trap was aimed at realizing the original aim of developing an atomic clock to surpass existing standards. In 1995 a fractional instability, in a sampling time of τ s, amounting to $3.3 \times 10^{-13}\tau^{-1/2}$ was achieved in a standard using the 40.5 GHz hyperfine resonance in ^{199}Hg$^+$ as reference [42]. More recent development work on standards in the optical region of the spectrum at NIST has achieved an uncertainty of 10 Hz in the 1.06×10^{15} Hz ($\lambda = 282$ nm) S-D clock transition on a single Hg$^+$ ion [43], and an Allan variance of $7 \times 10^{-15}\tau^{-1/2}$ for $\tau < 10$ s has been reported [44].

The ability to cool and observe individual ions in their lowest quantum vibrational state in a trap, and to manipulate their electronic states through resonance laser excitation, makes it possible to prepare and maintain such ions in what Schrödinger called *entangled states* [45]. These are quantum states in which more than one quantum degree of freedom, such as an ion's electronic state and center of mass vibrational state, cannot be represented as a product of electronic and vibrational wavefunctions, but rather as a general superposition of these states. Such entanglement introduces correlations between the states of different degrees of freedom, coherence which can persist as long as randomizing thermal perturbations do not destroy it. When the coherent motional states correspond to wave packets that are spatially well resolved, the result is a classical observable (position) being entangled with a quantum state called a *Schrödinger cat* state. The reference is of course to Schrödinger's *Gedanken* experiment in which a cat's state of being alive or dead (to make it a little more dramatic), is determined by whether or not a radioactive substance emits a particle that triggers the release of a deadly poison (cyanide). Until the particle has been observed to be emitted, the quantum description would require the radioactive substance be in a superposition of the states before and after emission. This correlates with a superposition of the states of the cat being alive and dead!

To "engineer" an entangled state involving two states of a trapped ion requires that each of the two states be associated with a different *coherent* vibrational state. An example of how this might be achieved [46] would be to laser cool an ion, in its ground electronic state $|g>$, to its lowest vibrational state, then excite it by a $\pi/2$ laser pulse resonant with the transition to an excited state $|e>$, putting it in a coherent superposition state $(|g>+|e>)/\sqrt{2}$. This is followed by applying pulses from two lasers, whose optical dipole interaction with the ion is modulated by the oscillatory motion in the trap. By appropriately choosing the polarizations of the lasers, the phases of the modulations of the two states $|g>$ and $|e>$ can be made to differ, resulting in an entangled state.

The interest in these entangled states arises from their application to quantum computing. A system under conditions where quantum effects are manifest, and having two eigenstates which can represent the Boolean states 0 and 1, are called *qubits*. A string of N such qubits is a quantum register, but unlike a classical register that can store only one N-bit binary number, the quantum register in which the qubits are entangled can in principle be used to simultaneously represent a superposition of *all* N-qubit states at once.

1.2 Principles of Particle Confinement

In considering the application of electric and magnetic fields to the problem of confining charged particles, it is useful to recall some salient properties of the motion of particles in such fields, particularly inhomogeneous high frequency electric fields and crossed electric and magnetic fields.

In the case of the former, some insight can be gained by assuming that the inhomogeneity of the field is so weak that the variation in the field intensity is negligible over the amplitude of the particle oscillation: the so called *adiabatic condition*. Thus consider a particle of mass M and charge Q moving in a weakly inhomogeneous electric field oscillating with an angular frequency Ω. Following Kapitza as quoted by Landau and Lifshitz [47] we consider the motion of the particle in an electric field having a static component $E_0(x)$ and a high frequency component $E_\Omega(x,t)$ such that, although $E_\Omega(x,t)$ is not necessarily small compared with $E_0(x)$, the amplitude of oscillation of the particle under the action of $E_\Omega(x,t)$ is assumed to be small. It is anticipated

1.2 Principles of Particle Confinement

that the motion will be an oscillation of small amplitude at the frequency Ω superimposed on a smooth average motion. Thus we attempt a solution of the form:

$$x(t) = X(t) + \xi(t) , \qquad (1.6)$$

where ξ is oscillatory at frequency Ω, so that the average of $x(t)$ over a period of the field is $X(t)$. Expanding the field in powers of ξ and retaining only first order terms we have for the equation of motion:

$$\frac{d^2 X}{dt^2} + \frac{d^2 \xi}{dt^2} = \frac{Q}{M} \left[E_0 + \xi \frac{dE_0}{dx} + E_\Omega(X) \cos \Omega t + \xi \frac{dE_\Omega(X)}{dX} \cos \Omega t \right] . \qquad (1.7)$$

The crucial step is to require the rapidly oscillating terms and the smoothly varying terms to separately satisfy the equation. Thus for the oscillating terms we have:

$$\frac{d^2 \xi}{dt^2} = \frac{Q}{M} E_\Omega(X) \cos \Omega t , \qquad (1.8)$$

which has the solution

$$\xi = -\frac{Q}{M} \frac{E_\Omega(X)}{\Omega^2} \cos \Omega t . \qquad (1.9)$$

Substituting this result in the equation of motion, and averaging over the oscillation period of the field, we find:

$$\frac{d^2 X}{dt^2} = \frac{Q}{M} E_0 - \frac{Q^2}{M^2 \Omega^2} < E_\Omega \frac{dE_\Omega}{dX} \cos^2 \Omega t > . \qquad (1.10)$$

It follows that the smooth or *secular* motion is determined by an effective potential given by

$$U_{eff} = U_0 + \frac{Q}{4M\Omega^2} E_\Omega^2(X) . \qquad (1.11)$$

Clearly this can be directly generalized to three dimensions, and since the phase of the high frequency field is not involved, it is possible to establish a three dimensional effective potential well in which to trap ions.

The motion of charged particles in crossed static electric and magnetic fields is also of interest in the trapping of particles, since a *static* electric field can be designed to trap particles along one axis, and a static magnetic field to trap particles in a plane perpendicular to that axis. To illustrate this, suppose a particle moves in a uniform magnetic field $\boldsymbol{B} = B\boldsymbol{k}$ and a diverging static electric field $\boldsymbol{E} = E_x \boldsymbol{i} + E_y \boldsymbol{j}$. The motion in the x–y-plane is governed by the equations

$$\frac{dv_x}{dt} = \frac{Q}{M} E_x + \omega_c v_y , \qquad \frac{dv_y}{dt} = \frac{Q}{M} E_y - \omega_c v_x , \qquad (1.12)$$

where $\omega_c = QB/M$ is the cyclotron frequency. If the electric field $\boldsymbol{E}$ is uniform, the solution of these equations can be put in the following form:

$$v_x = v_0 \cos(\omega_c t) + \frac{E_y}{B}, \quad v_y = v_0 \sin(\omega_c t) - \frac{E_x}{B}, \tag{1.13}$$

to show that the cyclotron motion in the magnetic field is superimposed on a constant drift velocity E/B perpendicular to $\boldsymbol{E}$. In the case where the magnetic field is so intense that the drift velocity is small compared with the velocity in the cyclotron orbit, one may picture the motion as consisting of a cyclotron rotation about a *guiding center* that moves with the $x-, y-$ velocity components E_y/B and $-E_x/B$. If the electric field is not uniform but is, for example, radial so that $E_x = kx$ and $E_y = ky$, where k is a constant, then under the assumed conditions the motion of the guiding center is a uniform rotation about the origin at the frequency k/B. This clearly indicates the possibility of constraining the motion of a particle in a divergent electric field by using a strong magnetic field.

For three-dimensional confinement of charged particles, a potential energy minimum must be established at some point in space, in order that the corresponding force be directed toward that point in all three dimensions. In general, the dependence of the magnitude of this force on the coordinates can have an arbitrary form; however, it is convenient to seek a binding force that is harmonic, since this simplifies the analytical description of the particle motion. Thus we assume

$$\boldsymbol{F} \sim -\boldsymbol{r}. \tag{1.14}$$

It follows from

$$\boldsymbol{F} = -\boldsymbol{\nabla} U, \tag{1.15}$$

where U is the potential energy, that in general the required function U is a quadratic form in the Cartesian coordinates x, y, z, as follows:

$$U = \gamma(Ax^2 + By^2 + Cz^2), \tag{1.16}$$

where A, B, C are some constants and γ can be a time-dependent function. If we attempt to achieve such confinement using an electrostatic field acting on an ion of charge Q, and write $U = Q\Phi$ where

$$\Phi = \frac{\Phi_0}{2d^2}(Ax^2 + By^2 + Cz^2), \tag{1.17}$$

we find that in order to satisfy Laplace's equation

$$\Delta\Phi = 0, \tag{1.18}$$

the coefficients must satisfy $A + B + C = 0$. For the interesting case of rotational symmetry around the z-axis, this leads to $A = B = 1$ and $C = -2$, giving us the quadrupolar form

$$\Phi = \frac{\Phi_0}{2d^2}(x^2 + y^2 - 2z^2) = \frac{\Phi_0}{2d^2}(r^2 - 2z^2), \tag{1.19}$$

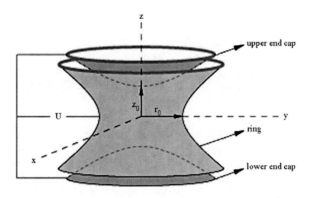

Fig. 1.2. General setup of the electrode configuration to create a quadrupole potential

where we have introduced the radius in cylindrical coordinates, $r^2 = x^2 + y^2$. From the difference in signs between the radial and axial terms, we see that the potential has a saddle point at the origin, having a minimum along one coordinate, but a maximum along the other. It is the basic premise of Earnshaw's theorem [48] that it is not possible to generate a minimum of the electrostatic potential in free space. Nevertheless, as we have sought to make plausible above, it *is* possible to achieve three-dimensional confinement: either by combining that static field with an axial magnetic field [12] forming the *Penning trap*, or by using an electric quadrupole field alternating at high frequency [49, 50] to form the *Paul trap*.

For particle confinement the electrodes must define a region which includes the saddle point at the origin. The electrodes consist, therefore, of three hyperbolic sheets of revolution, an hour glass-shaped cylinder (*"ring"*) and two coaxial bowl-shaped *"end caps"* (Fig. 1.2), which share the same asymptotic cone (with the slope $1/\sqrt{2}$, or an angle of 35.26° for a pure quadrupole).

It was originally common practice to shape the inside surfaces of trap electrodes to approximate, as far as possible, the hyperbolic equipotential surfaces, as used originally by Paul et al. as shown in Fig. 1.3. This was done mainly in order to simplify the theoretical prediction of the stability of entrapment of ions, and their oscillation frequencies, in order to identify them with certainty from the observed resonance spectrum.

However, it was early recognized that any electrode geometry which produces a saddle point in the equipotential surfaces, would have the potential field in the neighborhood of that point in fact precisely of the Paul form. Thus if ξ, η, ς are the Cartesian coordinates of a point near such a point taken as the origin, the Taylor expansion of the field to the second order, taking into account Laplace's equation and the rotational symmetry, is as follows:

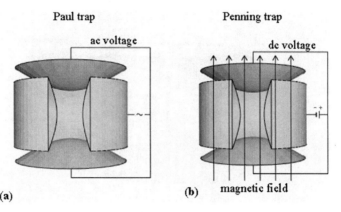

Fig. 1.3. Basic arrangement for Paul and Penning traps. The inner electrode surfaces are hyperboloids. (a) The dynamic stabilization in the Paul trap is given by the ac-voltage $V_0 \cos \Omega t$; (b) The static stabilization in the Penning trap is given by the dc-voltage $U = U_0$ and the axial magnetic field B

$$\Phi(x,y,z) = \Phi_0 + \frac{1}{2}\left(\frac{\partial^2 \Phi}{\partial x^2}\right)_0 (r^2 - 2z^2), \quad r^2 = x^2 + y^2. \tag{1.20}$$

On the basis of this it was realized, for example, that simply using axisymmetric electrodes consisting of a right circular cylinder with two plane end caps, the potential at least near the center of the trap, would have the same saddle point distribution as for hyperbolic electrodes. The relative ease with which cylindrical electrodes could be fabricated, and particularly the availability of analytical expressions for the microwave field modes in a such a cylindrical cavity, were strong inducements to use this electrode geometry. To ensure a reasonable trapping volume near the center, cylindrical traps tended to be of large dimensions, requiring higher operating voltages for a given well depth, and reduced electrical coupling between the trapped ions and the outside circuits. However, the use of additional compensation electrodes have largely overcome these drawbacks and made the cylindrical geometry adaptable even for precise frequency measurements.

If the radial distance from the center ($r = z = 0$) of a hyperbolic trap to ring electrode is called r_0, and the axial distance to an end cap is z_0, the equations for the hyperbolic electrode surfaces are:

$$r^2 - 2z^2 = r_0^2,$$
$$r^2 - 2z^2 = -2z_0^2. \tag{1.21}$$

If the potential difference between the ring and end caps is taken to be Φ_0, then $2d^2 = r_0^2 + 2z_0^2$. By choosing the ratio $r_0/z_0 = \sqrt{2}$, originally assumed in the Paul trap, we achieve electrical symmetry, in the sense that the potential on the ring and end caps will be equal and opposite with respect to the

Fig. 1.4. Electrodes of a charged particle trap

center of the trap. In general, any other ratio between r_0 and z_0 can be used to create the quadrupole potential [51], however, the choice $r_0/z_0 = \sqrt{2}$ allows the largest effective confinement space within the field defined by electrode surfaces. The size of the device ranges, in different applications, from several centimeters for the characteristic dimension d, to fractions of a millimeter. The trapped charged particles are constrained to a very small region of the trap, whose position can be centered by using a small additional dc-field.

Figure 1.4 shows a photograph of a trap as used in spectroscopic experiments. The trap design incorporates one end cap electrode in the form of a grid, to make the inside accessible for the optical detection of ions.

2 The Paul Trap

2.1 Theory of the Ideal Paul Trap

In an ideal Paul trap, an oscillating electric potential, usually in combination with a static component, $U_0 + V_0 \cos \Omega t$ is applied between the ring and the pair of end cap electrodes. This creates a potential of the form

$$\Phi = \frac{U_0 + V_0 \cos \Omega t}{2d^2}(2z^2 - x^2 - y^2) , \quad d = \sqrt{\frac{1}{2}r_0^2 + z_0^2} . \quad (2.1)$$

The shape of the potential of the ideal Paul trap, for $\Phi_0 = V_0 \cos \Omega t$ and $y = 0$ at two instants differing by half of the oscillation period, is illustrated in Fig. 2.1. The corresponding contour plot of equipotentials is shown in Fig. 2.2, where the arrows indicate the direction of the force at these instants.

Since the trapping field is inhomogeneous, the average force acting on the particle, taken over many oscillations of the field, is not necessarily zero. Depending on the amplitude and frequency of the field, the net force may be convergent toward the center of the trap leading to confinement, or divergent leading to the loss of the particle. Thus although the electric force alternately causes convergent and divergent motion of the particle in any given direction, it is possible by appropriate choice of field amplitude and frequency, to have a time averaged restoring force in all three dimensions towards regions of weak field, that is, towards the center of the trap [52], as required for confinement.

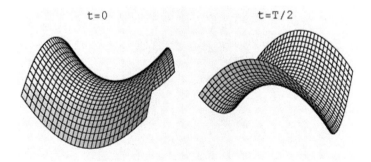

Fig. 2.1. Oscillation with period T of Paul's equipotential saddle-shaped surface

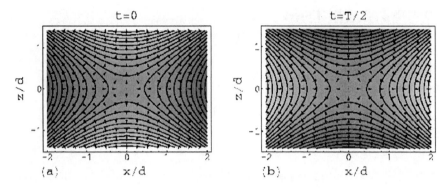

Fig. 2.2. Contour plots of the reduced Paul potential Φ/V_0 for $U_0 = 0$, $|x| \leq 2d$, and $|z| \leq d/\sqrt{2}$. (a) $t = 0$. (b) $t = T/2 = \pi/\Omega$

The dynamic stabilization can be demonstrated by Paul's analogue mechanical device, which consists of a small sphere put near the saddle point of a surface [53] shaped as in Fig. 2.1. If the device is kept stationary, the ball simply will roll off the surface. However, if the surface is rotated with an appropriate frequency around the normal axis, then the ball will remain on the surface for a long time making small oscillations around the rotation axis.

The conditions for stable confinement of an ion with mass M and charge Q in the Paul field may be derived by solving the equation of motion

$$\frac{d^2x}{dt^2} - \frac{Q}{Md^2}(U_0 + V_0 \cos \Omega t)x = 0 ,$$

$$\frac{d^2y}{dt^2} - \frac{Q}{Md^2}(U_0 + V_0 \cos \Omega t)y = 0 ,$$

$$\frac{d^2z}{dt^2} + \frac{2Q}{Md^2}(U_0 + V_0 \cos \Omega t)z = 0 . \quad (2.2)$$

Using the notation $u_1 = x$, $u_2 = y$, $u_3 = z$, and the dimensionless parameters

$$a_x = a_y = -\frac{4QU_0}{Md^2\Omega^2} , \quad q_x = q_y = \frac{2QV_0}{Md^2\Omega^2} , \quad (2.3)$$

$$a_z = \frac{8QU_0}{Md^2\Omega^2} , \quad q_z = -\frac{4QV_0}{Md^2\Omega^2} , \quad \tau = \frac{1}{2}\Omega t ,$$

we obtain a system of three differential equations of the homogeneous Mathieu type [54, 55]:

$$\frac{d^2u_j}{dt^2} + (a_j - 2q_j \cos 2\tau)u_j = 0 , \quad j = 1, 2, 3 , \quad (2.4)$$

where a_j and q_j are real parameters and τ is a real variable. In the general discussion of the solutions of these equations we shall drop the subscript j.

According to Floquet's theorem, solutions of the Mathieu equation have the form

$$w_1(\tau) = e^{\mu\tau}\Phi(\tau)\,, \quad w_2(\tau) = e^{-\mu\tau}\Phi(-\tau)\,, \tag{2.5}$$

where Φ is a π-periodic function and the Lyapunov characteristic exponent is $\mu = \alpha + i\beta$, where α and β are real functions of the parameters a, q. If $\alpha \neq 0$, or if $\alpha = 0$ and $\beta \neq n$, where n is an integer, the solutions w_1 and w_2 are linearly independent, and the general solution of the Mathieu equation can be written as a linear combination of w_1 and w_2.

If $\alpha \neq 0$, then the general solution contains a positive exponential factor and is unbounded as $\tau \to \infty$, and is classed as unstable. If $\alpha = 0$ and β is not an integer, then the general solution is bounded as $\tau \to \infty$; such a solution is called stable. A stable solution is periodic if and only if β is rational.

If however $\mu = in$, where n is an integer, then the general solution comprises one periodic solution w_1 (period π or 2π) and a second linearly-independent solution which grows linearly in τ for $\tau \to \infty$, provided $q \neq 0$ and is therefore classed as unstable. The value of μ and hence the stability or instability of the solutions depends only on the parameters a and q. This suggests a division of the a–q-plane into stable and unstable regions. The condition $\mu = in$ is met by values of (a, q) which lie on characteristic

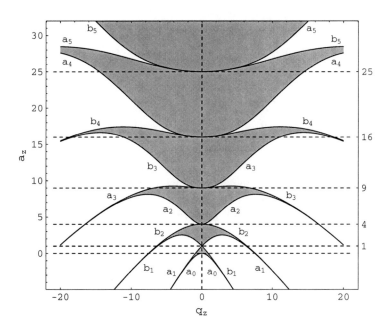

Fig. 2.3. Stability diagram for Mathieu's equation for axial direction (z) (the stable regions are *shaded*)

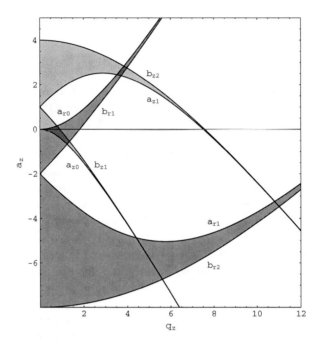

Fig. 2.4. The stability domains for the ideal Paul trap. *Light grey*: z-direction; *dark grey*: r-direction

curves a_n, b_n in the a–q-plane, forming the boundaries separating stable from unstable regions (Fig. 2.3).

On the a-axis, the Mathieu equation reduces to that of an harmonic oscillator with simple periodic solutions for which

$$a_n(0) = n^2, \quad b_{n+1}(0) = (n+1)^2, \tag{2.6}$$

for $n \geq 0$.

The values of a and q for which the solutions are stable simultaneously for the axial and radial directions, an obvious requirement for three-dimensional confinement, are found by using the relationship $a_z = -2a_r$, $q_z = 2q_r$ to make a composite plot of the boundaries of stability for both directions on the same set of axes (Fig. 2.4); the overlap regions lead to three dimensional confinement.

Most important for practical purposes is the stable region near the origin (Fig. 2.5) which has been exclusively used for ion confinement.

The stable solutions of the Mathieu equation can be expressed in the form of a Fourier series, thus

$$u_j(\tau) = A_j \sum_{n=-\infty}^{+\infty} c_{2n} \cos\left[(\beta_j + 2n)\tau\right] + B_j \sum_{n=-\infty}^{+\infty} c_{2n} \sin\left[(\beta_j + 2n)\tau\right], \tag{2.7}$$

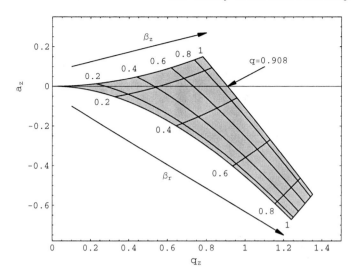

Fig. 2.5. The lowest stability domain of the Paul trap including lines of constant values for β_r and β_z

where A_j and B_j are constants depending on the initial conditions. For the stability parameter β one obtains a continued fraction expression:

$$\beta_j^2 = a_j + f_j(\beta_j) + f_j(-\beta_j) ,$$

$$f_j(\beta_j) = \cfrac{q_j^2}{(2+\beta_j)^2 - a_j - \cfrac{q_j^2}{(4+\beta_j)^2 - a_j - \cdots}} . \qquad (2.8)$$

For the coefficients c_{2n} which are the amplitudes of the Fourier components of the particle motion, we have the following recursion formula:

$$\frac{c_{2n,\,j}}{c_{2n\mp 2,\,j}} = -\cfrac{q_j}{(2n+\beta_j)^2 - a_j - \cfrac{q_j^2}{(2n\pm 2+\beta_j)^2 - a_j - \cdots}} . \qquad (2.9)$$

Examples of ion trajectories for different values of the stability parameters are shown in Fig. 2.6. Further examples and trajectories in the x–z-plane as well as phase space trajectories are given in Appendix B.

Under conditions similar to those in Fig. 2.6, Wuerker et al. [39] have taken photographs of single particle trajectories of microparticles in a Paul-type trap at low frequencies of the trapping field, showing the Lissajous-like shape of the trajectory of a stored particle (Fig. 2.7).

We note that because of the dependence of the stability parameter β on the square of the charge, positive and negative particles can be confined simultaneous. Figure 2.8 shows the first stable region for positive and negative ions of the same mass; in the overlapping region simultaneously storage can be achieved [56].

22   2 The Paul Trap

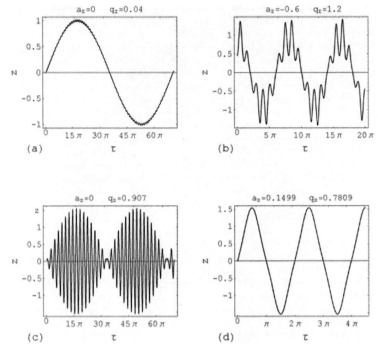

Fig. 2.6. The secular motion in an ideal Paul trap, for different values of a_z and q_z

Fig. 2.7. Observed Lissajous-like trajectory of a charged microparticle in a Paul trap [39]

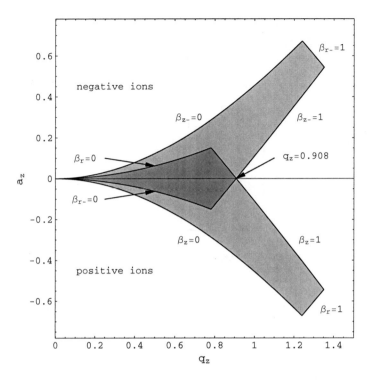

Fig. 2.8. Simultaneously stability (overlapping common area) for positive and negative ions in the quadrupole ion trap

2.2 Motional Spectrum in Paul Trap

From (2.7) we see that the motional spectrum of a stable confined particle contains the frequencies

$$\omega_{j,n} = (\beta_j + 2n)\Omega t/2 , \quad -\infty < n < \infty , \qquad (2.10)$$

with the fundamental frequencies given by $n = 0$, thus

$$\omega_{j,0} = \frac{1}{2}\beta_j \Omega . \qquad (2.11)$$

To experimentally demonstrate this spectrum, the motion is excited by an additional (weak) rf-detection field applied to the electrodes. When resonance occurs between the detection field frequency and one of the frequencies in the ion spectrum, the motion becomes excited and some ions may leave the trap, providing a signal commensurate with the number of trapped ions. Figure 2.9 shows an example where a cloud of stored N_2^+ ions is excited at resonances occurring at the frequencies predicted by theory.

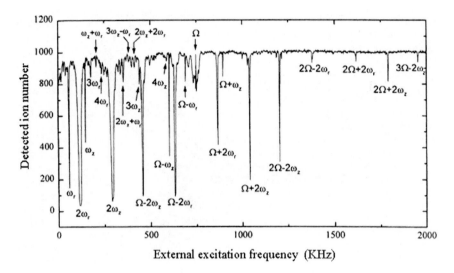

Fig. 2.9. The spectrum of motional resonances of an N_2^+ ion cloud in a quadrupole Paul trap [57]

2.3 Adiabatic Approximation

A useful approximation in the analysis of the motion of a charged particle in any inhomogeneous high frequency electric field, is obtained when the inhomogeneity is such that the amplitude of the field is nearly constant over the oscillation of the particle. Under this condition it has been shown in Chap. 1 that the particle motion averaged over the high frequency oscillation is governed by an effective potential energy U_eff

$$U_\text{eff} = \frac{Q^2 E_\Omega^2(\mathbf{r})}{4M\Omega^2} \ . \tag{2.12}$$

Moreover, this potential energy is just the mean kinetic energy of oscillation in the high frequency field. This approximation is called the *adiabatic approximation*, since the total kinetic energy remains constant as the particle moves through the high frequency field, simply exchanging between the high frequency oscillation and slower average motion.

In the case of the Paul trap, it is readily verified that for the adiabatic approximation to be valid, requires $a, q \ll 1$, and that the stability parameter β can be approximated by

$$\beta^2 \approx a + \frac{q^2}{2} \ . \tag{2.13}$$

In this approximation the coefficients c_{2n} becomes rapidly smaller with increasing n. For $n = 1$ we have $c_{-2} = c_{+2} = -(q_i/4)c_0$. The ion motion simplifies to

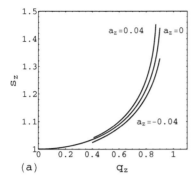
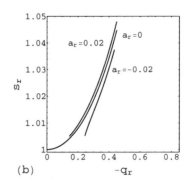

Fig. 2.10. The ratio $s = \beta/\beta_{ad}$ between exact value of β given by the Mathieu equation and its value resulting from the adiabatic approximation in an ideal Paul trap for some typical working values of a, and $0 \leq q \leq 0.9$

$$u_i(t) = G\left(1 - \frac{q_i}{2}\cos\Omega t\right)\cos\omega_i t, \qquad (2.14)$$

with $G = c_0\sqrt{A^2 + B^2}$, $\omega_i = \beta_i\Omega/2$. This can be considered as the motion of an oscillator of frequency ω whose amplitude is modulated with the trap's driving frequency Ω. Since it is assumed that $\beta \ll 1$ the oscillation at ω, usually called the *secular* or *macromotion*, is slow compared with the superimposed fast *micromotion* at Ω. Because of the large difference in the frequencies ω and Ω, the ion motion can be well separated in two components and the behavior of the slow motion at frequency ω can be considered as separate, while time averaging over the fast oscillation at Ω.

The validity of this approximation is illustrated in Fig. 2.10, where the ratio between the exact values for β and those obtained by the adiabatic approximation is plotted vs. the value of q.

2.3.1 Potential Depth

In the adiabatic approximation an expression for the depth of the confining potential can be easily derived. If we consider the secular motion only, the ion behaves in the axial direction as an harmonic oscillator of frequency ω_z. The equation of motion reads

$$\frac{d^2 z}{dt^2} = -\omega_z^2 z = -\frac{\beta_z^2 \Omega^2 z}{4}. \qquad (2.15)$$

For no dc voltage ($a = 0$) we substitute $q_z = 2QV_0/Mz_0^2\Omega^2$ and obtain

$$\frac{d^2 z}{dt^2} = -\frac{Q^2 V_0^2}{2M^2 z_0^4 \Omega^2} z, \qquad (2.16)$$

which can be written as

$$M\frac{d^2z}{dt^2} = -Q\frac{dD_z}{dz}.\qquad(2.17)$$

Here $dD_z/dz = QV^2z/(2Mz_0^4\Omega^2)$ can be interpreted as a field in the axial direction generated by a parabolic *pseudopotential* $\bar{D}_z$. The depth of this potential can be expressed by the stability parameter q as

$$\bar{D}_z = \int_0^{z_0} \frac{dD_z}{dz}dz = \frac{QV_0^2}{4Mz_0^2\Omega^2} = \frac{Mz_0^2\Omega^2 q_z^2}{16Q}.\qquad(2.18)$$

In a similar way one derives the potential depth in the radial direction

$$\bar{D}_r = \frac{QV_0^2}{4Mr_0^2\Omega^2},\qquad(2.19)$$

and for $r_0^2 = 2z_0^2$ we have

$$\bar{D}_r = \frac{\bar{D}_z}{2}.\qquad(2.20)$$

The effect of an additional dc voltage on the trap electrodes is to alter the depth of the radial and axial potentials in opposite directions. If the voltage U_0 is applied symmetrically to the trap electrodes (that is, the trap center is at zero potential) we have

$$\bar{D}'_z = \bar{D}_z + \frac{U_0}{2},\quad \bar{D}'_r = \bar{D}_r - \frac{U_0}{2}.\qquad(2.21)$$

If we use the definitions of a and q we find that for $a_z = -q_z^2/2$, which corresponds to the $\beta_z = 0$ line at the limit of the stability diagram, the depth of the potential becomes zero, as expected. Similarly we find zero radial potential depth at the $\beta_r = 0$ line.

2.3.2 Optimum Trapping Conditions

To achieve the optimum trapping conditions for a Paul trap, defined as those which yield the highest density of trapped particles, we are faced with conflicting requirements: On the one hand, the maximum trapped ion number increases with the potential depth D, since the number is limited by the condition that the space charge potential of the ions not exceed D, which in the adiabatic approximation increases as q^2 (see (2.18)). On the other hand the oscillation amplitudes $r_{\max}$ and $z_{\max}$ defined by

$$r_{\max}, z_{\max} = \sqrt{A^2 + B^2}\sum_n |c_{2n}|,\qquad(2.22)$$

also increase with q (see (2.9)), resulting in increasing ion loss for higher q, in a trap of a given size. Calculations on the maximum ion density [29, 58]

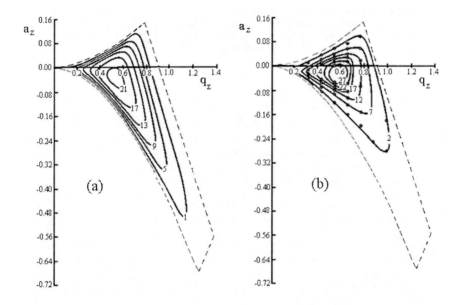

Fig. 2.11. Optimum trapping conditions. (**a**) Computed lines of equal ion density within the stability diagram. The numbers give relative densities; (**b**) experimental lines of equal ion density from laser-induced fluorescence [29]

show that the maximum ion number is obtained in a region around $q_z = 0.5$, $a_z = -0.02$ (Fig. 2.11a). This is confirmed experimentally by systematic variation of the trapping parameters and measurement of the relative trapped ion number by laser induced fluorescence (Fig. 2.11b).

2.4 Real Paul Traps

A single ion in a perfect quadrupole potential does not describe a real experimental situation. Truncation of the electrodes, misalignments or machining errors change the shape of the potential field. Some typical dimensions and operating conditions for a Paul trap can be seen in Table 2.1.

The traps designed to study ions or elementary particles clearly must operate under ultra high vacuum conditions; however, areas of research involving microparticles have been carried out even under standard atmospheric conditions of pressure and temperature [59–61].

The equations of motion as discussed above are valid only for a single confined particle. Simultaneous confinement of several particles requires the consideration of space charge effects.

Table 2.1. Typical dimensions and operating conditions for a Paul trap

parameter	value
size	1–5 cm
rf-amplitude	100–1000 V
rf-frequency	300–3000 kHz
dc-amplitude	± 20 V
maximum ion density	10^6 cm^{-3}
storage time	several hours
(uncooled) ion temperature	10 000 K
trap depth well	several tens of eV

2.4.1 Models for Ion Clouds

If more than one ion is confined, the Coulomb interaction between ions gives rise to long range effects in addition to strong scattering events; the former is accounted for by an average space charge potential, which adds to the applied potentials, while the latter is expected to statistically affect the ion energy and the approach to thermal equilibrium.

At first we will assume that the trapping potential is ideal, and consider only the ensemble behaviour of a cloud of ions trapped in such a field; the effects of trap imperfections will be discussed later.

For a low density cloud it seems reasonable to assume that the particles move in somewhat modified orbits, mainly independent of each other except for rare Coulomb scattering events. These collisions ultimately serve to establish a thermal equilibrium between the particles. They may be considered as small perturbations to the *average* particle motion, provided the time av-

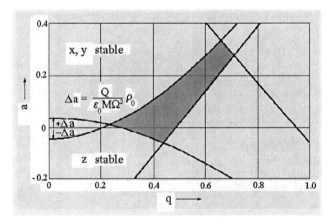

Fig. 2.12. The shift of the first stability domain due to an homogeneous space charge ρ_0 [56]

erage of the Coulomb interaction potential is small compared to the average energy of the individual particle. Then the particle cloud may be described as an ideal gas of noninteracting particles in thermal equilibrium [62]. With this assumption the probability of finding an ion with velocity **v** and position ρ in the range $d^3\mathbf{v}$ and $d^3\rho$, respectively is

$$f(\rho, \mathbf{v}) d^3\rho\, d^3\mathbf{v} \sim \exp[-E(\rho, \mathbf{v})/k_B T] d^3\rho\, d^3\mathbf{v} \ . \tag{2.23}$$

For ions in the trap potential well, the energy is given by

$$E(r, z, \mathbf{v}) = \frac{1}{2} M v^2 + \frac{r^2}{r_0^2} Q D_r + \frac{z^2}{z_0^2} Q D_z \ , \tag{2.24}$$

where $r^2 = \rho^2 - z^2$. The distribution function f then becomes

$$f(r, z, \mathbf{v}) = \frac{N}{r_0^2 z_0} \frac{(8Q^3 M^3 D_r^2 D_z)^{1/2}}{(2\pi k_B T)^3}$$

$$\times \exp\left(-\frac{Mv^2}{2k_B T} - \frac{r^2 Q D_r}{r_0^2 k_B T} - \frac{z^2 Q D_z}{z_0^2 k_B T}\right) , \tag{2.25}$$

where N is the total number of ions. Integrating over ρ gives the usual Maxwellian distribution of velocities. To obtain the density distribution $n(\rho)$ we integrate over **v** to obtain

$$n(r, z) = \frac{1}{r_0^2 z_0} \frac{1}{(\pi k_B T)^{3/2}} (Q^3 D_r^2 D_z)^{1/2} \exp\left[-\left(\frac{r}{\Delta r}\right)^2 - \left(\frac{z}{\Delta z}\right)^2\right] , \tag{2.26}$$

with $\Delta r = r_0 (k_B T / Q D_r)^{1/2}$ and $\Delta z = z_0 (k_B T / Q D_z)^{1/2}$.

The Gaussian density distribution can be experimentally verified by observing the scattering of a laser beam scanned across the stored ion cloud (Fig. 2.13) [63, 64]. The intensity of fluorescence light emitted from the ions

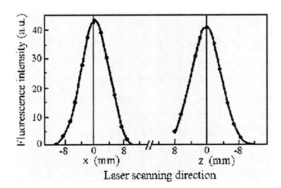

Fig. 2.13. Vertical and horizontal laser scan through the ion trap. Experimental points fitted Gaussian [64]

after laser excitation is proportional to the ion number inside the laser beam profile. From the width of the distribution a value for the mean kinetic energy of the cloud can be derived. When no particular damping mechanism is applied one finds experimentally to a good approximation $\bar{E} = (1/10)\,QD$. The fact that the density distribution can be well described by a Gaussian, which implies a Maxwellian velocity distribution, justifies the idea that a temperature can be ascribed to the ion cloud. The equilibrium temperature for a given well depth is determined by a balance between the heating effects produced by the time-dependent trapping field, and energy loss mechanisms such as collisions with a lighter background gas, or escape of fast ions from the trap.

Similar results can be obtained in a model of the ion cloud where the particle motion is influenced by randomly fluctuating forces, such as those arising from the fluctuating electric fields of neighbouring ions, or from collisions with background molecules. This leads to a description of the ion motion as Brownian motion of a parametric oscillator [65]. The model includes the contribution of both the secular and micromotion to the velocity distribution within the adiabatic approximation, leading to the expression for the average kinetic energy using (2.21):

$$\frac{1}{2}M\bar{v}^2 = \frac{1}{2}\left(\omega^2 + \frac{\Omega^2 q^2}{8}\right)\bar{x}^2 \ . \tag{2.27}$$

The first term on the right hand side is the contribution of an harmonic oscillator with frequency ω corresponding to the macromotion, while the second term accounts for the micromotion. The model thus provides the explicit dependence of the mean kinetic energy of the trapped particle cloud on the operating parameters of the trap.

A calculation of the average energy for different parameters and the width of the spatial distribution (Fig. 2.14) has been partially verified by experiments [66, 67].

An important result in ion cloud models is the prediction of shifts in the oscillation frequencies of any given ion due to the average space charge field acting on it from all the other ions. Note however that the motion of the center-of-charge of the ion cloud cannot be affected by the internal Coulomb interactions. In a simplified model, the stored ion cloud can be considered as charged plasma of constant density and spherical shape [68]. This would be true if the cloud temperature approaches zero; however, above zero it is still a good approximation near the cloud center, because there the density does not vary to first order in the radial distance. In this case the electric potential created by the space charge is quadratic like the confining potential. The ion motion remains harmonic, however at a shifted frequency. The shift can be expressed in units of the plasma frequency

$$\omega_p = (Q^2 n/\varepsilon_0 M)^{1/2} \ , \tag{2.28}$$

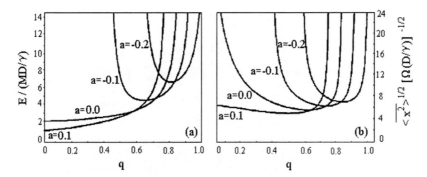

Fig. 2.14. The time averaged $\bar{E}$ in units of MD/γ (a) and spatial width $\overline{\langle x^2 \rangle}^{1/2}$ (b) as a function of the trap potential parameter q, for a damping coefficient $\gamma/\Omega = 10^{-7}$ corresponding to usual experimental conditions [65]

where n is the ion density. For the axial motion we can write the shifted frequency as

$$\omega'_z = \omega_z(1 - \omega_p^2/3\omega_z^2)^{1/2} , \qquad (2.29)$$

and similarly for the radial frequency. In this approximation the shift is linear with the ion density. The maximum trapped ion density is given by

$$\omega_p^2/\omega_z^2 = 3 . \qquad (2.30)$$

If we assume for example a potential depth of 10 eV, we obtain as limiting density $n_{\max} = 5 \cdot 10^7 \, \text{cm}^{-3}$.

A more refined model assumes a Maxwellian distribution of charge density at a finite temperature T, thus:

$$n(r) = \frac{C}{k_B T} \exp\left[-\frac{1}{2} M \omega_s^2 r^2 + Q\Phi_{sc}(r) \right] , \qquad (2.31)$$

where C is a normalization constant and k_B the Boltzmann constant. Here ω_s is defined by

$$\omega_s = 2QV_0/M\Omega(r_0^2 + 2z_0^2) , \qquad (2.32)$$

and $\Phi_{sc}(r)$ is the space charge potential that is itself determined by the density through the Poisson equation. Assuming spherical symmetry, it can be shown [70] that to a first approximation the potential is given by:

$$\Phi_s(r) = (QN/4\pi\varepsilon_0 r) erf(r/R) , \qquad (2.33)$$

where R is the value of r at which the cloud density is reduced to $1/e$ of its center value, and can be roughly considered as the cloud radius. The ion oscillation frequency, shifted by space charge, now depends on the distance from the trap center. The numerical results agree with those previously

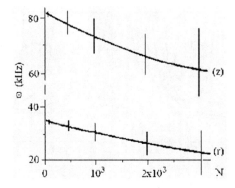

Fig. 2.15. Calculated variation of fundamental frequencies in axial and radial directions versus ion number N; $a = 0$, $q = 0.2$ [69]. Copyright (2003) by the American Physical Society

obtained by a statistical model using similar density distributions [69] and graphically represented in Fig. 2.15. The shift averaged over the total cloud would be experimentally observed. The axial resonance, taken with high resolution, exhibits two components [71]: the center-of-mass frequency *(collective*

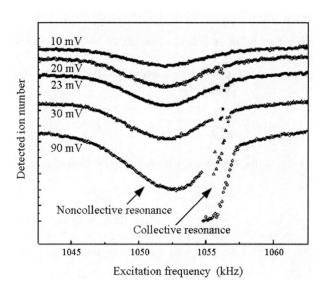

Fig. 2.16. Axial resonance of a stored ion cloud showing excitation of the center-of-mass (collective) and individual (noncollective) ion oscillations for five different excitation voltages. The initial number of ions for the five different curves is the same. The *curves* are vertically shifted for clarity of presentation [71]. Copyright (2003) by the American Physical Society

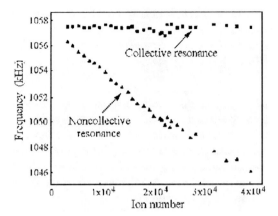

Fig. 2.17. Observed variation of fundamental frequencies versus ion number [71]. Copyright (2003) by the American Physical Society

resonance), unaffected by space charge, and the individual ion oscillation frequency *(noncollective resonance)* (Fig. 2.16). The latter is shifted linearly downwards as the ion number increases (Fig. 2.17). From a comparison of the calculated values with the observed frequency shifts, the total number of stored ions can be derived.

2.5 Instabilities in an Imperfect Paul Trap

Deviations of the trap potential from the ideal quadrupolar form can be treated by a series expansion in spherical harmonics, thus

$$\Phi(\rho,\theta) = (U_0 + V_0 \cos \Omega t) \sum_{n=2}^{\infty} c_n \left(\frac{\rho}{d}\right)^n P_n(\cos\theta) , \qquad (2.34)$$

where $P_n(\cos\theta)$ are the Legendre polynomials of order n. For rotational symmetry the odd coefficients c_n vanish. The terms beyond the quadrupole (c_2) may be looked on as perturbing potentials, the lowest of which is the octupole (c_4), followed by the dodecapole (c_6); their dependence on (r, z) coordinates is given in App. C.1. As a consequence of the presence of the higher order terms in the trapping potential, the motional eigenfrequencies are shifted with respect to the pure quadrupole field, and moreover in an amplitude dependent way. These shifts are of particular importance in the case of Penning traps, which are used for very high resolution mass spectrometry. Therefore we will discuss these shifts in detail in the Sec. 3.4.

Regarding the potential Φ as the sum of the pure quadrupole part Φ_2 and the higher order terms as a perturbation $\tilde{\Phi}$, the equations of motion for a single particle in an imperfect Paul trap now become coupled inhomogeneous differential equations:

$$\frac{d^2 u_i}{d\tau^2} + (a_i - 2q_i \cos 2\tau) u_i = -(a_i - 2q_i \cos 2\tau) \frac{\partial \tilde{\Phi}}{\partial u_i} \,, \qquad (2.35)$$

where $u_i = x, y, z$, $a_i = a_x, a_y, a_z$ and $q_i = q_x, q_y, q_z$, respectively.

This set of equations cannot be solved analytically. Kotowski [72] and later Wang, Wanczek and Franzen [73] have shown that, under certain conditions that would otherwise give stability in a perfect quadrupole field, the amplitude of the ion motion can increase to infinity with time, that is, the motion becomes unstable. These conditions can be expressed in terms of the stability parameters β_r and β_z thus:

$$n_r \beta_r + n_z \beta_z = 2k \,, \qquad (2.36)$$

where n_r, n_z, k are integers. Multiplying this expression by the frequency Ω of the trapping field we obtain

$$n_r \omega_r + n_z \omega_z = k\Omega \,, \qquad (2.37)$$

where $\omega_r = \beta_r \Omega / 2$ and $\omega_z = \beta_z \Omega / 2$ are the radial and axial frequencies, respectively. This relationship states that if a linear combination of harmonics of the ion macrofrequencies coincides with a harmonic of the high frequency trapping field, an ion will gain energy from that field until it gets lost from the trap. The amount of energy gain per unit time is different for different combinations of the numbers n_z and n_r. The sum $|n_z| + |n_r| = 2n$ defines n, the resonance order. The maximum resonance order corresponds to the order n of the perturbing potential in the series expansion, whose magnitude is determined by the shape of the imperfections actually present in the trap. The instabilities should occur along certain lines within the stability diagram, as graphically represented in Fig. 2.18 for the lowest orders. If we take an octupole perturbation ($n = 4$) as an example, there are lines of instabilities arising from this contribution defined by: $4\beta_z = 2$, $3\beta_z + \beta_r = 2$, $2\beta_z + 2\beta_r = 2$, $\beta_z + 3\beta_r = 2$, $4\beta_r = 2$. All these lines meet in one point.

Experimental proof of the instabilities has been obtained by measurements of the number of trapped ions at different operating points. Early evidence was obtained by Guidugli and Traldi [74] who used the term "black holes" for the observed strong instabilities. A high resolution scan of a limited region of the stability diagram was performed by Eades et al. [75]; however, a complete account for most of the stability diagram was given by Alheit et al. [200], who observed instabilities resulting from very high orders of perturbing potentials (Fig. 2.19). From this figure it is evident that strong instabilities occur at the high-q region of the stability diagram, due to hexapole and octupole terms in the expansion of the potential, which are the highest orders expected in a reasonably well machined trap. This makes it very difficult to

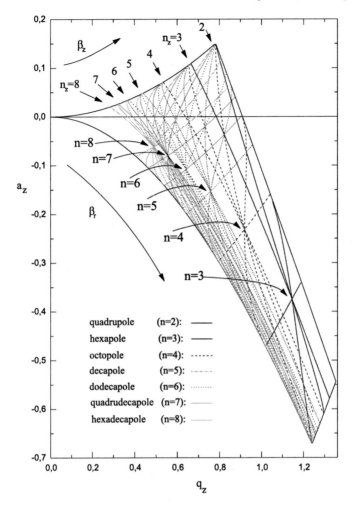

Fig. 2.18. Theoretical lines of instabilities in the first region of stability of the Paul trap for perturbations of order $n = 3$ to $n = 8$

obtain long storage times at high amplitudes of the trapping voltage (for a given frequency). In fact, the most stable conditions are obtained for small q values near the $a = 0$-axis; here the instability condition on the frequencies is met only for a very high resonance order n which would occur only if the field was highly imperfect, with the multipole expansion significant to very high orders.

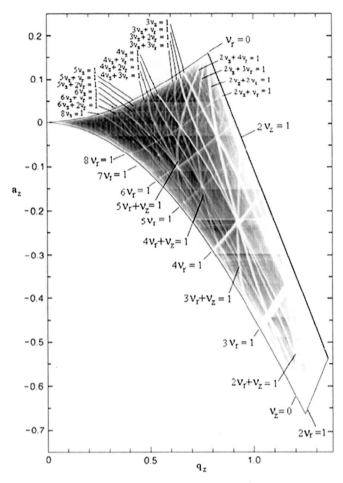

Fig. 2.19. Experimental lines of instabilities in the first region of stability of a real Paul trap taken with H_2^+ ions. The unstable lines are assigned according to (2.36), where $\Omega/2\pi$ is normalized to 1. The intensity of *grey* is proportional to the trapped ion number

2.6 The Role of Collisions in a Paul Trap

The presence of neutral background gas particles in a Paul trap, whether introduced intentionally or as inevitable residual gas in the vacuum system, will result in collisions between the ions and the background particles, that will abruptly change the stable orbits of the ions, and determine statistically the evolution of the mean energy and storage time of the ion population. For all ion energies commonly of interest (excluding the extreme of low energy, where $\lambda/a \sim 1$, with λ de Broglie wavelength of the ion and a the range of the ion-particle interaction), the collision may be described in terms of classical particle trajectories. Further it is expected that in cases where the ions are produced by electron impact with the parent atom, the most frequently collisions are likely to result in resonant charge exchange, with little change in linear momentum. The effect of a collision is to suddenly put the ion in a different orbit, which after a succession of such impulses may cause the ion to leave the trap, limiting its lifetime in it. However, repeated collisions may, on the *average*, increase or decrease the mean ion energy, depending on the relative mass of the colliding particles, as was first pointed out by Major and Dehmelt [76]. Early attempts at a statistical analysis of the effects of collisions are found in [77, 78]. Using the adiabatic approximation, in which the motion is separated into a micromotion superimposed on a secular motion, it was shown explicitly that elastic collisions lead to a characteristic time of decay or growth of the mean ion energy, a time which is a function of the mass ratio of the colliding particles. There is a damping of the ion motion when the ion mass is very much larger than the mass of the background particles, just as the free swinging of a massive pendulum is damped by collisions with air molecules. The opposite is true of a light ion scattering from more massive particles, where the ion will on the average gain energy from the time varying trapping field. This is analogous to the rf-heating in a conductor in which electrons may be thought of as scattering from fixed centers in a solid. Collisional cooling will be discussed later in the context of a different mechanism to reduce the stored ion kinetic energy.

Of equal practical importance is the effect of collisions on the mean lifetime of the ion population in the trap. Clearly, if through collisions with a lighter gas there is on average a loss of kinetic energy by the ion population, their statistical energy distribution will have a reduced number at the energy needed to escape, and the lifetime is increased. Figure 2.20 gives evidence for the extended storage times achieved with increasing pressure of a *light* buffer gas. The effect of heavy collision partners on the storage time has been investigated in some detail by Moriwaki and Shimizu [80].

In order to derive a more quantitative description of the ion motion in a Paul trap in the presence of a background of very light particles, the effect of the collisions can be modeled as a viscous damping force proportional to the ion velocity, thus $\boldsymbol{F} = -D\boldsymbol{v}$. If we define a damping constant as $b = D/M\Omega$, then the equation of motion of an ion in a Paul trap reads

$$\frac{d^2u}{d\tau^2} + 2b\frac{du}{d\tau} + (a_u - 2q_u \cos 2\tau)u = 0 \ . \tag{2.38}$$

The transformation $u = w \exp(-b\tau)$ leads to the equation

$$d^2w/d\tau^2 + (a - b^2 - 2q \cos 2\tau)w = 0 \ . \tag{2.39}$$

This is again the Mathieu equation with the coefficient a replaced by $(a - b^2)$; the solution would thus seem to be essentially the same as in the case of no damping. However, Hasegawa and Uehara [81] and also Nasse and

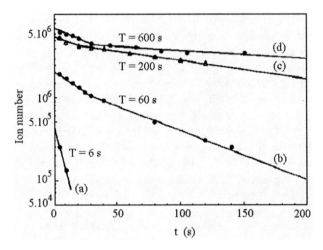

Fig. 2.20. Increase of mean storage time of Tl$^+$ ions in the presence of neutral Helium buffer gas: **(a)** without He; **(b)** $p_{He} = 2.10^{-5}$ Pa; **(c)** $p_{He} = 5.10^{-4}$ Pa; **(d)** $p_{He} = 10^{-3}$ Pa. The given value of the storage time T are deduced from the slopes of the second parts of the curves [79]

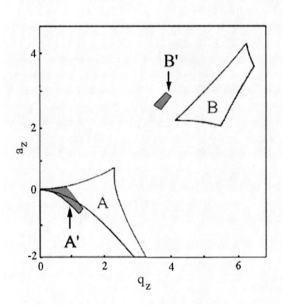

Fig. 2.21. Stability diagram in the a_z–q_z-plane. Two main stable regions A and B for $b = 1.0$ and the corresponding regions A' and B' for $b = 0$ (*grey*) are shown [81]

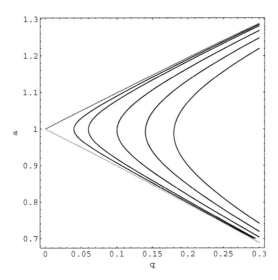

Fig. 2.22. Limits of the stability diagram in axial direction near $q = 0$ for various values of the damping constant b: 0.00, 0.02, 0.03, 0.05, 0.07 and 0.09 (*from left to right*) [83]

Foot [82] have shown that this is not the case; the exponential factor in the solution affects its character. The result of their calculation is that the stability region is not only enlarged, as one might assume intuitively, but also shifted (Fig. 2.21). The shift of the boundaries of the stable region for several damping factor b calculated numerically from (2.38) is shown in more detail in the Fig. 2.22 [83].

From Fig. 2.21 [81] it is evident that at some operating points the motion of a trapped particle, which is stable for $b = 0$, actually becomes unstable, when damping is introduced. This effect, however, becomes significant only under pressure conditions exceeding 1.0 Pa, while under more typical high or ultra-high vacuum conditions, it can be neglected. More significantly, it appears that damping reduces the effect of trap imperfections on the stability of the ions, as discussed above. At background pressures of the order of 10^{-3} Pa no indication of instabilities for small values of the trapping parameters has been reported.

2.7 Quantum Dynamics in Paul Traps

2.7.1 Quantum Parametric Oscillator

The quantum motion of a charged particle in a quadrupole Paul trap can be described as three parametric oscillators. A review of quantum dynamics in Paul traps has been given in [84, 85].

The quantum parametric oscillator is a one-dimensional time-dependent harmonic oscillator described by the Schrödinger equation

$$i\hbar \frac{\partial \Psi}{\partial t} = -\frac{\hbar^2}{2M}\frac{\partial^2 \Psi}{\partial x^2} + \frac{1}{2}k(t)x^2\Psi , \qquad (2.40)$$

where k is a T-periodic function of t.

A systematic study of this equation has been made by a number of authors [86–89]; in the context of charged particle traps, the quantum parametric oscillator has been considered in [90–97].

In order to describe the quasienergy spectrum of (2.40), we introduce the dimensionless variables

$$q = \sqrt{\frac{M\Omega}{2\hbar}}\, z , \qquad \tau = \Omega t/2 , \qquad (2.41)$$

and obtain the Schrödinger equation of the parametric oscillator

$$i\frac{d\psi}{d\tau} = -\frac{1}{2}\frac{\partial^2 \psi}{\partial q^2} + \frac{1}{2}g(\tau)q^2\psi , \qquad (2.42)$$

where $g(\tau) = 4k(t)/(M\Omega^2)$. Here g is a π-periodic function of τ. The exact solutions of (2.42) were constructed by Husimi [86] using the Gaussian wave packet

$$\psi(q,\tau) = \pi^{-1/4}\exp\left[-\frac{1}{2}(aq^2 - 2bq + c)\right], \qquad (2.43)$$

where a, b, and c are complex functions of τ. Inserting (2.43) into (2.42) we obtain

$$i\frac{da}{d\tau} = a^2 - g(\tau) , \quad i\frac{db}{d\tau} = ab , \quad i\frac{dc}{d\tau} = b^2 - a . \qquad (2.44)$$

The first equation (2.44) is of Riccati type and is satisfied by

$$a = -\frac{i}{w}\frac{dw}{d\tau} , \qquad (2.45)$$

where w is a stable complex solution of the Hill equation

$$\frac{d^2 w}{d\tau^2} + g(\tau)w = 0 . \qquad (2.46)$$

The Wronskian of w and w^* is

$$w^*\frac{dw}{d\tau} - w\frac{dw^*}{d\tau} = 2i\delta , \qquad (2.47)$$

where $\delta > 0$. Then we can write $w = \rho\exp(i\gamma)$, where $\rho = |w|$ and $\gamma = \arg w$. Moreover, $a = a_1 + ia_2$, where

$$a_1 = \frac{\delta}{\rho^2} = \frac{d\gamma}{d\tau} , \quad a_2 = -\frac{1}{\rho}\frac{d\rho}{d\tau} . \qquad (2.48)$$

According to [87], the last two equations of (2.44) have solutions

$$b = \sqrt{2a_1}\,\alpha \exp(-i\gamma)\,, \quad c = \alpha^2 \exp(-2i\gamma) - \frac{1}{2}\ln a_1 + i\gamma\,, \tag{2.49}$$

where α is a complex parameter and

$$\frac{d\gamma}{d\tau} = \frac{\delta}{\rho^2}\,, \quad \exp(-2i\gamma) = \frac{w^*}{w}\,. \tag{2.50}$$

The Gaussian packet (2.43) can be written as

$$\psi(q,\tau,\alpha) = \left(\frac{a_1}{\pi}\right)^{1/4} \exp\left[-\frac{1}{2}(a_1 q^2 + i\gamma)\right] \tag{2.51}$$
$$\times \exp\left[\sqrt{2a_1}\,\alpha q \exp(-i\gamma) - \frac{1}{2}\alpha^2 \exp(-2i\gamma)\right].$$

Using the generating function for the Hermite polynomials H_n (D.12) we get

$$\psi(q,\tau,\alpha) = \sum_{n=0}^{\infty} \frac{\alpha^n}{\sqrt{n!}}\psi_n(q,\tau)\,, \tag{2.52}$$

where

$$\psi_n(q,\tau) = a_1^{1/4}\varphi_n(\sqrt{a_1}\,q)\exp\left\{-\frac{i}{2}[a_2 q^2 + (2n+1)\gamma]\right\}\,. \tag{2.53}$$

Thus we obtain a complete set of orthonormal solutions ψ_n, $n = 0, 1, \ldots$, of the Schrödinger equation (2.42), and φ_n is the normalized Hermite function.

According to Floquet's theorem for the regions of stability of the Hill equation (2.46), we can write $w = v\exp(i\beta\tau)$, where v is a π-periodic function of τ and the characteristic exponent is given by

$$\beta = \frac{\delta}{\pi}\int_0^\pi \frac{1}{\rho^2}d\tau\,. \tag{2.54}$$

The quasienergy functions can be rewritten

$$\psi_n(q,\tau) = \left(\frac{\sqrt{\delta}}{\rho}\right)^{1/2}\varphi_n\left(\frac{\sqrt{\delta}}{\rho}q\right)\exp\left[-i\left(n+\frac{1}{2}\right)\beta\tau\right]\exp\left(\frac{i}{2\rho}\frac{d\rho}{d\tau}q^2\right)\,, \tag{2.55}$$

and

$$\psi_n(q,\tau+\pi) = \exp\left[-i\pi\beta\left(n+\frac{1}{2}\right)\right]\psi_n(q,\tau)\,. \tag{2.56}$$

Thus the functions ψ_n have the scaled quasienergies $\varepsilon_n = (n+1/2)\beta$, and the quasienergy spectrum is discrete. Unstable classical solutions yield quasienergy functions with continuous spectra [87].

We define
$$\psi^\alpha(q,t) = \exp\left(-\frac{1}{2}|\alpha|^2\right)\psi(q,w,\alpha) . \tag{2.57}$$

The expectation values of the operators q and $p = -i\partial/\partial q$ are
$$\bar{q} = \langle\psi^\alpha|q|\psi^\alpha\rangle = \frac{1}{\sqrt{2}}(\alpha w^* + \alpha^* w) ,$$
$$\bar{p} = \langle\psi^\alpha|p|\psi^\alpha\rangle = \frac{1}{\sqrt{2}}\left(\alpha\frac{dw^*}{d\tau} + \alpha^*\frac{dw}{d\tau}\right) . \tag{2.58}$$

The position variance σ_{qq}, the momentum variance σ_{pp}, the covariance σ_{qp}, and the correlation coefficient r_{qp} of the operators q and $p = -i\partial/\partial q$ are given by
$$\sigma_{qq} = \langle q^2\rangle - \langle q\rangle^2 , \quad \sigma_{pp} = \langle p^2\rangle - \langle p\rangle^2 ,$$
$$\sigma_{qp} = \frac{1}{2}\langle qp + pq\rangle - \langle q\rangle\langle p\rangle , \quad r_{qp} = \frac{\sigma_{qp}}{\sqrt{\sigma_{qq}\sigma_{pp}}} , \tag{2.59}$$

where $\langle\rangle$ means the expectation value. The Schrödinger uncertainty relation [98] can be written in the form
$$\sigma_{qq}\sigma_{pp} - \sigma_{qp}^2 \geq \frac{1}{4} . \tag{2.60}$$

The inequality (2.60) implies the Heisenberg uncertainty relation
$$\sigma_{qq}\sigma_{pp} \geq \frac{1}{4} . \tag{2.61}$$

Using (2.55), (2.57), and (2.59), we find
$$\sigma_{qq}^{(0)} = \sigma_{qq}^{[\alpha]} = \frac{\rho^2}{2\delta} , \quad \sigma_{qp}^{(0)} = \sigma_{qp}^{[\alpha]} = \frac{\rho}{2\delta}\frac{d\rho}{d\tau} , \tag{2.62}$$

$$\sigma_{pp}^{(0)} = \sigma_{pp}^{[\alpha]} = \frac{1}{2\delta}\left|\frac{dw}{d\tau}\right|^2 = \frac{\delta}{2\rho^2} + \frac{1}{2\delta}\left(\frac{d\rho}{d\tau}\right)^2 , \tag{2.63}$$

$$\frac{\sigma_{qq}^{(n)}}{\sigma_{qq}^{(0)}} = \frac{\sigma_{pp}^{(n)}}{\sigma_{pp}^{(0)}} = \frac{\sigma_{qp}^{(n)}}{\sigma_{qp}^{(0)}} = n + \frac{1}{2} , \tag{2.64}$$

$$r_{qp}^{(n)} = r_{qp}^{[\alpha]} = \left[1 + \frac{\delta^2}{\rho^2}\left(\frac{d\rho}{d\tau}\right)^{-2}\right]^{-1/2} . \tag{2.65}$$

Here the superscripts (n) and $[\alpha]$ denote the expectation values in the states represented by ψ_n and ψ^α, respectively. The correlation coefficient (2.65) is independent of n and α. The variances, covariances and correlation coefficients given by (2.61)–(2.63) are π-periodic functions of τ. In terms of these the Gaussian solution ψ^α can be written as

$$\psi^\alpha(q,\tau) = \left(2\pi\sigma_{qq}^{(0)}\right)^{-1/2} \exp\left(-\frac{i}{2}\gamma\right) \tag{2.66}$$

$$\times \exp\left[-\frac{1}{4\sigma_{qq}^{(0)}}\left(1 - i\sigma_{pq}^{(0)}\right)(q-\bar{q})^2 + i\bar{p}\left(q - \frac{\bar{q}}{2}\right)\right].$$

For ψ^α the Schrödinger uncertainty relation is satisfied with the equality sign.

The quantum driven parametric oscillator is described by the Schrödinger equation

$$i\frac{\partial\phi}{\partial\tau} = \left[-\frac{1}{2}\frac{\partial^2}{\partial q^2} + \frac{1}{2}g(\tau)q^2 - f(\tau)\right]\phi. \tag{2.67}$$

The solution of (2.67) is

$$\phi(q,\tau) = \psi(q - w_0, \tau)\exp(i\lambda), \tag{2.68}$$

where ψ is a solution of (2.42), w_0 is a solution of the equation

$$\frac{d^2w}{d\tau^2} + g(\tau)w = f(\tau), \tag{2.69}$$

$$\lambda = i(q - w_0)\frac{dw_0}{d\tau} + i\int_0^\tau L(\tau')d\tau', \tag{2.70}$$

and L is the classical Lagrangian

$$L(\tau) = \frac{1}{2}\left(\frac{dw_0}{d\tau}\right)^2 - \frac{1}{2}g(\tau)w_0^2 + f(\tau)w_0. \tag{2.71}$$

2.7.2 Quantum Dynamics in Ideal Paul Trap

The Schrödinger equation for a charged particle of mass M confined in a quadrupole Paul trap can be written as

$$i\hbar\frac{\partial\Psi}{\partial t} = -\frac{\hbar^2}{2M}\Delta\Psi + \frac{1}{2}\left[k_1(t)x^2 + k_2(t)y^2 + k_3(t)z^2\right], \tag{2.72}$$

where k_1, k_2, and k_3 are time-periodic functions of period $T = 2\pi/\Omega$ with $k_1 + k_2 + k_3 = 0$.

For convenience we introduce the dimensionless variables

$$x_1 = \sqrt{\frac{M\Omega}{2\hbar}}\,x, \quad x_2 = \sqrt{\frac{M\Omega}{2\hbar}}\,y, \quad x_3 = \sqrt{\frac{M\Omega}{2\hbar}}\,z, \quad \tau = \frac{1}{2}\Omega t. \tag{2.73}$$

A solution of (2.72) is

$$\Psi(x,y,z,t) = \left(\frac{M\Omega}{2\hbar}\right)^{3/2}\psi^{(1)}(x_1,\tau)\psi^{(2)}(x_2,\tau)\psi^{(3)}(x_3,\tau), \tag{2.74}$$

where
$$i\frac{\partial \psi^{(j)}}{\partial \tau} = -\frac{1}{2}\frac{\partial^2 \psi^{(j)}}{\partial x_j^2} + \frac{1}{2}g_j(\tau)x_j^2 \psi^{(j)} , \tag{2.75}$$

$$g_j(\tau) = \frac{4}{M\Omega^2}k_j(t) , \quad j = 1, 2, 3 .$$

The classical equations of motion of a charged particle in a quadrupole Paul trap reduce to the Hill equations

$$\frac{d^2 w_j}{d\tau^2} + g_j(\tau)w_j = 0 , \tag{2.76}$$

where the functions g_j are π-periodic.

It is convenient to introduce the stable complex solutions w_j of (2.76) with the Wronskians

$$w_j^* \frac{dw_j}{d\tau} - w_j \frac{dw_j^*}{d\tau} = 2i\delta_j , \tag{2.77}$$

where $\delta_j > 0$. According to Floquet's theorem, we can write

$$w_j = \rho_j \exp(i\gamma_j) = v_j \exp(i\beta_j \tau) , \tag{2.78}$$

where $\rho_j = |w_j|$, $\gamma_j = \arg w_j$, the characteristic exponents β_j are positive and the functions v_j are π-periodic. Using (2.74) and (2.55), we obtain the following quasienergy solutions of (2.72):

$$\psi_{n_1 n_2 n_3}(x, y, z, t) = \left(\frac{M\Omega}{2\hbar}\right)^{3/4} \psi_{n_1}(x_1, \tau)\psi_{n_2}(x_2, \tau)\psi_{n_3}(x_3, \tau) , \tag{2.79}$$

where n_1, n_2, and n_3 are nonnegative integers. Here

$$\psi_{n_j}(q_j, \tau) = \left(\frac{\delta}{\pi}\right)^{1/4} (2^{n_j} n_j! \rho_j)^{-1/2} \exp\left[-i\left(n_j + \frac{1}{2}\right)\gamma_j\right] \tag{2.80}$$
$$\times H_n\left(\frac{\sqrt{\delta_j}}{\rho_j}x_j\right)\exp\left[-\left(\frac{\delta_j}{2\rho_j^2} - \frac{i}{2\rho_j}\frac{d\rho_j}{d\tau}\right)x_j^2\right] ,$$

where H_n are Hermite polynomials. The quasienergy of $\psi_{n_1 n_2 n_3}$ is

$$E_{n_1 n_2 n_3} = \frac{1}{2}\hbar\Omega \sum_{j=1}^{3}\beta_j\left(n_j + \frac{1}{2}\right) . \tag{2.81}$$

The set of quasienergy functions (2.79) is complete and orthonormal.

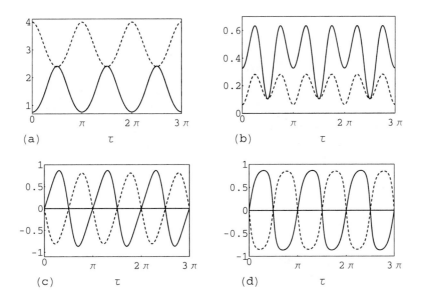

Fig. 2.23. Variances and correlation coefficients for the quasienergy ground state in an ideal Paul trap with the Mathieu parameters $a = 0.01$ and $q = 0.5$. The radial curves are *dashed*, while the axial ones are *full*. (a) The position variances $\sigma_{x_1 x_1}$ and $\sigma_{x_3 x_3}$; (b) the momentum variances $\sigma_{p_1 p_1}$ and $\sigma_{p_3 p_3}$; (c) the covariances $\sigma_{x_1 p_1}$ and $\sigma_{x_3 p_3}$; (d) the correlation coefficient $r_{x_1 p_1}$ and $r_{x_3 p_3}$

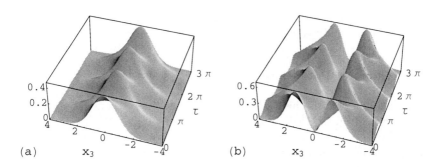

Fig. 2.24. The axial probability distributions $P_n(x_3, \tau)$ for an ideal Paul trap with the Mathieu parameters $a = 0.01$ and $q = 0.5$. (a) $n = 0$; (b) $n = 1$

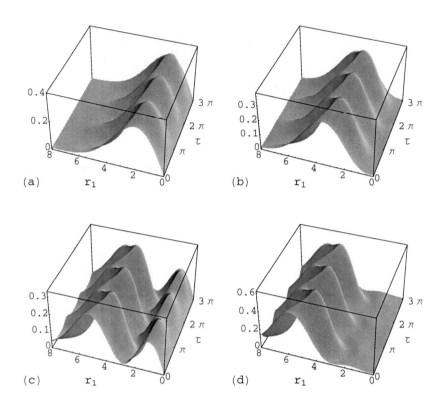

Fig. 2.25. The radial probability distributions $P_{nl}(r_1, \tau)$ for an ideal Paul trap with the Mathieu parameters $a = 0.01$ and $q = 0.5$. **(a)** $n = l = 0$; **(b)** $n = 0$ and $l = 1$; **(c)** $n = 1$ and $l = 0$; **(d)** $n = l = 1$

2.7.3 Effective Potentials

The first quantum-mechanical treatment of the time-dependent trapping potential was given by Cook, Shankland, and Wells in [99]. We present a general effective potential approach to the quantum motion of charged particles in rapidly oscillating fields. Consider the Schrödinger equation of a particle of mass M

$$i\hbar \frac{\partial \Psi}{\partial t} = -\frac{\hbar^2}{2M} \boldsymbol{\Delta} \Psi + U(\mathbf{r}) \Psi + V(\mathbf{r}, t) \Psi , \qquad (2.82)$$

where V is a time-periodic function of period $T = 2\pi/\Omega$ and $\langle V \rangle = 0$. Here $\langle \rangle$ denotes the time-average over a period. We write the solution to (2.82) in the form

$$\Psi = \exp\left(-\frac{i}{\hbar} W\right) \varphi , \qquad (2.83)$$

2.7 Quantum Dynamics in Paul Traps

where W is a function of $\mathbf{r}$ and t such that

$$\frac{\partial W}{\partial t} = V, \quad \langle W \rangle = 0 . \tag{2.84}$$

Substituting (2.83) into (2.82) we obtain

$$i\hbar \frac{\partial \varphi}{\partial t} = -\frac{\hbar^2}{2M} \triangle \varphi + U(\mathbf{r}) \varphi + S\varphi , \tag{2.85}$$

$$S = \frac{i\hbar}{M} (\nabla W) \nabla + \frac{i\hbar}{2M} \Delta W + \frac{1}{2M} (\nabla W)^2 . \tag{2.86}$$

We define the time-independent effective potential by $V^{\text{eff}} = \langle S \rangle$. Then

$$V^{\text{eff}} = \frac{(\nabla W)^2}{2M} . \tag{2.87}$$

The expression (2.87) is equivalent to the classical result of [100].

The solution of the Schrödinger equation (2.82) for high frequency Ω and small V may be approximated by

$$\Psi^{\text{eff}}(\mathbf{r}, t) = \exp\left(-\frac{i}{\hbar} W\right) \Phi , \tag{2.88}$$

where Φ is a solution of the Schrödinger equation in which the time-dependent potential V is replaced by the static potential V^{eff}:

$$i\hbar \frac{\partial \Phi}{\partial t} = -\frac{\hbar^2}{2M} \triangle \Phi + \left[U(\mathbf{r}) \Phi + V^{\text{eff}}(\mathbf{r}) \right] \Phi . \tag{2.89}$$

In reference [99] it is assumed that $V(\mathbf{r}, t) = \cos(\Omega t) v(\mathbf{r})$. Then

$$W(\mathbf{r}, t) = \frac{\sin(\Omega t)}{\Omega} v(\mathbf{r}) , \quad V^{\text{eff}} = \frac{(\nabla v)^2}{4M\Omega^2} . \tag{2.90}$$

If Ω is a high frequency, the dominant effect of the oscillating potential is contained in the phase factor of (2.88) and Φ is a slowly function of time [99].

In order to study the limitations of the effective potential approximation, we consider the axial Schrödinger equation in an ideal Paul trap:

$$i\hbar \frac{\partial \Psi}{\partial t} = -\frac{\hbar^2}{2M} \frac{\partial^2 \Psi}{\partial z^2} + \frac{1}{2} [c_0 + c_1 \cos(\Omega t)] z^2 \Psi . \tag{2.91}$$

By (2.88)-(2.90), we obtain

$$\Psi^{\text{eff}}(z, t) = \exp\left[-\frac{ic_1}{2\hbar\Omega} \sin(\Omega t) z^2\right] \Phi , \quad V^{\text{eff}} = \frac{c_1^2}{4M\Omega^2} z^2 , \tag{2.92}$$

2 The Paul Trap

$$i\hbar \frac{\partial \Phi}{\partial t} = -\frac{\hbar^2}{2M} \frac{\partial^2}{\partial z^2} + \frac{1}{2}\left(c_0 + \frac{c_1^2}{2M\Omega^2}\right) z^2 \Phi . \tag{2.93}$$

Using

$$\xi = \sqrt{\frac{M\Omega}{2\hbar}}\, z , \quad \tau = \frac{1}{2}\Omega t , \tag{2.94}$$

we obtain

$$i\frac{\partial \phi}{\partial \tau} = -\frac{1}{2}\frac{\partial^2 \phi}{\partial \xi^2} + \frac{1}{2}\left(\beta^{\text{eff}}\right)^2 \xi^2 \phi , \tag{2.95}$$

where

$$a = \frac{4c_0}{M\Omega^2} , \quad q = -\frac{2c_1}{M\Omega^2} , \quad \beta^{\text{eff}} = \sqrt{a + \frac{1}{2}q^2} . \tag{2.96}$$

Then

$$\psi_n^{\text{eff}}(\xi, \tau) = \exp\left[\frac{i}{2}q \sin(2\tau)\xi^2\right] \phi_n(\xi, \tau) , \tag{2.97}$$

$$\phi_n(\xi, \tau) = \sqrt{\beta^{\text{eff}}}\, \varphi_n\left(\sqrt{\beta^{\text{eff}}}\,\xi\right) \exp\left[-i\left(n+\frac{1}{2}\right)\beta^{\text{eff}}\tau\right] \exp\left[\frac{i}{2}q \sin(2\tau)^2 \xi^2\right] . \tag{2.98}$$

Here the Hermite functions φ_n are defined by (D.16).

Consider the complex solution

$$w = \text{Ce}(a, q, \tau) + i\text{Se}(a, q, \tau) \tag{2.99}$$

of the Mathieu equation

$$\frac{d^2 w}{d\tau^2} + [a - 2q \cos(\Omega t)] w = 0 , \tag{2.100}$$

where $\text{Ce}(a, q, \tau)$ and $\text{Se}(a, q, \tau)$ are the even and odd Mathieu functions, respectively. The quasienergy solutions of (2.95) are

$$\psi_n(\xi, \tau) = \left(\frac{\sqrt{\delta}}{\rho}\right)^{1/2} \varphi_n\left(\frac{\sqrt{\delta}}{\rho}\xi\right) \exp\left[-i\left(n+\frac{1}{2}\right)\gamma\right] \exp\left(\frac{i}{2\rho}\frac{d\rho}{d\tau}\xi^2\right) , \tag{2.101}$$

where $\rho = |w|$, $\gamma = \arg w$, and

$$\frac{d\gamma}{d\tau} = \frac{\delta}{\rho^2} , \quad \delta = \frac{1}{2i}\left(w^* \frac{dw}{d\tau} - w \frac{dw^*}{d\tau}\right) . \tag{2.102}$$

The second order approximation $\psi_n^{(2)}$ of ψ_n is given by (2.101) for

$$\beta = a + \frac{q^2}{2(1-a)} + \cdots , \quad \delta = \beta\left[1 - \frac{q^2}{2(1-\beta^2)^2}\right] + \cdots , \tag{2.103}$$

2.7 Quantum Dynamics in Paul Traps   49

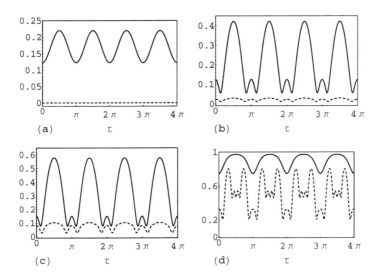

Fig. 2.26. The error function f^{eff} between the effective and quasienergy ground states (*full lines*) and the error function $f^{(2)}$ between the quasienergy ground state and the corresponding second order approximation (*dashed lines*) for the Mathieu parameters a and q of the ideal Paul trap. (**a**) $a = 0$ and $q = 0.1$; (**b**) $a = 0.1$ and $q = 0.5$; (**c**) $a = -0.1$ and $q = 0.7$; (**d**) $a = -0.67$ and $q = 1.24$

$$\rho^2 = 1 - \frac{\cos(2\tau)}{1-\beta^2}q + \frac{3\cos(4\tau)}{4(4-\beta^2)(1-\beta^2)}q^2 + \cdots, \quad (2.104)$$

We consider two pure states $\boldsymbol{\Psi} = |\Psi\rangle\langle\Psi|$ and $\boldsymbol{\Phi} = |\Phi\rangle\langle\Phi|$, where Ψ and Φ are normalized state vectors. Then the state $\boldsymbol{\Phi}$ is an approximation of the state $\boldsymbol{\Psi}$ with the error [101]

$$\|\boldsymbol{\Phi} - \boldsymbol{\Psi}\| = \left[1 - |\langle\Phi,\Psi\rangle|^2\right]^{1/2}. \quad (2.105)$$

By (2.105), we have $0 \le \|\boldsymbol{\Phi} - \boldsymbol{\Psi}\| \le 1$. Moreover, the distance between states $\boldsymbol{\Psi}$ and $\boldsymbol{\Phi}$ is $d(\boldsymbol{\Phi},\boldsymbol{\Psi}) = \arcsin\|\boldsymbol{\Phi} - \boldsymbol{\Psi}\|$ [101]. We introduce the following π-periodic functions of τ:

$$f^{\text{eff}} = \left[1 - |\langle\psi_0^{\text{eff}},\psi_0\rangle|^2\right]^{1/2}, \quad f^{(2)} = \left[1 - |\langle\psi_0^{(2)},\psi_0\rangle|^2\right]^{1/2}. \quad (2.106)$$

In Fig. 2.26 the error functions f^{eff} and $f^{(2)}$ are plotted against τ for the first stability region of the ideal Paul trap. As the f^{eff} decreases, the accuracy of the effective potential approximation increases. Combescure [90] has shown some limitations to the effective potential approximation. However, the bases $\{\psi_n^{\text{eff}}\}$ and $\{\psi_n^{(2)}\}$ may be a starting point for the study of small anharmonic perturbations of a quadrupole electromagnetic trap.

3 The Penning Trap

3.1 Theory of the Ideal Penning Trap

The Penning trap uses static electric and magnetic fields to confine charged particles. The ideal Penning trap is formed by the superposition of a homogeneous magnetic field $\boldsymbol{B} = (0, 0, B_0)$ and an electric field $\boldsymbol{E} = -\boldsymbol{\nabla}\Phi$ derived from the potential

$$\Phi = \frac{U_0}{2d^2}(2z^2 - x^2 - y^2) ,\tag{3.1}$$

whose equipotential surfaces are hyperboloids of revolution. As in the case of the previously discussed Paul trap (Fig. 1.2), a conducting surface following the equation $2z^2 - x^2 - y^2 = -r_0^2$ is the ring electrode, and $2z^2 - x^2 - y^2 = 2z_0^2$ the two end caps. U_0 is the static voltage applied between them, so that $d = \sqrt{z_0^2 + r_0^2/2}$.

A particle of mass M, charge Q, and velocity $\boldsymbol{v} = (v_x, v_y, v_z)$ moving in the fields $\boldsymbol{E}$ and $\boldsymbol{B}$ experiences a force

$$\boldsymbol{F} = -Q\boldsymbol{\nabla}\Phi + Q(\boldsymbol{v} \times \boldsymbol{B}) .\tag{3.2}$$

Since the magnetic field is along the z-axis, the z-component of the force is purely electrostatic, and therefore to confine the particle in the z-direction we must have $QU_0 > 0$. The x- and y-components of $\boldsymbol{F}$ are a combination of a dominant restraining force due to the magnetic field, characterized by the *cyclotron frequency*

$$\omega_c = \frac{|QB_0|}{M} ,\tag{3.3}$$

and a repulsive electrostatic force, that tries to push the particle out of the trap in the radial direction.

Newton's equations of motion in Cartesian coordinates are as follows:

$$\frac{d^2x}{dt^2} - \omega_0 \frac{dy}{dt} - \frac{1}{2}\omega_z^2 x = 0 ,\tag{3.4}$$

$$\frac{d^2y}{dt^2} + \omega_0 \frac{dx}{dt} - \frac{1}{2}\omega_z^2 y = 0 ,\tag{3.5}$$

$$\frac{d^2z}{dt^2} + \omega_z^2 z = 0 ,\tag{3.6}$$

where
$$\omega_0 = \frac{QB_0}{M}, \quad \omega_z = \sqrt{\frac{2QU_0}{Md^2}}. \qquad (3.7)$$

It is apparent that $|\omega_0|$ equals the free particle cyclotron frequency ω_c, and the motion in the z-direction is a simple harmonic oscillation with an *axial frequency* ω_z, decoupled from the transverse motion in the x- and y-directions.

To describe the motion in the x, y-plane we introduce the complex variable $u = x + iy$. The radial equations of motion (3.4) and (3.5) then reduce to

$$\frac{d^2 u}{dt^2} + i\omega_0 \frac{du}{dt} - \frac{1}{2}\omega_z^2 u = 0. \qquad (3.8)$$

The general solution is found by setting $u = \exp(-i\omega t)$ to obtain the algebraic condition

$$\omega^2 - \omega_0 \omega + \frac{1}{2}\omega_z^2 = 0. \qquad (3.9)$$

The solutions ω_+ and ω_- of (3.9) may be written as $\mathrm{sgn}(\omega_0)\omega_\pm$, where

$$\omega_+ = \frac{1}{2}(\omega_c + \omega_1), \quad \omega_- = \frac{1}{2}(\omega_c - \omega_1), \quad \omega_1 = \sqrt{\omega_c^2 - 2\omega_z^2}. \qquad (3.10)$$

Here ω_+ is the *modified cyclotron frequency* and ω_- is the *magnetron frequency*. In order that the motion be bounded, the roots of (3.9) must be real, leading to the trapping condition

$$\omega_c^2 - 2\omega_z^2 > 0. \qquad (3.11)$$

Using (3.7), (3.3), and (3.11), the conditions for stable confinement of charged particles in the ideal Penning trap can be expressed in terms of the applied fields as follows:

$$\frac{|Q|}{M} B_0^2 > \frac{4|U_0|}{d^2}, \quad QU_0 > 0. \qquad (3.12)$$

It determines the minimum magnetic field required to balance the radial component of the applied electric field.

Several useful relations exist between the eigenfrequencies of the trapped particle:

$$\omega_+ + \omega_- = \omega_c, \qquad (3.13)$$
$$2\omega_+\omega_- = \omega_z^2, \qquad (3.14)$$
$$\omega_+^2 + \omega_-^2 + \omega_z^2 = \omega_c^2. \qquad (3.15)$$

The general solution of (3.8) is

$$u = R_+ \exp[-i(\omega_+ t + \alpha_+)] + R_- \exp[-i(\omega_- t + \alpha_-)], \qquad (3.16)$$

hence the parametric equations for a trajectory of the charged particle can be written as

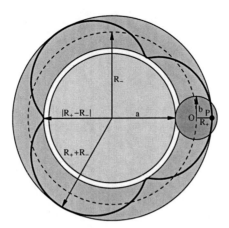

Fig. 3.1. Generation of an epitrochoid in the radial plane of the ideal Penning trap for $\omega_+ = 4\omega_-$, $R_+ = 3R_-/16$, $\alpha_+ = \alpha_- = 0$. The epitrochoid is the locus of a point P that is rigidly attached to a circle of radius $b = \omega_- R_-/\omega_+$ which rolls without slippage on the outside of a fixed circle of radius $a = \omega_1 R_-/\omega_+$. The distance from P to the center O of the rolling circle is equal to the cyclotron radius R_+. The *dashed magnetron circle* of radius R_- is generated by O. The epitrochoid lies in an annulus bounded by circles of radii $R_+ + R_-$ and $|R_+ - R_-|$

$$x = R_+ \cos(\omega_+ t + \varphi_+) + R_- \cos(\omega_- t + \varphi_-) , \qquad (3.17)$$

$$y = -\frac{Q}{|Q|}[R_+ \sin(\omega_+ t + \varphi_+) + R_- \sin(\omega_- t + \varphi_-)] , \qquad (3.18)$$

$$z = R_z \cos(\omega_z t + \varphi_z) . \qquad (3.19)$$

The cyclotron radius R_+, the magnetron radius R_-, the axial amplitude R_z, and the phases φ_+, φ_-, φ_z are determined by the initial conditions. By insertion of (3.17) and (3.18) into the radial variable $r = \sqrt{x^2 + y^2}$, we obtain

$$r = [R_+^2 + R_-^2 + 2R_+ R_- \cos(\omega_1 t + \varphi_+ - \varphi_-)]^{1/2} . \qquad (3.20)$$

From (3.20), follows $|R_+ - R_-| \leq r \leq R_+ + R_-$, and thus the trajectory given by (3.17)–(3.19) is bounded by the cylinder

$$|R_+ - R_-| \leq r \leq R_+ + R_- , \quad |z| < R_z . \qquad (3.21)$$

The projection on to the x–y-plane of a particle trajectory whose parametric equations are given by (3.17) and (3.18) is a figure which can be described as an epitrochoid (see also Appendix B).

The $x - y$ projection of the motion is periodic if ω_+/ω_- is rational; otherwise if ω_+/ω_- is irrational, the radial motion is quasiperiodic.

Some trajectories with the initial conditions $x(0) = x_0 \neq 0$, $y(0) = 0$, and $v_x(0) = 0$ are illustrated in Figs. 3.2 and 3.3.

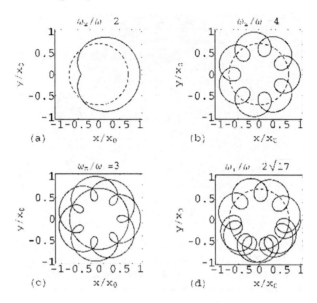

Fig. 3.2. Some radial projections of trajectories with $R_- = 2.5R_+$. Periodic orbits for (a) $\omega_+/\omega_- = 2$; (b) $\omega_+/\omega_- = 8$; (c) $\omega_+/\omega_- = 9/2$; (d) quasiperiodic orbit for $\omega_+/\omega_- = 2\sqrt{17}$

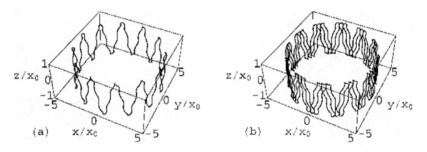

Fig. 3.3. Trajectories in three dimensions with the initial conditions $R_- = 50R_+$, $R_z = 0.2R_+$. (a) Periodic orbit for $\omega_+/\omega_z = 6$; (b) quasiperiodic orbit for $\omega_+/\omega_z = \sqrt{35}$

The energy content of the different motions can be derived using the Hamiltonian formalism. We choose the symmetric gauge with the vector potential $\boldsymbol{A} = B_0(-y/2,\ x/2,\ 0)$. Then the Hamiltonian is

$$H = \frac{1}{2M}(\boldsymbol{p} - Q\boldsymbol{A})^2 + Q\Phi \qquad (3.22)$$
$$= \frac{1}{2M}\boldsymbol{p}^2 + \frac{1}{8}M\omega_1^2(x^2 + y^2) + \frac{1}{2}M\omega_z^2 z^2 - \frac{Q}{2|Q|}\omega_c L_z\ ,$$

where $\boldsymbol{p} = M\boldsymbol{v}$ is the momentum vector and the axial component of the canonical angular momentum is

$$L_z = xp_y - yp_x \ . \tag{3.23}$$

Here H and L_z are constants of the motion. By insertion of (3.17)–(3.19) into (3.23), we obtain for the axial angular momentum

$$L_z = \frac{Q}{2|Q|} M\omega_1 (R_-^2 - R_+^2) \ , \tag{3.24}$$

and for the total energy of the charged particle we have

$$H = E_k + E_p = \frac{1}{2} M R_z^2 \omega_z^2 + M\omega_1 (R_+^2 \omega_+ - R_-^2 \omega_-) \ . \tag{3.25}$$

Note that the part of the total energy arising from the magnetron motion has a negative sign indicating that an increase in the magnetron radius leads to a decrease in the energy.

The instantaneous particle kinetic energy E_k and the potential energy E_p are given by

$$E_k = \frac{1}{2} M v^2 = \frac{1}{2} M R_z^2 \omega_z^2 \sin^2(\omega_z t + \alpha_z)$$
$$+ \frac{1}{2} M [R_+^2 \omega_+^2 + R_-^2 \omega_-^2 + 2 R_+ R_- \omega_+ \omega_- \cos(\omega_1 t + \alpha_+ - \alpha_-)] \ , \tag{3.26}$$

$$E_p = Q\Phi = \frac{1}{2} M R_z^2 \omega_z^2 \cos^2(\omega_z t + \alpha_z) \tag{3.27}$$
$$- \frac{1}{4} M \omega_z^2 [R_+^2 + R_-^2 + 2 R_+ R_- \cos(\omega_1 t + \alpha_+ - \alpha_-)] \ .$$

The time-averaged kinetic and potential energies are

$$\bar{E}_k = \frac{1}{4} M R_z^2 \omega_z^2 + \frac{1}{2} M R_+^2 \omega_+^2 + \frac{1}{2} M R_-^2 \omega_-^2 \ , \tag{3.28}$$

$$\bar{E}_p = \frac{1}{4} M \omega_z^2 (R_z^2 - R_+^2 - R_-^2) \ . \tag{3.29}$$

It is convenient to consider the following canonical transformation to the action and angle variables J_+, J_-, J_z, φ_+, φ_-, and φ_z [62]:

$$x = \sqrt{\frac{2}{M\omega_1}} \left(\sqrt{J_+} \cos \varphi_+ + \sqrt{J_-} \cos \varphi_- \right) \ , \tag{3.30}$$

$$p_x = \sqrt{\frac{M\omega_1}{2}} \left(-\sqrt{J_+} \sin \varphi_+ + \sqrt{J_-} \sin \varphi_- \right) \ , \tag{3.31}$$

$$y = -\frac{|Q|}{Q}\sqrt{\frac{2}{M\omega_1}}\left(\sqrt{J_+}\sin\varphi_+ + \sqrt{J_-}\sin\varphi_-\right), \tag{3.32}$$

$$p_y = \frac{|Q|}{Q}\sqrt{\frac{M\omega_1}{2}}\left(-\sqrt{J_+}\cos\varphi_+ + \sqrt{J_-}\cos\varphi_-\right), \tag{3.33}$$

$$z = \sqrt{\frac{2}{M\omega_z}}\sqrt{J_z}\cos\varphi_z, \quad p_z = -\sqrt{2M\omega_z}\sqrt{J_z}\sin\varphi_z. \tag{3.34}$$

According to (3.22) and (3.30)–(3.34), the Hamiltonian of the ideal Penning trap can be written as

$$H = \omega_+ J_+ - \omega_- J_- + \omega_z J_z. \tag{3.35}$$

The action variables are constants of the motion characterized by

$$J_\pm = \frac{1}{2}M\omega_1 R_\pm^2, \quad J_z = \frac{1}{2}M\omega_z R_z^2, \tag{3.36}$$

where R_+, R_-, and R_z are the orbital amplitudes in (3.17)–(3.19).

3.2 Motional Spectrum in Penning Trap

The motional spectrum of an ion in a Penning trap contains the fundamental frequencies ω_+, ω_-, and ω_z as shown in the solution of the equations of motion.

These frequencies can be measured using resonant excitation of the ion motion by applying an additional dipole rf-field to the trap electrodes. For the axial frequency this has to be applied between the two trap end caps, while for excitation of ω_+ and ω_- in the radial plane, the ring electrode has to be split in two segments. An indication of resonant excitation is a rapid loss of stored ions, due to the gain in energy from the rf-field, which enables them to escape. The electrostatic field inside the trap is expected to be a superposition of many multipole components, hence many combinations of the fundamental frequencies and their harmonics become visible, depending on the amplitude of the exciting field. Figure 3.4 shows the number of trapped electrons as a function of the frequency of an excitation rf-field which enters the apparatus by an antenna placed near the trap. Identification of the resonances is through their different dependence on the electrostatic field strength. The axial frequency ω_z depends on the square root of the trapping voltage (3.7), while ω_+ and ω_- vary linearly with the voltage. Figure 3.5 shows the positions of measured resonances at different operating voltages applied to the trap.

Of particular interest is the combination $\omega_+ + \omega_-$, since it is independent of the trapping voltage, and equals the cyclotron frequency of the free ion $\omega_c = (Q/M)B$, whose measurement can serve to calibrate the magnetic field at the ion position.

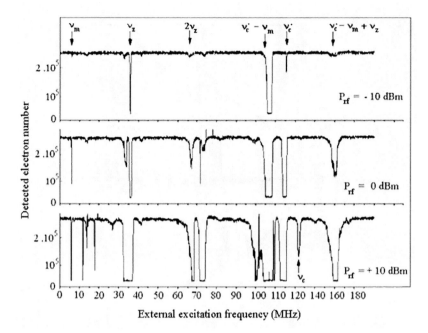

Fig. 3.4. Motional resonances of an electron cloud in a Penning trap taken at different amplitudes of an exciting radio-frequency field [102]

The frequency ω_c can be obtained in three different ways: direct excitation of the sideband $\omega_+ + \omega_-$, measurement of the fundamental frequency ω_+ at different trapping voltages U_0 and extrapolating to $U_0 = 0$, or using (3.15). The last is particularly useful since it is, to first order, independent of perturbations arising from field imperfections, which may shift the individual eigenfrequencies, as will be discussed in a next section.

3.3 Real Penning Traps

In recent years there has been a wide proliferation of the use of the essential Penning field configuration, with adaptations to accommodate a variety of applications. These range from the slowing and bunching of beams of radionuclides to high energy elementary particles and their precise mass determination, to the study of nuclear decay times, mass resolution of isobars, and precise g-factor measurements. In each case, the electrode design must meet often conflicting aspects such as purity of the quadrupole field, but in an open electrode structure that allows compatibility with ion beam or laser optics. Thus in real Penning traps the conditions are considerably more complicated than the ideal description in the previous section. The hyperbolic

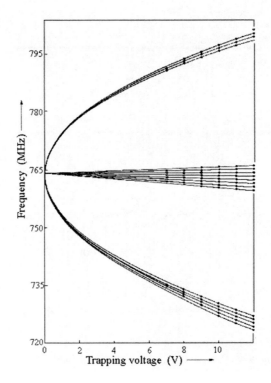

Fig. 3.5. Measured motional frequencies of electrons in a Penning trap. The *center part* shows the frequencies $\omega_+ \pm n\omega_-$, $n = 0, 1, 2, \ldots$ having a linear dependence on the trapping voltage U_0. The *upper* and *lower branches* are resonances at $\omega_+ \pm k\omega_z \pm n\omega_-$, which depend on $U_0^{1/2}$ [103]

surfaces are necessarily truncated, and errors in construction and alignment may cause the electrodes to deviate from the ideal shape and the trap axis to be tilted with respect to the direction of the magnetic field. Moreover, the actual behavior of the trapped particles, if there are more than one, will be affected by the long range Coulomb interaction, which ultimately sets a limit on the number of particles that can be confined by a given magnetic field intensity. This limit, called the *Brillouin limit*, a designation borrowed from plasma physics, is given by

$$n_{\text{lim}} = \frac{B^2}{2\mu_0 M c^2}, \tag{3.37}$$

where μ_0 is the permeability of free space. Naturally, as this limit is approached the individual particle picture is no longer valid: one is then dealing with a plasma.

The departures of the field from the pure quadrupole introduce nonlinearity in the equations of motion, and coupling between the degrees of freedom. The main effect of these imperfections on the behaviour of the trapped ions are a shift of the eigenfrequencies and reduction in the storage capability of the trap, as will be discussed in the following sections.

3.4 Shift of the Eigenfrequencies

3.4.1 Electric Field Imperfections

As in the case of the previously discussed Paul trap, imperfections in the electrostatic field can be treated by a multipole expansion of the cylindrically symmetric potential, thus

$$\Phi = U_0 \sum_{n=2}^{\infty} c_n \left(\frac{\rho}{d}\right)^n P_n \left(\frac{z}{\rho}\right) , \qquad (3.38)$$

where $\rho = (x^2 + y^2 + z^2)^{1/2}$. The size of the shifts in the spectrum of the ions depends on the contribution of the higher order multipole perturbations in the series expansion (3.38) characterizing the field, and also on the ion oscillation amplitude. The shifts have been calculated by different authors [62,104–106], by solving the equations of motion in the perturbed potential. Under the assumption that the odd orders of the expansion (3.38) vanish because of symmetry about the mid-plane of the trap, which is rather easy to ensure in practice, the shifts for the different motions are given as functions of the coefficients c_n and the amplitudes R_+, R_-, R_z of the oscillations in the different modes.

We consider the Hamiltonian

$$H' = H + V , \qquad (3.39)$$

where the anharmonic perturbation is given by

$$V = QU_0 \sum_{n=2}^{\infty} \frac{c_{2n}}{d^{2n}} H_{2n}(r,z) , \qquad (3.40)$$

and the polynomials $H_{2n}(r,z) = \rho^{2n} P_{2n}(\cos\theta)$ are given by (C.8):

$$H_{2n}(r,z) = (2n)! \sum_{k=0}^{n} \frac{(-4)^{-k} r^{2k} z^{2n-2k}}{(2n-2k)!(k!)^2} . \qquad (3.41)$$

The angle average of a function F in action-angle variables is defined by

$$\bar{F}(J) = \frac{1}{(2\pi)^3} \int_0^{2\pi}\int_0^{2\pi}\int_0^{2\pi} F(J,\varphi)\, d\varphi_+\, d\varphi_-\, d\varphi_z . \qquad (3.42)$$

According to the principle on averaging [107], the frequency shifts can be approximated from

$$\overline{H}' = \omega_+ J_+ - \omega_- J_- + \omega_z J_z + \overline{V} , \quad (3.43)$$

by

$$\Delta\omega_\pm = \pm\frac{\partial \overline{V}}{\partial J_\pm} , \quad \Delta\omega_z = \frac{\partial \overline{V}}{\partial J_z} . \quad (3.44)$$

Using (3.30)–(3.34), and (3.10) we find

$$r^2 = \frac{2}{M\omega_1}\left[J_+^2 + J_-^2 + 2\sqrt{J_+ J_-}\cos(\varphi_+ - \varphi_-)\right] , \quad (3.45)$$

$$z^2 = \frac{2}{M\omega_z} J_z \cos^2\varphi_z . \quad (3.46)$$

By (3.46), we have:

$$z^{2m} = \left(\frac{1}{2M\omega_z}J_z\right)^m \sum_{j=0}^{2m} \frac{(2m)!\exp[i(2m-2j)\varphi_z]}{j!(2m-j)!} , \quad (3.47)$$

$$\overline{z^{2(n-k)}} = \left(\frac{1}{2M\omega_z}J_z\right)^{n-k} \frac{(2n-2k)!}{[(n-k)!]^2} . \quad (3.48)$$

By (3.45), we find:

$$r^{2k} = \left(\frac{2}{M\omega_1}\right)^k \sum_{p=0}^{k} \frac{k!}{(k-p)!}(J_+ + J_-)^{k-p}(J_+ J_-)^{p/2} S_p , \quad (3.49)$$

$$S_p = \sum_{j=0}^{p} \frac{1}{j!(p-j)!}\exp[i(p-2j)\varphi] . \quad (3.50)$$

Using (3.49), (3.50), and the relations $\overline{S_p} = 0$ for p odd and $\overline{S_{2q}} = (q!)^{-2}$ for p even, $p = 2q$, we obtain

$$\overline{r^{2k}} = \left(\frac{2}{M\omega_1}\right)^k \sum_{q=0}^{[k/2]} \frac{k!}{(k-2q)![(q)!]^2}(J_+ + J_-)^{k-2q}(J_+ J_-)^q ,$$

$$= \left(\frac{2}{M\omega_1}\right)^k \sum_{q=0}^{[k/2]} \sum_{q=0}^{p} \frac{k!}{[(q)!]^2} \frac{1}{(j-q)!(k-q-j)!} J_+^{k-j} J_-^j . \quad (3.51)$$

By (3.51) and the identity

$$\sum_{q} \frac{j!(k-j)!}{(j-q)!(k-j-q)![(q)!]^2} = \frac{k!}{j!(k-j)!} , \quad (3.52)$$

3.4 Shift of the Eigenfrequencies

we have

$$\overline{r^{2k}} = \left(\frac{2}{M\omega_1}\right)^k \sum_{j=0}^{k} \frac{(k!)^2}{[j!(k-j)!]^2} J_+^{k-j} J_-^j \ . \tag{3.53}$$

Then we obtain a compact explicit form for the even part of (3.38):

$$\overline{V} = \sum_{n=2}^{\infty} \overline{V}_{2n} \ , \tag{3.54}$$

$$\overline{V}_{2n} = \frac{c_{2n}(2n)!}{4(2Md^2)^{n-1}} \sum_{k=0}^{n} \sum_{j=0}^{k} \frac{(-1)^k J_+^{k-j} J_-^j J_z^{n-k}}{\omega_1^k \omega_z^{n-k-2} [j!(k-j)!(n-k)!]^2} \ .$$

According to (3.44), (3.54) and (3.36), we obtain [108]

$$\Delta^{(2n)}\omega_\pm = \pm \frac{\partial \overline{V}_{2n}}{\partial J_\pm} \ , \quad \Delta^{(2n)}\omega_z = \frac{\partial \overline{V}_{2n}}{\partial J_z} \ , \tag{3.55}$$

$$\Delta^{(2n)}\omega_\pm = \pm \frac{c_{2n}(2n)!\omega_z^2}{4^n d^{2n-2}\omega_1} \sum_{k=1}^{n} \sum_{j=1}^{k} \frac{(-1)^k R_\pm^{2(j-1)} R_\mp^{2(k-j)} R_z^{2(n-k)}}{j[(j-1)!(k-j)!(n-k)!]^2} \ , \tag{3.56}$$

$$\Delta^{(2n)}\omega_z = \frac{c_{2n}(2n)!\omega_z}{4^n d^{2n-2}} \sum_{k=0}^{n-1} \sum_{j=0}^{k} \frac{(-1)^k R_+^{2(k-j)} R_-^{2j} R_z^{2(n-k-1)}}{(n-k)[j!(k-j)!(n-k-1)!]^2} \ . \tag{3.57}$$

Restriction to the octupole term with strength c_4 in the series expansion gives the results:

$$\Delta^{(4)}\omega_\pm = \pm \frac{3c_4\omega_z^2}{4d^2\omega_1}\left(R_\pm^2 + 2R_\mp^2 - 2R_z^2\right) \ , \tag{3.58}$$

$$\Delta^{(4)}\omega_z = \frac{3c_4\omega_z}{4d^2}\left(R_z^2 - 2R_+^2 - 2R_-^2\right) \ . \tag{3.59}$$

The next order component is the dodecapole c_6 term with the shifts

$$\Delta^{(6)}\omega_\pm = \mp \frac{15c_6\omega_z^2}{16d^4\omega_1}\left[R_\pm^4 + 6R_+^2 R_-^2 + 3R_\mp^4 - 6R_z^2(R_\pm^2 + 2R_\mp^2) + 3R_z^4\right] \ , \tag{3.60}$$

$$\Delta^{(6)}\omega_z = \frac{15c_6\omega_z}{16d^4}\left[3(R_+^4 + 4R_+^2 R_-^2 + R_-^4) - 6R_z^2(R_+^2 + R_-^2) + R_z^4\right] \ . \tag{3.61}$$

The hexadecapole c_8 contribution is given by

$$\Delta^{(8)}\omega_\pm = \pm \frac{35c_8\omega_z^2}{32d^6\omega_1}\left[R_\pm^6 + 18R_\pm^4 R_\mp^2 + 18R_\pm^2 R_\mp^4 + 4R_\mp^6 - 4R_z^6 \right. \tag{3.62}$$
$$\left. + 18R_z^4(R_\pm^2 + 2R_\mp^2) + 12R_z^2(R_\pm^4 + 6R_+^2 R_-^2 + 3R_\mp^4)\right] \ ,$$

$$\Delta^{(8)}\omega_z = \frac{35c_8\omega_z}{32d^6}\left[4(R_\pm^6 + 9R_\pm^4 R_\mp^2 + 9R_\pm^2 R_\mp^4 + R_\mp^6) \right. \tag{3.63}$$
$$\left. + 12R_z^4(R_\pm^2 + R_\mp^2) - 18R_z^2(R_\pm^4 + 4R_+^2 R_-^2 + R_\mp^4) - R_z^6\right] \ .$$

Of particular importance is the sideband $\omega_c = \omega_+ + \omega_-$ because it is most often used in high precision mass spectrometry using Penning traps. For such applications where accuracy is paramount, a correction characterized by c_4, c_6 and c_8 is needed. We have

$$\Delta\omega_c = \frac{3\omega_z^2}{4d^2\omega_1}(R_-^2 - R_+^2)\left[c_4 + \frac{5c_6}{2d^2}(R_+^2 + R_-^2 - 3R_z^2)\right. \tag{3.64}$$
$$\left. + \frac{35c_8}{8d^4}\left(R_+^4 + 3R_+^2 R_-^2 + R_-^4 - 8R_z^2 R_+^2 - 8R_z^2 R_-^2 + 6R_z^4\right)\right].$$

3.4.2 Magnetic Field Inhomogeneities

An inhomogeneity in the superimposed magnetic field of the Penning trap also leads to a shift of the field-dependent frequencies.

Assume
$$\boldsymbol{A'} = \frac{1}{2}(1+\Lambda)\,\boldsymbol{B}\times\boldsymbol{\rho}\,, \tag{3.65}$$

where $\boldsymbol{\rho} = (x, y, z)$ and $\boldsymbol{B} = (0, 0, B_0)$. The perturbation Λ describes magnetic field inhomogeneities. The magnetic field is given by

$$\boldsymbol{B'} = \boldsymbol{\nabla}\times\boldsymbol{A'} = B_0\left(-\frac{x}{2}\frac{\partial\Lambda}{\partial z},\ -\frac{y}{2}\frac{\partial\Lambda}{\partial z},\ 1+\Lambda+\frac{1}{2}r\frac{\partial\Lambda}{\partial r}\right). \tag{3.66}$$

The Hamiltonian can be written as

$$H' = \frac{1}{2M}(\boldsymbol{p} - Q\boldsymbol{A'})^2 + Q\Phi = H + W\,, \tag{3.67}$$

where H is given by (3.25) and the perturbation is

$$W = W_1 + W_2\,,\quad W_1 = \frac{1}{4}\Lambda\left(M\omega_c^2 r^2 - 2\omega_0 L_z\right)\,,\quad W_2 = \frac{1}{8}M\omega_c^2\Lambda^2\,. \tag{3.68}$$

As in the electric field case, a magnetostatic field in a current-free region can be represented by a harmonic function, which can be expanded in the following multipole series:

$$B = B_0\sum_{n=1}^{\infty} b_{2n}\left(\frac{\rho}{d}\right)^{2n} P_{2n}(\cos\theta) = B_0\left(\Lambda + \frac{1}{2}r\frac{\partial\Lambda}{\partial r}\right), \tag{3.69}$$

$$\Lambda = \sum_{n=1}^{\infty} \frac{b_{2n}}{d^{2n}} K_{2n}(r, z)\,, \tag{3.70}$$

$$K_{2n}(r, z) = (2n)!\sum_{k=0}^{n}\frac{(-4)^{-k} r^{2k} z^{2n-2k}}{k!(k+1)!(2n-2k)!}.$$

Using (3.48), (3.51), (3.68), (3.71), and the relation $L_z = (J_- - J_+)Q/|Q|$, we have

$$\overline{W_1} = M\omega_c \sum_{n=1}^{\infty} \frac{b_{2n}(2n)!}{2^{2n+1}d^{2n}} \overline{W_1}^{(2n)} , \qquad (3.71)$$

$$\overline{W_1}^{(2n)} = \sum_{k=0}^{n} \frac{(-1)^k R_z^{2n-2k} \left[2\left(\omega_+ R_+^{2k+2} + \omega_- R_-^{2k+2}\right) + S^{(2n,k)} \right]}{k!(k+1)![(n-k)!]^2} ,$$

$$S^{(2n,k)} = \sum_{m=1}^{k} [(k+1)\omega_+ - m\omega_1] \frac{(k+1)[k!]^2 R_+^{2k-2m+2} R_-^{2m}}{[m!(k+1-m)!]^2} .$$

The frequency shifts are given by

$$\Delta^{(2n)}\omega_\pm = \pm \frac{\partial \overline{W_1}}{\partial J_\pm}, \qquad \Delta^{(2n)}\omega_z = \frac{\partial \overline{W_1}}{\partial J_z}. \qquad (3.72)$$

Using (3.72) and (3.72), we obtain [108]

$$\Delta\omega_z^{(2n)} = \frac{2}{M\omega_z} \sum_{k=0}^{n-1} \frac{(-1)^k R_z^{2n-2k-2} \left[2\left(\omega_+ R_+^{2k+2} + \omega_- R_-^{2k+2}\right) + \overline{W_1}^{(2n,k)} \right]}{k!(k+1)!(n-k-1)!(n-k)!} , \qquad (3.73)$$

$$\Delta^{(2n)}\omega_\pm = \pm \frac{4}{M\omega_1} \sum_{k=0}^{n} \frac{(-1)^k R_z^{2n-2k} \left[2(k+1)\omega_\pm R_\pm^{2k} + \overline{W}_\pm^{(2n,k)} \right]}{k!(k+1)![(n-k)!]^2} ,$$

$$\overline{W}_\pm^{(2n,k)} = \sum_{m=0}^{k-1} [(k+1)\omega_+ - (m+1)\omega_1] \frac{k!(k+1) R_+^{2k-2m} R_-^{2m}}{m!(m+1)![(k-m)!]^2} . \qquad (3.74)$$

In the case of mirror symmetry in the trap's mid-plane, the odd coefficients vanish. If we retain only the b_2 and b_4 terms in the expansion, the frequency shifts using (3.72) are as follows:

$$\Delta^{(2)}\omega_z = \frac{\omega_c b_2}{2\omega_1 d^2} \left(\omega_+ R_+^2 + \omega_- R_-^2\right) , \qquad (3.75)$$

$$\Delta^{(2)}\omega_\pm = \frac{\omega_c b_2}{2\omega_1 d^2} \left(\omega_\pm R_z^2 - \omega_\pm R_\pm^2 - \omega_c R_\mp^2\right) , \qquad (3.76)$$

$$\Delta^{(4)}\omega_z = \frac{3\omega_c b_4}{4\omega_z d^4} \left[\left(\omega_+ R_+^2 + \omega_- R_-^2\right) R_z^2 - \omega_+ R_+^4 - \omega_- R_-^4 - 2\omega_c R_+^2 R_-^2 \right] , \qquad (3.77)$$

$$\Delta^{(4)}\omega_\pm = \frac{3\omega_c b_4}{16\omega_1 d^4} \left[\omega_\pm R_\pm^4 + (\omega_\mp + \omega_c) R_\mp^4 + 2\left(\omega_\pm + \omega_c\right) R_+^2 R_-^2 \right]$$

$$- \frac{3\omega_c b_4}{16\omega_1 d^4} R_z^2 \left(4\omega_\pm R_\pm^2 + \omega_c R_\mp^2 - \omega_\pm R_z^2 \right) . \qquad (3.78)$$

Using (3.76) and (3.78), we obtain the cyclotron frequency shift

$$\begin{aligned}\Delta\omega_c &= \Delta^{(2)}\omega_+ + \Delta^{(2)}\omega_- + \Delta^{(4)}\omega_+ + \Delta^{(4)}\omega_- \\ &= \frac{\omega_c b_2}{2\omega_1 d^2}\left(\omega_1 R_z^2 + \omega_- R_+^2 - \omega_+ R_-^2\right) \\ &\quad + \frac{3\omega_c b_4}{16\omega_1 d^4}\left[2\omega_1 R_+^2 R_-^2 - \omega_c R_-^4 - \omega_c R_+^4\right] \\ &\quad + \frac{3\omega_c b_4}{16\omega_1 d^4} R_z^2 \left[4\omega_- R_+^2 - 4\omega_+ R_-^2 + \omega_1 R_z^2\right] .\end{aligned} \quad (3.79)$$

3.4.3 Distortions and Misalignments

If the ring electrode of the trap shows a slight ellipticity, the potential remains harmonic as in the case of the ideal trap, however with different strength in the radial coordinates x and y. In the principal axis coordinate system the potential energy thus has the form [104]

$$W = \frac{1}{4}M\omega_z^2[2z^2 - x^2 - y^2 - \varepsilon(x^2 - y^2)] , \quad (3.80)$$

where ω_z is defined by (3.7) and ε is an ellipticity parameter.

In addition, assume the magnetic field is misaligned with respect to the z-axis, thus

$$\mathbf{B} = B_0\left(\sin\theta\cos\varphi, \sin\theta\sin\varphi, \cos\theta\right) , \quad (3.81)$$

where θ and φ are the spherical polar coordinates, and $0 < \theta < \pi$. Using (3.2), (3.80), and (3.81), we obtain Newton's equations of motion for a particle of mass M and charge Q confined in a general quadrupole Penning trap:

$$\frac{d^2x}{dt^2} = \omega_0\left(\cos\theta\frac{dy}{dt} - \sin\theta\sin\varphi\frac{dz}{dt}\right) + \frac{1}{2}\omega_z^2(1+\varepsilon)x , \quad (3.82)$$

$$\frac{d^2y}{dt^2} = \omega_0\left(-\cos\theta\frac{dx}{dt} + \sin\theta\cos\varphi\frac{dz}{dt}\right) + \frac{1}{2}\omega_z^2(1-\varepsilon)y , \quad (3.83)$$

$$\frac{d^2z}{dt^2} = \omega_0\sin\theta\left(\sin\varphi\frac{dx}{dt} - \cos\varphi\frac{dy}{dt}\right) - \omega_z^2 z , \quad (3.84)$$

where $\omega_0 = QB_0/M$.

Putting $x = a\exp(-i\omega t)$, $y = b\exp(-i\omega t)$, and $z = c\exp(-i\omega t)$, we obtain a linear system of three homogeneous equations in a, b, and c. There exists nontrivial solutions only if the determinant of the system is zero:

$$\begin{vmatrix} \omega^2 + \frac{1}{2}\omega_z^2(1+\varepsilon) & -i\omega\omega_0\cos\theta & i\omega\omega_0\sin\theta\sin\varphi \\ i\omega\omega_0\cos\theta & \omega^2 + \frac{1}{2}\omega_z^2(1-\varepsilon) & -i\omega\omega_0\sin\theta\cos\varphi \\ -i\omega\omega_0\sin\theta\sin\varphi & i\omega\omega_0\sin\theta\cos\varphi & \omega^2 - \omega_z^2 \end{vmatrix} = 0 . \quad (3.85)$$

3.4 Shift of the Eigenfrequencies

The characteristic equation (3.85) can be written as

$$\omega^6 - \omega_c^2 \omega^4 + \omega_z^2 \left[\omega_c^2(1-\eta) - \frac{1}{4}\omega_z^2(3+\varepsilon^2)\right]\omega^2 - \frac{1}{4}\omega_z^6(1-\varepsilon^2) = 0 , \quad (3.86)$$

where the misalignment parameter η is defined by

$$\eta = \frac{1}{2}\sin^2\theta(3+\varepsilon\cos 2\varphi) . \quad (3.87)$$

It is convenient to introduce the parameters β and λ:

$$\beta = \frac{\omega_z^2}{2\omega_c^2}, \quad \lambda = 2\beta(1-\eta) - \beta^2(3+\varepsilon^2) . \quad (3.88)$$

Replacing ω in (3.86) by the new unknown

$$u = \omega^2 - \frac{1}{3} , \quad (3.89)$$

we obtain the canonical cubic equation

$$u^3 + pu + q = 0 , \quad (3.90)$$

where the parameters p and q are given by

$$p = \lambda - \frac{1}{3}, \quad q = \frac{p}{3} - 2\beta^3(1-\varepsilon^2) + \frac{1}{27} . \quad (3.91)$$

All three roots of (3.90) are real and distinct when the discriminant $D = 27q^2 + 4p^3$ is negative. The motional eigenfrequencies given by (3.86) are denoted by $\pm\omega'_+$, $\pm\omega'_-$, and $\pm\omega'_z$. The prime denotes the frequencies under the influence of the perturbations. Some important relations between the perturbed motional frequencies exist [109]. The roots of the polynomial equation (3.86) are related to its coefficients by Viète's formulas:

$$\omega'^2_+ + \omega'^2_- + \omega'^2_z = \omega_c^2 , \quad (3.92)$$

$$\omega'^2_+ \omega'^2_- + \omega'^2_+ \omega'^2_z + \omega'^2_+ \omega'^2_z = \omega_c^4 \lambda , \quad (3.93)$$

$$\omega'^2_+ \omega'^2_- \omega'^2_z = \frac{1}{4}\omega_z^6(1-\varepsilon^2) . \quad (3.94)$$

The solutions of (3.82)–(3.84) are stable when the roots of (3.86) are real and distinct. According to (3.93) and (3.94), the stability conditions are [110]

$$27q^2 + 4p^3 < 0 , \quad \lambda > 0 , \quad |\varepsilon| < 1 . \quad (3.95)$$

The invariance theorem (3.92) makes possible high precision determination of the free ion cyclotron frequency in a Penning trap, since harmonic trap distortions and magnetic field misalignments are experimentally inevitable, and their cancellation to first order is therefore of primary importance. Note that (3.92) is independent of η and ε.

If $\eta = 0$, then the positive roots of (3.86) are the eigenfrequencies in the absence of misalignment, defined as $\omega'_\pm = \tilde{\omega}_\pm$ and $\tilde{\omega}_z$, where

$$\tilde{\omega}_\pm = \frac{1}{\sqrt{2}} \left(\omega_c^2 - \omega_z^2 \pm \sqrt{\omega_c^4 - 2\omega_c^2\omega_z^2 + \varepsilon^2\omega_z^4} \right)^{1/2}, \quad \tilde{\omega}_z = \omega_z . \tag{3.96}$$

According to condition (3.90) for $\eta \neq 0$, the general roots of (3.86) cannot be expressed in terms of the coefficients by means of radicals of real numbers. The eigenfrequencies have been calculated by Brown and Gabrielse [109] by treating it as an eigenvalue problem for small η and independently by Kretzschmar [62] using perturbation theory.

Introducing $w = \omega^2/\omega_c^2$ in (3.86) we obtain the cubic equation

$$w^3 - w^2 + \lambda w - 2\beta^3(1 - \varepsilon^2) = 0 . \tag{3.97}$$

Using (3.96) and (3.97), we find

$$\omega'_\pm = \omega_c \sqrt{w_\pm}, \quad \omega'_z = \omega_c \sqrt{w_z}, \tag{3.98}$$

$$\tilde{\omega}_\pm = \omega_c \sqrt{\tilde{w}_\pm}, \quad \tilde{\omega}_z = \omega_c \sqrt{\tilde{w}_z}, \tag{3.99}$$

where w_+, w_-, and w_z are the roots of (3.97) and

$$\tilde{w}_\pm = \frac{1}{2} \left[1 - 2\beta \pm \sqrt{1 - 4\beta + 4\varepsilon^2\beta} \right], \quad \tilde{w}_z = 2\beta . \tag{3.100}$$

If η is small, we consider the power-series expansions:

$$w_\sigma = \sum_{n=0}^{\infty} w_{\sigma,n} \xi_\sigma^n , \tag{3.101}$$

where $\sigma = +, -, z$, and

$$\xi_\sigma = 2\beta\eta \left[3\tilde{w}_\sigma^2 - 2\tilde{w}_\sigma + 2\beta - \beta^2(3 + \varepsilon^2) \right]^{-1} . \tag{3.102}$$

Using (3.97), (3.101), and (3.102), we obtain the recursive relations

$$w_{\sigma,0} = \tilde{w}_\sigma , \tag{3.103}$$

$$w_{\sigma,n+1} = w_{\sigma,n} + (1 - 3w_{\sigma,0}) \sum_{p=1}^{n} w_{\sigma,p} w_{\sigma,n+1-p}$$

$$- \sum_{p=1}^{n-1} \sum_{q=1}^{n-p} w_{\sigma,p} w_{\sigma,q} w_{\sigma,n+1-p-q} . \tag{3.104}$$

The solution of (3.104) can be written as

$$w_\sigma = \tilde{w}_\sigma [1 + \xi_\sigma + (1 - \tilde{w}_\sigma)(1 + 3\tilde{w}_\sigma)\xi_\sigma^2 \\ + (1 + 3\tilde{w}_\sigma - 8\tilde{w}_\sigma^2 - 12\tilde{w}_\sigma^3 + 18\tilde{w}_\sigma^4)\xi_\sigma^3 + \cdots] . \quad (3.105)$$

According to (3.105), the analytical expressions for the frequency shifts are given by [110]

$$\omega_\sigma' - \tilde{\omega}_\sigma = \frac{1}{2}\tilde{\omega}_\sigma \xi_\sigma \left[1 + \frac{1}{4}\left(3 + 8\tilde{w}_\sigma - 12\tilde{w}_\sigma^2\right)\xi_\sigma \right. \quad (3.106)$$

$$\left. + \frac{1}{8}\left(5 + 16\tilde{w}_\sigma - 52\tilde{w}_\sigma^2 - 96\tilde{w}_\sigma^3 + 144\tilde{w}_\sigma^4\right)\xi_\sigma^2 + \cdots \right] ,$$

where $\tilde{\omega}_\sigma$, $\tilde{w}_\sigma$ and ξ_σ, are given by (3.96), (3.98), (3.102), respectively.

Then the eigenfrequencies for small field misalignments can be expressed as

$$\omega_\pm' - \tilde{\omega}_\pm = \frac{\omega_c \eta \beta \left(1 - 2\beta \pm \sqrt{1 - 4\beta + 4\beta^2 \varepsilon^2}\right)}{1 - 4\beta + 4\varepsilon^2 \beta^2 \pm (1 - 6\beta)\sqrt{1 - 4\beta + 4\beta^2 \varepsilon^2}} + O(\eta^2) , \quad (3.107)$$

$$\omega_z' - \omega_z = -\frac{\omega_z \eta}{2 - (9 - \varepsilon^2)\beta} + O(\eta^2) . \quad (3.108)$$

Using (3.107) and (3.108) for $\omega_c \gg \omega_z$ and $|\sin\theta| \ll 1$, we obtain the approximations

$$\omega_\pm' \approx \omega_\pm + \frac{1}{2}\omega_- \sin^2\theta (3 + \varepsilon\cos 2\varphi) ,$$

$$\omega_z' \approx \omega_z - \frac{1}{4}\omega_z \sin^2\theta (3 + \varepsilon\cos 2\varphi) . \quad (3.109)$$

3.4.4 Space Charge Shift

As discussed already for the Paul trap, the space charge potential of a trapped ion cloud shifts the eigenfrequencies of individual ions. From a simple model of a homogeneous charge distribution in the cloud, we are led to a decrease in the axial frequency proportional to the square root of the ion number.

More important for high precision spectroscopy is the space charge shift in the frequencies when we are dealing with small numbers of ions, where the simple model is no longer adequate. The observed shift depends on the mean inter-ion distance, and consequenly on the ion temperature. For protons cooled to 4K, van Dyck et al. [111] have observed shifts of opposite signs in the perturbed cyclotron and magnetron frequencies, amounting to 0.5 ppb per stored ion (Fig. 3.6). The shift increases with the charge of the ions. The electrostatic origin of the shifts suggests that the sum of the perturbed cyclotron and magnetron frequencies, which equals the free ion cyclotron frequency, should be independent of the ion number. In fact, the shifts observed in the sideband $\omega_+ + \omega_- = \omega_c$ by van Dyck are consistent with zero.

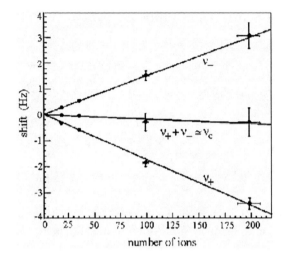

Fig. 3.6. The observed shifts in the perturbed cyclotron frequency ν_+, in the magnetron frequency ν_-, and in their sum approximately equal to the trap independent cyclotron frequency ν_c vs the number of trapped protons [111]. Copyright (2003) by the American Physical Society

3.4.5 Image Charges

An oscillating ion induces in the trap electrodes image charges, which create an electric field that reacts on the stored ion and shifts its motional frequencies. The shift has been calculated by van Dyck and coworkers [111] and by J.V. Porto [112], using a simple model. If the trap is replaced by a conducting spherical shell of radius a, the image charge creates an electric field given by:

$$\boldsymbol{E} = \frac{1}{4\pi\varepsilon_0} \frac{Qa}{(a^2 - r^2)^2} \boldsymbol{r} \ . \tag{3.110}$$

Since this field is added to the trap field, it causes a shift in the axial frequency amounting to

$$\Delta\omega_z = -\frac{1}{4\pi\varepsilon_0} \frac{Q^2}{2Ma^3\omega_z} \ . \tag{3.111}$$

It scales linearly with the ion number n, and is significant only for small trap sizes. Since it is of purely electrostatic origin, the electric field-independent combination frequency $(\omega_+ + \omega_-) = \omega_c$ is not affected.

3.5 Instabilities of the Ion Motion

Another consequence of the presence of higher order multipoles in the field is that ion orbit instability can occur at certain operating points, where it would otherwise be stable in a perfect quadrupole field. Using perturbation theory to solve the equations of motions when the trap potential is written as a series expansion in spherical harmonics (3.38), Kretzschmar [62] has shown that the solution exhibits singularities for operating points at which

$$n_+\omega_+ + n_-\omega_- + n_z\omega_z = 0 \; , \tag{3.112}$$

where n_+, n_-, n_z are integers. The ion trajectory at such points is unstable and the ion is lost from the trap. Indications of such instabilities have been

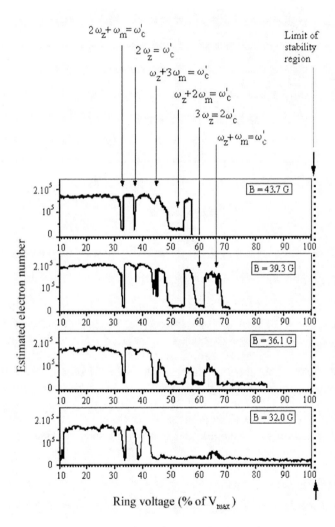

Fig. 3.7. Observed instabilities of an electron cloud vs. the trapping potential for different values of the magnetic field [113]

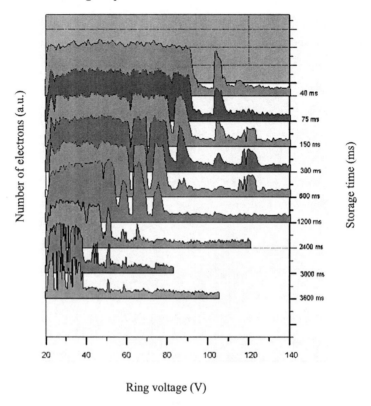

Fig. 3.8. Observed number of trapped electrons vs. trapping potential at different storage times. The shortest available observation time in the given experimental setup was 40 ms [113]

obtained by Yu et al., [68], and Schweikhard et al., [114]. In Fig. 3.7 the results are shown of measurements obtained for different trap voltages and various values of the magnetic field, on a cloud of trapped electrons in which the combined effect of space charge and trap imperfections leads to the loss of the particles. The observed instabilities which become more discernible for extended storage time (Fig. 3.8), can be assigned to operating conditions predicted from (3.112) for $n_+ = 1$. These results show that in practice it is very difficult to obtain stable operating conditions when the trapping voltage exceeds about half the value allowed by the stability criterion (3.12).

3.6 Tuning the Trap

The size of the amplitudes c_n of higher order perturbing potentials has been calculated by Beaty [115] for various trap designs or truncations of the ideal hyperbolic electrodes. The values for the most important contributions, c_4

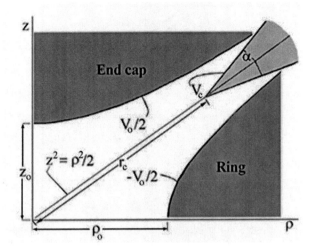

Fig. 3.9. Model used for calculating the electrostatic potential in an asymptotically symmetric hyperbolic Penning trap, with $\alpha = 30°$ and $r_c/d \simeq 2.2$; endcap, ring, and compensation (guard) electrodes are at $V_0/2$, $-V_0/2$, and V_c, respectively [116]. Copyright (2003) by the American Physical Society

and c_6, are typically of the order 10^{-3}. They can be made much smaller by potentials applied to additional compensation electrodes placed between the ring and end cap electrodes. Calculations on the influence of the shape and position of those electrodes on the magnitudes of the higher order multipoles have been carried out by Gabrielse [116].

Introducing the tunability T_n of the trap defined as

$$T_n = U_0 \frac{dc_n}{dV_c} \; , \qquad (3.113)$$

one can calculate the dependence of this dimensionless quantity on the radius and the shape of the compensation electrode as defined in [116] (Fig. 3.9). In Fig. 3.10 is shown the calculated tunability T_2 for a flat correction electrode ($\alpha = 180°$) and for $\alpha = 30°$, as function of the ratio of the distance r_c of the compensation electrode to the characteristic trap dimension d. The tunabilities of c_4, c_6 and c_8 which are also functions of r_c/d have ratios to T_2 which become constant when r_c/d is larger than 2, independently of α. The most conveniently tunable trap would show no change in the quadrupole part of the trapping potential as the voltage on the compensation electrode is "tuned" to reduce the higher orders. This "orthogonality" condition, calculations show, holds for a trap characterized by

$$r_0 = 1.16 z_0 \; , \quad r_0/d > 2 \; , \qquad (3.114)$$

rather than the generally adopted asymptotically symmetric proportion ($r_0 = \sqrt{2} z_0$).

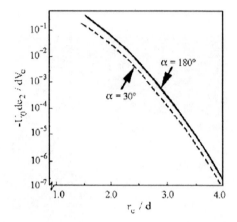

Fig. 3.10. Tunability $U_0 dc_2/dV_c$ for asymptotically symmetric, compensated Penning traps [116]. Copyright (2003) by the American Physical Society

3.7 Quantum Dynamics in Ideal Penning Trap

3.7.1 Spinless Particle Dynamics

We consider the nonrelativistic quantum Hamiltonian of a spinless particle of mass M and electric charge Q confined in a time-independent electromagnetic field

$$H = \frac{1}{2M}(\boldsymbol{p} - Q\boldsymbol{A})^2 + Q\Phi \,, \tag{3.115}$$

where $\boldsymbol{p} = -i\hbar\boldsymbol{\nabla} = (p_x, p_y, p_z)$ is the momentum operator, $\boldsymbol{A}$ is the vector potential and Φ is the electric potential.

For an ideal Penning trap characterized by a static homogeneous magnetic field $\boldsymbol{B} = (0, 0, B_0)$ directed along the positive z-axis and an axially symmetric quadrupole electric potential Φ, we have

$$\boldsymbol{A} = \frac{1}{2}\boldsymbol{B} \times \boldsymbol{\rho}, \quad \Phi = \frac{U_0}{2d^2}\left(2z^2 - x^2 - y^2\right) \,, \tag{3.116}$$

where $\boldsymbol{\rho} = (x, y, z)$ is the position operator. Then the Hamiltonian (3.115) can be written as [117]:

$$H_0 = H_{xy} + H_z \,, \tag{3.117}$$

$$H_{xy} = \frac{1}{2M}\left(p_x^2 + p_y^2\right) + \frac{1}{8}M\omega_1^2 r^2 - \frac{1}{2}\omega_0 L_z \,, \tag{3.118}$$

$$H_z = \frac{1}{2M}p_z^2 + \frac{1}{2}M\omega_z^2 z^2 \,, \tag{3.119}$$

where $L_z = xp_y - yp_x$ is the axial component of the angular momentum operator and $r = (x^2 + y^2)^{1/2}$. The frequencies ω_0, ω_1, and ω_z, are as defined in (3.7) and (3.10), where $\omega_c = |\omega_0|$ is the cyclotron frequency:

3.7 Quantum Dynamics in Ideal Penning Trap

$$\omega_z = \sqrt{\frac{2QU_0}{Md^2}}, \quad \omega_1 = \sqrt{\omega_c^2 - 2\omega_z^2}, \quad \omega_0 = \frac{QB_0}{M}. \tag{3.120}$$

Stable confinement requires $QU_0 > 0$ and $\omega_c > \sqrt{2}\omega_z$. According to (3.118) and (3.119), H_z is the Hamiltonian of a quantum one-dimensional harmonic oscillator and H_{xy} is the Hamiltonian of a quantum two-dimensional isotropic harmonic oscillator, plus a term proportional to L_z.

The analytical solution of the Schrödinger equation for (3.118) has been obtained by Fock [118] and Darwin [119]. Explicit expressions for eigenstates, coherent states and propagators have been given in [86,120,121]. Applications of the spectral properties of the Hamiltonian (3.117) to the Penning trap have been presented in [85, 103, 104, 117, 122].

In order to obtain an analytical expression for the wave functions in the coordinate representation, it is convenient to use the cylindrical coordinates (r, θ, z) with $x = r\cos\theta$ and $y = r\sin\theta$. The explicit form of the eigenfunctions can be obtained by the separation of variables [86], thus

$$\Psi = r^{-1/2} R(r) A(z) \exp(im\theta), \tag{3.121}$$

with $L_z\Psi = m\hbar\Psi$, where the quantum number m takes integer values. This leads to the stationary Schrödinger equations for the axial and radial directions:

$$-\frac{\hbar^2}{2M}\frac{d^2 A}{dz^2} + \frac{1}{2}M\omega_z^2 z^2 A = E'A, \tag{3.122}$$

$$-\frac{\hbar^2}{2M}\frac{d^2 R}{dr^2} + \frac{\hbar^2}{2Mr^2}\left(m^2 - \frac{1}{4}\right)R + \frac{1}{8}M\omega_1^2 r^2 R = E''R, \tag{3.123}$$

$$E'' = E + \frac{1}{2}m\hbar\omega_0 - E'. \tag{3.124}$$

It is convenient to introduce the axial and radial dimensionless coordinates η and ξ:

$$\eta = \frac{z}{z_1}, \quad \xi = \frac{r}{r_1},$$

$$z_1 = \sqrt{\frac{\hbar}{M\omega_z}}, \quad r_1 = \sqrt{\frac{2\hbar}{M\omega_1}}. \tag{3.125}$$

Then (3.122) is reduced to the standard equation of a quantum harmonic oscillator

$$-\frac{1}{2}\frac{d^2 A}{d\xi^2} + \frac{1}{2}\xi^2 A = \frac{E'}{\hbar\omega_z}A, \tag{3.126}$$

with the solution

$$A = \left(\sqrt{\pi}2^k k!\right)^{-1/2} H_k(\xi) \exp\left(-\frac{1}{2}\xi^2\right), \quad E' = \left(k + \frac{1}{2}\right)\hbar\omega_z, \tag{3.127}$$

where l is a nonnegative integer and H_l is the Hermite polynomial. Moreover, (3.123) is reduced to the standard differential equation

$$-\frac{1}{2}\frac{d^2 R}{d\eta^2} + \frac{1}{2}\eta^2 R + \left(m^2 - \frac{1}{4}\right)\frac{1}{\eta^2}R = \frac{2E''}{\hbar\omega_1}R, \qquad (3.128)$$

with the solution

$$R = \sqrt{2\frac{j!}{(j+|m|)!}}\,\eta^{|m|+1/2}L_j^{|m|}(\eta^2)\exp\left(-\frac{1}{2}\eta^2\right),$$

$$E'' = \frac{1}{2}\hbar\omega_1(2j + |m| + 1), \qquad (3.129)$$

where j and $m = |m'|$ are nonnegative integers, and $L_j^{|m|}$ the associated Laguerre polynomial. Complete solution of the stationary Schrödinger equation $H_0\Phi_{jml} = E_{jml}\Phi_{jml}$ can be written as

$$E_{nlk} = \frac{1}{2}\hbar\omega_1(2j + |m| + 1) + \hbar\omega_z\left(k + \frac{1}{2}\right) \qquad (3.130)$$

$$= \hbar\omega_+\left(n + \frac{1}{2}\right) - \hbar\omega_-\left(l + \frac{1}{2}\right) + \hbar\omega_z\left(k + \frac{1}{2}\right),$$

$$\Psi_{nlk} = \frac{1}{\pi^{3/4}r_1 z_1^{1/2}}\left(\frac{j!}{2^k k!\,(j+|m|)!}\right)^{1/2} \qquad (3.131)$$

$$\times \left(\frac{r}{r_1}\right)^{|m|} L_j^{|m|}\left(\frac{r^2}{r_1^2}\right) H_k\left(\frac{z}{z_1}\right) \exp\left(-\frac{r^2}{2r_1^2} - \frac{z^2}{2z_1^2} + im\theta\right),$$

where n, l, and k are the modified cyclotron quantum number, the magnetron quantum number, and the axial quantum number, respectively. The modified cyclotron frequency ω_+ and the magnetron frequency ω_- are defined by $\omega_\pm = (\omega_c \pm \omega_1)/2$. The quantum numbers satisfy the relations

$$n = j, \quad l = j + m \quad \text{for} \quad \omega_0 m > 0,$$
$$n = j + m, \quad l = j \quad \text{for} \quad \omega_0 m < 0. \qquad (3.132)$$

The negative sign in (3.130) reveals the inverse energy of the magnetron oscillator.

The complete orthonormal system of eigenstate vectors of (3.117) are those of the well known harmonic oscillators. A level diagram of the energy states is given in Fig. 3.11. The probability distribution $|\Phi_{nlk}|^2$ for some low lying energy states are given in Fig. 3.12.

3.7 Quantum Dynamics in Ideal Penning Trap 75

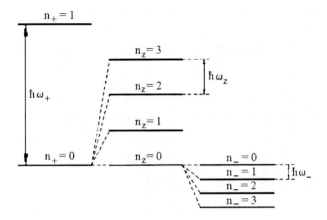

Fig. 3.11. Energy diagram of harmonic oscillator levels of a spinless particle in an ideal Penning trap

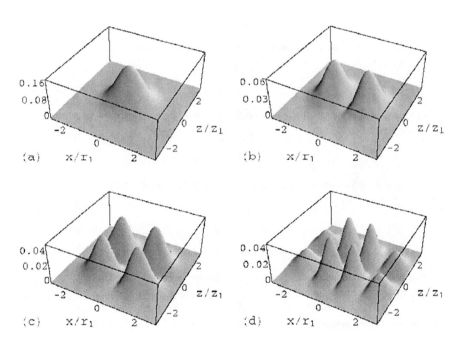

Fig. 3.12. The probability distributions $r_1^2, z_1 |\Phi_{jml}|^2$ of a charged particle in the ideal Penning trap for $0 \leq r \leq 3r_1$ and $0 \leq |z| \leq 3z_1$. **(a)** $j = m = l = 0$. **(b)** $j = l = 1$ and $m = 0$. **(c)** $m = 1$ and $j = l = 0$. **(d)** $j = m = l = 1$

We consider the following creation and annihilation operators:

$$a_z = \frac{1}{\sqrt{2}z_1}\left(z + \frac{i}{M\omega_z}p_z\right) , \quad a_z^\dagger = \frac{1}{\sqrt{2}z_1}\left(z - \frac{i}{M\omega_z}p_z\right) , \quad (3.133)$$

$$a_\pm = \frac{1}{2r_1}\left[x \pm i\frac{Q}{|Q|}y + \frac{2i}{M\omega_1}\left(p_x \pm i\frac{Q}{|Q|}p_y\right)\right] , \quad (3.134)$$

$$a_\pm^\dagger = \frac{1}{2r_1}\left[x \mp i\frac{Q}{|Q|}y - \frac{2i}{M\omega_1}\left(p_x \mp i\frac{Q}{|Q|}p_y\right)\right] . \quad (3.135)$$

The operators (3.133)–(3.135) commute with each other except for the cases

$$\left[a_+, a_+^\dagger\right] = \left[a_-, a_-^\dagger\right] = [a_z, a_z^\dagger] = 1 . \quad (3.136)$$

By (3.133)–(3.135), we have

$$H_0 = \hbar\omega_+\left(a_+^\dagger a_+ + \frac{1}{2}\right) - \hbar\omega_-\left(a_-^\dagger a_- + \frac{1}{2}\right) + \hbar\omega_z\left(a_z^\dagger a_z + \frac{1}{2}\right) , \quad (3.137)$$

$$L_z = \frac{Q}{|Q|}\hbar\left(a_-^\dagger a_- - a_+^\dagger a_+\right) . \quad (3.138)$$

The complete orthonormal system of eigenstate vectors of (3.137) are those of the well known harmonic oscillators

$$\Psi_{nlk} = (n!l!k!)^{-1/2}(a_+^\dagger)^n(a_-^\dagger)^l(a_z^\dagger)^k \Psi_0 , \quad (3.139)$$

where n, l, and k are nonnegative integers and the normalized vacuum vector Ψ_0 is characterized by

$$a_+\Psi_0 = a_-\Psi_0 = a_z\Psi_0 = 0 . \quad (3.140)$$

Then Ψ_{nlk} is a simultaneous eigenvector of the commuting operators $a_+^\dagger a_+$, $a_-^\dagger a_-$, $a_z^\dagger a_z$, H, and L_z:

$$\Psi_{nlk} = a_+^\dagger a_+ \Psi_{nlk} , \quad a_-^\dagger a_- \Psi_{nlk} = k\Psi_{nlk} , \quad a_z^\dagger a_z \Psi_{nlk} = l\Psi_{nlk} ,$$
$$H\Psi_{nlk} = E_{nlk}\Psi_{nlk} , \quad L_z\Psi_{nlk} = m\hbar\Psi_{nlk} , \quad (3.141)$$

where the quantum number m is given by $m = (n-l)Q/|Q|$.

In order to find the relation between the quantum motion solutions and the classical trajectories we work in the Heisenberg representation. The evolution equations are

$$\frac{d^2\hat{x}}{dt} - \omega_c\frac{Q}{|Q|}\frac{d\hat{y}}{dt} - \frac{1}{2}\omega_z^2\hat{x} = 0 ,$$

$$\frac{d^2\hat{y}}{dt} + \omega_c\frac{Q}{|Q|}\frac{d\hat{x}}{dt} - \frac{1}{2}\omega_z^2\hat{y} = 0 ,$$

$$\frac{d^2\hat{z}}{dt} + \omega_z^2\hat{z} = 0 . \quad (3.142)$$

3.7 Quantum Dynamics in Ideal Penning Trap

Then the expectation values $\langle \hat{x} \rangle$, $\langle \hat{y} \rangle$, and $\langle \hat{z} \rangle$ of the coordinate operators in the time-independent normalized state $\tilde{\Psi}$ satisfy the classical motion equation (3.4)–(3.6). The quantum expectation values of the coordinate and momentum operators evolve according to the classical dynamics. A general wave-packet spreads as time evolves. However, under time evolution of the Schrödinger states for the harmonic oscillator [123], the Gaussian shape is preserved.

We introduce the new coordinate operators and the new momentum operators:

$$Q_\pm = \sqrt{\frac{\hbar}{2}}\left(a_\pm + a_\pm^\dagger\right), \quad P_\pm = i\sqrt{\frac{\hbar}{2}}\left(a_\pm^\dagger - a_\pm\right),$$

$$Q_z = \sqrt{\frac{\hbar}{2}}\left(a_\pm + a_\pm^\dagger\right), \quad P_z = i\sqrt{\frac{\hbar}{2}}\left(a_\pm^\dagger - a_\pm\right). \quad (3.143)$$

Now consider the following Schrödinger Gaussian wave packets centered at the phase space point $(Q_{+0}, Q_{-0}, Q_{z0}, P_{+0}, P_{-0}, P_{z0})$:

$$\tilde{\Psi} = \pi^{-3/4}\exp(i\varphi)$$
$$\times \exp\left[-\frac{1}{2}(Q_+ - Q_{+0})^2 - \frac{1}{2}(Q_- - Q_{-0})^2 - \frac{1}{2}(Q_z - Q_{z0})^2\right],$$
$$\varphi = P_{+0}Q_+ + P_{-0}Q_- + P_{z0}Q_z . \quad (3.144)$$

Then (3.144) are minimum-uncertainty states:

$$\Delta Q_+ \Delta P_+ = \Delta Q_- \Delta P_- = \Delta Q_z \Delta P_z = 1/2, \quad (3.145)$$

where $\Delta A = (\langle A^2 \rangle - \langle A \rangle^2)^{1/2}$.

The coherent state $\tilde{\Psi}$ may be rewritten in Glauber's form [92]:

$$\tilde{\Psi} = \exp\left[-\frac{1}{2}(|\alpha_+|^2 + |\alpha_-|^2 + |\alpha_z|^2)\right]\exp\left(\alpha_+ a_+^\dagger + \alpha_- a_-^\dagger + \alpha_z a_z^\dagger\right)\Psi_0, \quad (3.146)$$

where

$$\alpha_\pm = \frac{1}{\sqrt{2}}(Q_{\pm 0} + iP_{\pm 0}), \quad \alpha_z = \frac{1}{\sqrt{2}}(Q_{z0} + iP_{z0}). \quad (3.147)$$

The explicit relation of the quantum evolution of the coordinate operators to the classical trajectories can be found by integration of the Heisenberg equations

$$\frac{d\hat{a}_\pm}{dt} = \mp i\omega_\pm \hat{a}_\pm, \quad \frac{d\hat{a}_z}{dt} = -i\omega_z \hat{a}_z. \quad (3.148)$$

It follows that

$$\hat{a}_\pm = \exp(\mp i\omega_\pm t)a_\pm, \quad \hat{a}_z = \exp(-i\omega_z t)a_z. \quad (3.149)$$

By (3.149), we obtain the time evolution of the coordinate operators

$$\hat{x} + i\frac{Q}{|Q|}\hat{y} = r_1\left[\exp(-i\omega_+ t)a_+ + \exp(-i\omega_- t)a_-^\dagger\right],$$

$$\hat{z} = \frac{z_1}{\sqrt{2}}\left[\exp(-i\omega_z t)a_z + \exp(-i\omega_z t)a_z^\dagger\right]. \quad (3.150)$$

Using (3.150), we find the parametric equations of a classical trajectory (3.17)–(3.19)

$$\langle\hat{x}\rangle + i\frac{Q}{|Q|}\langle\hat{y}\rangle = R_+ \exp\left[-i(\omega_+ t + \varphi_+)\right] + R_- \exp\left[-i(\omega_- t + \varphi_-)\right],$$

$$\langle\hat{z}\rangle = R_z \cos(\omega_z t + \varphi_z), \quad (3.151)$$

where the amplitudes R_+, R_-, R_z, and the phases φ_+, φ_-, φ_z are given in terms of expectation values for the annihilation operators:

$$R_\pm \exp(\mp i\varphi_\pm) = r_1\langle a_\pm\rangle, \quad R_z \exp(-i\varphi_z) = \sqrt{2}z_1\langle a_z\rangle. \quad (3.152)$$

According to the canonical quantization of action variables with respect the quantum numbers n, l, and k, we have

$$\hbar\omega_+\left(n + \frac{1}{2}\right) = \frac{1}{2}M\omega_1^2 \tilde{R}_+^2, \quad \hbar\omega_-\left(l + \frac{1}{2}\right) = \frac{1}{2}M\omega_1^2 \tilde{R}_-^2,$$

$$\hbar\omega_z\left(k + \frac{1}{2}\right) = \frac{1}{2}M\omega_z^2 \tilde{R}_z^2, \quad (3.153)$$

where $\tilde{R}_+$, $\tilde{R}_-$, and $\tilde{R}_z$ are orbital amplitudes.

3.7.2 Spin Motion

The spectral properties of particles with spin 1/2 trapped in an ideal Penning trap have been presented in [122] and [104].

The Hamiltonian of a particle of mass M, electric charge Q, spin s, and magnetic moment $\boldsymbol{\mu}$ can be written as

$$H_s = \frac{1}{2M}(\boldsymbol{p} - Q\boldsymbol{A})^2 + Q\Phi - \boldsymbol{\mu}\boldsymbol{B}. \quad (3.154)$$

The magnetic moment is given by

$$\boldsymbol{\mu} = \frac{g}{2}\frac{|Q|}{M}\boldsymbol{S}, \quad (3.155)$$

where g is the gyromagnetic factor of the charged particle and $\boldsymbol{S}$ is the spin operator.

3.7 Quantum Dynamics in Ideal Penning Trap

For an ideal Penning trap, the Hamiltonian (3.154) is

$$H_s = \frac{1}{2M}\left(p_x^2 + p_y^2 + p_z^2\right) + \frac{1}{8}M\omega_1^2 + \frac{1}{2}M\omega_z^2 - \frac{1}{2}\omega_0 L_z - \frac{|g|}{2}\omega_0 S_z , \quad (3.156)$$

where S_z is the axial component of the spin operator.

Let us consider the normalized spinors φ_{m_s}, $m_s = -s, -s+1, \ldots, s$, such that

$$S_z \varphi_{m_s} = \hbar m_s \varphi_{m_s} . \quad (3.157)$$

We introduce the tensor products

$$\Psi_{nlkm_s} = \Psi_{nlk} \varphi_{m_s} . \quad (3.158)$$

Then Ψ_{nlkm_s} is an eigenstate vector of (3.154) with the energy

$$E_{nlkm_s} = \hbar\omega_+\left(n+\frac{1}{2}\right) - \hbar\omega_-\left(l+\frac{1}{2}\right) + \hbar\omega_z\left(k+\frac{1}{2}\right) - \hbar\omega_s m_s . \quad (3.159)$$

Here n, l, k are nonnegative integers and $|m_s| \le s$. The spin frequency is defined by $\omega_s = |g|\omega_0/2$.

For particles of spin $s = 1/2$, (3.156) is the Pauli Hamiltonian and the spin operator for particles of spin $s = 1/2$ is given by

$$\mathbf{S} = \frac{1}{2}\hbar\boldsymbol{\sigma} , \quad (3.160)$$

where $\boldsymbol{\sigma} = (\sigma_x, \sigma_y, \sigma_z)$ is the triplet of Pauli matrices:

$$\sigma_x = \begin{pmatrix} 0 & 1 \\ 1 & 0 \end{pmatrix}, \quad \sigma_y = \begin{pmatrix} 0 & -i \\ i & 0 \end{pmatrix}, \quad \sigma_z = \begin{pmatrix} 1 & 0 \\ 0 & -1 \end{pmatrix} . \quad (3.161)$$

In this case $m_s = \pm 1/2$ and the corresponding spinors can be written as

$$\varphi_{1/2} = \begin{pmatrix} 1 \\ 0 \end{pmatrix}, \quad \varphi_{-1/2} = \begin{pmatrix} 0 \\ 1 \end{pmatrix} . \quad (3.162)$$

We consider a relativistic particle of rest mass M, spin $1/2$, charge Q, and anomalous magnetic moment $a\mu_1$, where $\mu_1 = Q\hbar/2M$. The magnetic moment is $\mu = (1+a)\mu_1$. In the presence of the electric field $\mathbf{E}$ and the magnetic field $\mathbf{B}$, the Dirac Hamiltonian H_D is [124]

$$H_D = Mc^2\beta + H_+ + H_- , \quad (3.163)$$

$$H_+ = Q\Phi - a\mu_1 \boldsymbol{\Sigma} \cdot \mathbf{B} , \quad H_- = c\boldsymbol{\alpha} \cdot \boldsymbol{\Pi} + ia\mu_1\beta\boldsymbol{\alpha} \cdot \mathbf{E} , \quad (3.164)$$

where $\boldsymbol{\Pi} = \mathbf{p} - Q\mathbf{A}$ is the kinetic momentum operator, a is the magnetic moment anomaly, $\mu_1 = \hbar Q/2M$, and

$$\beta = \begin{pmatrix} \sigma_0 & 0 \\ 0 & -\sigma_0 \end{pmatrix}, \quad \boldsymbol{\alpha} = \begin{pmatrix} 0 & \boldsymbol{\sigma} \\ \boldsymbol{\sigma} & 0 \end{pmatrix}, \quad \boldsymbol{\Sigma} = \begin{pmatrix} \boldsymbol{\sigma} & 0 \\ 0 & \boldsymbol{\sigma} \end{pmatrix} , \quad (3.165)$$

where σ_0 is the 2×2 unit matrix.

In the nonrelativistic limit, the Dirac Hamiltonian reduces to the Pauli Hamiltonian. In order to obtain the relativistic corrections in this limit, we consider the Foldy–Wouthuysen transformation [125–128], which to the order of $1/(Mc^2)$ is given by:

$$H_{FW}^{(1)} = \exp\left(\frac{1}{2Mc^2}H_-\beta\right) H_D \exp\left(\frac{1}{2Mc^2}\beta H_-\right)$$
$$= Mc^2\beta + H_+^{(1)} + H_-^{(1)}, \qquad (3.166)$$

where $\beta H_\pm^{(1)} = \pm H_\pm^{(1)}\beta$. The $(1/Mc^2)^3$ order can be written as

$$H_{FW}^{(2)} = \exp\left(\frac{1}{2Mc^2}H_-^{(1)}\beta\right) H_D \exp\left(\frac{1}{2Mc^2}\beta H_-^{(1)}\right)$$
$$= Mc^2\beta + H_+ + \frac{1}{2Mc^2}\beta H_-^2$$
$$- \frac{1}{(2Mc^2)^2}\left[H_-, [H_-, H_+] - \frac{1}{(2Mc^2)^3}\beta H_-^4\right] + \cdots . \quad (3.167)$$

Then
$$H_{FW} = Mc^2 + H_P + H_{RC} + \cdots, \qquad (3.168)$$

where the nonrelativistic Hamiltonian is given by

$$H_P = \frac{1}{2M}\boldsymbol{\Pi}^2 + Q\Phi - (1+a)\mu_1\boldsymbol{\sigma}\cdot\boldsymbol{B}, \qquad (3.169)$$

and the relativistic corrections to the Hamiltonian are given by:

$$H_{RC} = -\frac{1}{2Mc^2}\left(\frac{1}{2M}\boldsymbol{\Pi}^2 - \mu_1\boldsymbol{\sigma}\cdot\boldsymbol{B}\right)^2 \qquad (3.170)$$
$$+ \frac{\mu_1}{4Mc^2}(1+2a)\boldsymbol{\sigma}\cdot(\boldsymbol{\Pi}\times\boldsymbol{E} - \boldsymbol{E}\times\boldsymbol{\Pi})$$
$$+ \frac{a\mu_1}{8M^2c^2}\left[(\boldsymbol{\sigma}\cdot\boldsymbol{\Pi})(\boldsymbol{\Pi}\cdot\boldsymbol{B} + \boldsymbol{B}\cdot\boldsymbol{\Pi}) + (\boldsymbol{\Pi}\cdot\boldsymbol{B} + \boldsymbol{B}\cdot\boldsymbol{\Pi})(\boldsymbol{\sigma}\cdot\boldsymbol{\Pi})\right].$$

The Pauli Hamiltonian given by (3.154) with $s = 1/2$ is equivalent to H_P.

The relativistic corrections to the Hamiltonian for an ideal Penning trap can be written as

$$H_{RC} = \frac{1}{2Mc^2}(H_0 - Q\Phi - \mu_1 B_0\sigma_3)^2$$
$$- \frac{\mu_1}{4Mc^2}(1+2a)\boldsymbol{\sigma}\cdot(\boldsymbol{E}\times\boldsymbol{\Pi}) + \frac{a\mu_1 B_0}{2M^2c^2}p_z\boldsymbol{\sigma}\cdot\boldsymbol{\Pi}, \qquad (3.171)$$

where H_0 is given by (3.168).

According to the first-order perturbation theory, the relativistic correction to E_{nlkm_s} is

$$\Delta E_{nlkm_s} = (\Psi_{nlkm_s}, H_{FW}\Psi_{nlkm_s}) \,. \tag{3.172}$$

Using (3.171) and

$$\mathbf{E} = -\frac{U_0}{d^2}(x, y, -2z) \,, \quad \mathbf{\Pi} = (p_x + \frac{1}{2}M\omega_0 y, p_y - \frac{1}{2}M\omega_0 x, p_z) \,, \tag{3.173}$$

we obtain

$$\Delta E_{nlkm_s} = -\frac{\omega_z^4}{4Mc^2\omega_1^2}\left(I_+ I_- + \frac{\hbar^2}{4}\right) - \frac{\omega_z^2}{16Mc^2}\left(I_z^2 + \frac{\hbar^2}{4}\right)$$
$$- \frac{1}{2Mc^2}\left(\frac{\omega_+^2}{\omega_1}I_+ + \frac{\omega_-^2}{\omega_1}I_- + \frac{1}{2}\omega_z I_z - \hbar\omega_0 m_s\right)^2$$
$$- \frac{\hbar\omega_z^2}{2Mc^2\omega_1}m_s(1+2a)(\omega_+ I_+ + \omega_- I_-) + \frac{\hbar\omega_0\omega_z}{Mc^2}am_s I_z \,.$$
$$\tag{3.174}$$

By (3.169), we have $g = -2$ for $a = 0$.

3.8 Quantum Dynamics in Real Penning Traps

3.8.1 Electric Field Perturbations

We apply the theory of stationary perturbations to the Penning trap Hamiltonians with respect to the oscillator basis (3.132).

It is assumed that the Hamiltonian of the Penning trap can be written as

$$H' = H_s + V \,, \tag{3.175}$$

where H_s is the unperturbed Hamiltonian (3.130) and V is an analytic function of coordinates.

The expectation value with respect to Ψ_{nlkm_s} is denoted by $\langle \ \rangle$. Then

$$\langle H \rangle = E_+ + E_- + E_z - \hbar\omega_s m_s + \langle V \rangle \,, \tag{3.176}$$

where the energies in the cyclotron, magnetron, and axial motions are written as

$$E_+ = \hbar\omega_+\left(n + \frac{1}{2}\right) \,, \quad E_- = -\hbar\omega_-\left(l + \frac{1}{2}\right) \,, \quad E_z = \hbar\omega_z\left(k + \frac{1}{2}\right) \,.$$
$$\tag{3.177}$$

Using the first-order perturbations, we introduce the frequency shifts

$$\delta\omega_\pm = \pm\omega_\pm \frac{\partial \langle V \rangle}{\partial E_\pm}, \quad \delta\omega_z = \omega_z \frac{\partial \langle V \rangle}{\partial E_z}, \tag{3.178}$$

where $\langle V \rangle$ is extended to an analytic function of E_+, E_-, E_z [109].

We consider an anharmonic perturbation consisting of octupole and dodecapole terms:

$$V = \frac{1}{2} M \omega_z^2 \left[\frac{c_4}{d^2} H_4(r, z) + \frac{c_6}{d^4} H_6(r, z) \right]. \tag{3.179}$$

According to (3.153), it is convenient to introduce the variables

$$R_\pm^2 = \pm \frac{2}{M \omega_1 \omega_\pm} E_\pm, \quad R_z^2 = \frac{2}{M \omega_z^2} E_z. \tag{3.180}$$

Using (3.125) and the relations

$$r^2 = r_1^2 (a_+ + a_-^\dagger)(a_- + a_+^\dagger), \quad z = \frac{z_1}{\sqrt{2}} (a_z + a_z^\dagger), \tag{3.181}$$

$$\left\langle \left(a_+^\dagger\right)^{n''} \left(a_-^\dagger\right)^{l''} \left(a_z^\dagger\right)^{k''} a_+^{n'} a_-^{l'} a_z^{k'} \right\rangle = \frac{n! l! k! \delta_{n'n''} \delta_{l'l''} \delta_{k'k''}}{(n-n')! (l-l')! (k-k')!}, \tag{3.182}$$

we obtain

$$\langle z^2 \rangle = \frac{1}{2} R_z^2, \quad \langle z^4 \rangle = \frac{3}{8} \left(R_z^4 + z_1^4 \right), \tag{3.183}$$

$$\langle r^2 \rangle = R_+^2 + R_-^2, \quad \langle r^4 \rangle = R_+^4 + R_-^4 + 4 R_+^2 R_-^2 + \frac{1}{2} r_1^4, \tag{3.184}$$

$$\langle r^6 \rangle = R_+^6 + R_-^6 + 3 R_+^2 R_-^2 \left(R_+^2 + R_-^2 \right) \left(3 R_+^2 R_-^2 + \frac{7}{2} r_1^4 \right). \tag{3.185}$$

Using (C.13), (C.15), and (3.182)–(3.185), we obtain:

$$\langle H_4(r, z) \rangle = \frac{3}{8} \left[R_z^4 - 4 R_z^2 \left(R_+^2 + R_-^2 \right) + R_+^4 + R_-^4 + 4 R_+^2 R_-^2 \right] + \frac{3}{16} \left[r_1^4 + z_1^4 \right], \tag{3.186}$$

$$\langle H_6(r, z) \rangle = \frac{5}{16} [R_z^4 - 9 R_z^2 \left(R_+^2 + R_-^2 \right) + 9 R_z^2 (R_+^4 + R_-^4 + 4 R_+^2 R_-^2) \\ - R_+^6 - R_-^6 - 93 R_+^2 R_-^2 \left(R_+^2 + R_-^2 \right) \\ - (9 z_1^4 + \frac{7}{2} r_1^4) \left(R_+^2 + R_-^2 \right) + (5 z_1^4 + \frac{9}{2} r_1^4) R_z^2]. \tag{3.187}$$

By (3.178), (3.180), (3.182)–(3.185), (3.186), and (3.187), we find the frequency shifts:

$$\Delta\omega_{\pm} = \pm\frac{3c_4\omega_z^2}{4d^2\omega_1}\left(R_{\pm}^2 + 2R_{\mp}^2 - 2R_z^2\right)$$
$$\mp\frac{15c_6\omega_z^2}{16d^4\omega_1}\left[R_{\pm}^4 + 6R_{+}^2R_{-}^2 + 3R_{\mp}^4 - 6R_z^2\left(R_{\pm}^2 + 2R_{\mp}^2\right) + 3R_z^4\right]$$
$$\pm\frac{15c_6\omega_z^2}{96d^4\omega_1}\left(18z_1^4 + 7r_1^4\right),\tag{3.188}$$

$$\Delta\omega_z = \frac{3c_4\omega_z}{4d^2}\left(R_z^2 - 2R_{+}^2 - 2R_{-}^2\right)$$
$$+\frac{15c_6\omega_z}{16d^4}\left[3\left(R_{+}^4 + 4R_{+}^2R_{-}^2 + R_{-}^4\right) - 6R_z^2\left(R_{+}^2 + R_{-}^2\right) + R_z^4\right]$$
$$+\frac{15c_6\omega_z^2}{96d^4\omega_1}\left(10z_1^4 + 9r_1^4\right).\tag{3.189}$$

By (3.188), we obtain the cyclotron frequency shift:

$$\Delta\omega_c = \Delta\omega_{+} + \Delta\omega_{-} = \frac{3\omega_z^2}{4d^2\omega_1}\left(R_{-}^2 - R_{+}^2\right)\left[c_4 + \frac{5c_6}{2d^2}\left(R_{+}^2 + R_{-}^2 - 3R_z^2\right)\right].\tag{3.190}$$

The shifts (3.188)–(3.190) agree with the classical result (3.58)–(3.61) for negligible r_1, and z_1.

3.8.2 Magnetic Field Perturbations

We consider the vector potential (3.65). According to (3.66) and (3.65), the Hamiltonian (3.154) can be written as

$$H'_s = \frac{1}{2M}(\mathbf{p} - Q\mathbf{A}')^2 + Q\Phi - \boldsymbol{\mu}\mathbf{B}.\tag{3.191}$$

$$H'_s = H_s + \frac{1}{8}M\omega_c^2(2\Lambda + \Lambda^2)r^2 - \frac{\omega_0}{2}\Lambda L_z$$
$$-\frac{|g|\omega_0}{2}\left[\Phi_m S_z - \frac{1}{2}\frac{\partial\Lambda}{\partial z}(xS_x + yS_y)\right].\tag{3.192}$$

Using (3.192) and the relation $\omega_0\langle L_z\rangle = \hbar\omega_c(l-n)$, the first-order perturbation is given by

$$\Delta E_{nlkm_s} = \langle H'_s - H_s\rangle = \frac{1}{8}M\omega_1^2\langle r^2(2\Lambda + \Lambda^2)\rangle \tag{3.193}$$
$$+\frac{1}{2}\hbar\omega_c(l-k)\langle\Lambda\rangle - \frac{|g|}{2}\hbar\omega_0 m_s\langle\Phi_m\rangle.\tag{3.194}$$

We retain only the b_2 and b_4 terms in the expansion (3.69)

$$\Lambda = \frac{b_2}{d^2}\left(z^2 - \frac{1}{4}r^2\right) + \frac{b_4}{d^4}\left(z^4 - \frac{3}{2}r^2z^2 + \frac{1}{8}r^4\right), \quad (3.195)$$

$$\Phi_m = \frac{b_2}{d^2}\left(z^2 - \frac{1}{2}r^2\right) + \frac{b_4}{d^4}\left(z^4 - 3r^2z^2 + \frac{3}{8}r^4\right). \quad (3.196)$$

Using (3.180), (3.182)–(3.185), (3.193)–(3.196), and (3.178) for $\langle V \rangle = \langle H'_s - H_s \rangle$, we find the cyclotron frequency shift

$$\delta\omega_c = \Delta\omega_c + \frac{3\omega_c b_4}{8d^4}\left(z_1^4 + \frac{1}{6}r_1^4\right) + \frac{|g|}{2d^2}\hbar\omega_0 m_s \left[b_2 + \frac{b_4}{d^2}\left(\tilde{R}_-^2 - \tilde{R}_+^2\right)\right], \quad (3.197)$$

$$\Delta\omega_c = \frac{\omega_c b_2}{2\omega_1 d^2}\left(\omega_1 R_z^2 + \omega_- R_+^2 - \omega_+ R_-^2\right) + \frac{|g|b_2}{2d^2}\hbar\omega_0 m_s$$
$$+ \frac{3\omega_c b_4}{16\omega_1 d^4}\left[2\omega_1 R_+^2 R_-^2 - \omega_c R_+^4 - \omega_c R_-^4\right]$$
$$+ \frac{3\omega_c b_4}{16\omega_1 d^4} R_z^2 \left[4\omega_- R_+^2 - 4\omega_+ R_-^2 + \omega_1 R_z^2\right]$$
$$+ \frac{3\omega_c b_4}{8d^4}\left(z_1^4 + \frac{1}{6}r_1^4\right) + \frac{|g|b_4}{2d^4}\hbar\omega_0 m_s \left(R_-^2 - R_+^2\right). \quad (3.198)$$

Note that (3.198) is given by the classical result (3.79) for negligible r_1, z_1, and g.

3.8.3 The General Hamiltonian

The linear terms of the electric potential can be eliminated by suitable translations of coordinates and momenta. Then the general Hamiltonian (3.115) of a charged particle in a nonlinear Penning trap can be written as

$$H = H' + Q\Phi, \quad (3.199)$$

where Φ is the anharmonic part of the electric potential, and

$$H' = \frac{1}{2M}p^2 + \frac{Q^2}{8M}\left[\rho^2 B^2 - (\rho B)^2\right] - \frac{Q}{2M}LB - \mu B + W. \quad (3.200)$$

Here L is the angular momentum operator and the harmonic part of the electric potential energy is given by (3.80). Ellipticity and misalignment are included in H'.

If the stability conditions (3.95) are satisfied, then the harmonic Hamiltonian (3.199) can be written as:

$$H' = \hbar\omega'_+ \left(p'^2_+ + q'^2_+\right) - \hbar\omega'_- \left(p'^2_- + q'^2_-\right) + \hbar\omega'_z \left(p'^2_z + q'^2_z\right) - \mu B, \quad (3.201)$$

3.8 Quantum Dynamics in Real Penning Traps

where the operators $q'_\pm$, q'_z, $p'_\pm$, and p'_z are linear combinations of x, p_x, y, p_y, z, and p_z. These operators commute with each other except for the cases

$$\left[q'_+, p'^\dagger_+\right] = \left[q'_-, p'^\dagger_-\right] = \left[q'_z, p'^\dagger_z\right] = i\hbar \ . \qquad (3.202)$$

The characteristic frequencies ω'_+, ω'_-, and ω'_z are the positive roots of (3.86).

4 Other Traps

4.1 Combined Traps

4.1.1 Equations of Motion

A trap in which a radio-frequency potential is applied to the electrodes in addition to a uniform magnetic field $\mathbf{B} = (0, 0, B)$ directed along the symmetry axis, is a hybrid device commonly called a "combined trap", since it combines the properties of the static Penning and dynamic Paul traps. Operation of a trap in the combined mode displays interesting features that may prove useful in certain applications. Of particular note, as will be shown below, is the stable trapping over a wider range of parameters than with either the pure Penning trap or the Paul trap. Furthermore, it can be used to trap charged particles of different sign and widely differing mass and charge at the same time [129]. Experimental demonstration of these properties has been obtained by Walz et al. [130], who confined electrons and positive ions simultaneously in a trap operating in the combined mode.

The equations of motions for a charged particle of charge-to-mass ratio Q/M in such a combined trap are [56, 129, 131]:

$$\frac{d^2 x}{dt^2} - \frac{Q}{Md^2}(U_0 + V_0 \cos \Omega t)x - \omega_c \frac{dy}{dt} = 0,$$

$$\frac{d^2 y}{dt^2} - \frac{Q}{Md^2}(U_0 + V_0 \cos \Omega t)y + \omega_c \frac{dx}{dt} = 0,$$

$$\frac{d^2 z}{dt^2} + \frac{2Q}{Md^2}(U_0 + V_0 \cos \Omega t)z = 0, \qquad (4.1)$$

where $\omega_c = QB/M$. The transformation to rotating coordinates such that

$$x = \tilde{x} \cos \frac{\omega_c}{2} t - \tilde{y} \sin \frac{\omega_c}{2} t, \quad y = \tilde{x} \sin \frac{\omega_c}{2} t + \tilde{y} \cos \frac{\omega_c}{2} t, \qquad (4.2)$$

uncouples the equations for x and y in (4.1), which, written in terms of dimensionless parameter $\tau = \Omega t/2$, become

$$\frac{d^2 \tilde{x}}{d\tau^2} + \left(\frac{\omega_c^2}{\Omega^2} + a_x - 2q_x \cos 2\tau\right) \tilde{x} = 0,$$

$$\frac{d^2\tilde{y}}{d\tau^2} + \left(\frac{\omega_c^2}{\Omega^2} + a_y - 2q_r \cos 2\tau\right)\tilde{y} = 0 ,$$

$$\frac{d^2 z}{d\tau^2} + (a_z - 2q_z \cos 2\tau) z = 0 , \qquad (4.3)$$

where

$$a_x = a_y = -\frac{4QU_0}{Md^2\Omega^2} , \quad q_x = q_y = \frac{2QV_0}{Md^2\Omega^2} ,$$

$$a_z = \frac{8QU_0}{Md^2\Omega^2} , \quad q_z = -\frac{4QV_0}{Md^2\Omega^2} . \qquad (4.4)$$

These, it will be noted, still have the form of Mathieu equations

$$\frac{d^2 u}{d\tau^2} + (a_u - 2q_u \cos 2\tau) u = 0 , \qquad (4.5)$$

where $u = r, z$. However, although the stability parameters a_u and q_u for the axial motion are identical to those of a Paul trap (as defined in (2.3)), since the magnetic field has no influence on the motion in this direction, the value of a'_r for the radial motion is modified thus

$$a'_r = a_r + g^2 , \qquad (4.6)$$

where a_r is the Mathieu parameter (2.3) in the absence of the magnetic field, and

$$g = \frac{\omega_c}{\Omega} . \qquad (4.7)$$

Thus the effect of the magnetic field on the stability parameters is equivalent to the addition of a dc voltage U_B:

$$U_B = \frac{Qd^2 B^2}{4M} . \qquad (4.8)$$

As a consequence, the areas of radial stability in the a–q-plane are shifted, whereas the axial ones remain the same.

In Fig. 4.1 the regions of stability in an ideal combined trap are shown. It can be seen that, in the presence of a magnetic field, the area of the lowest stability domain is enlarged. This is particularly interesting in the case of low mass particles like electrons, when even at low magnetic fields, ω_c^2/Ω^2 may be much larger than unity.

Since the transformed equations of motion are of the Mathieu type, the solutions are written as for a Paul trap:

$$u(\tau) = A_u \sum_{-\infty}^{+\infty} c_{2n} \cos\left(\beta_u' + 2n\right)\tau + B_u \sum_{-\infty}^{+\infty} c_{2n} \sin\left(\beta_u' + 2n\right)\tau , \qquad (4.9)$$

where β_u' can be approximated by $\beta_u' = (a_u' + \frac{1}{2}q_u^2)^{1/2}$ as long as $a', q \ll 1$. In order to transform back this solution, from the rotating reference frame to

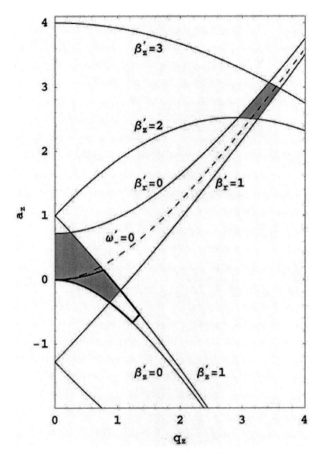

Fig. 4.1. The regions of stability for charged particles in an ideal combined trap (*grey area*) with $g = 0.36$, and the comparison with the lowest stability domain in the ideal Paul trap (*thick boundary*). The *dashed line* indicates trapping parameters for magnetron-free operation

the laboratory one, we have to substitute the solution (4.9) into the equations (4.2) to obtain:

$$\begin{aligned}
x &= [R_- \cos(\Omega_- t + \alpha_-) + R_+ \cos(\Omega_+ t + \alpha_+)] \, C_r(t) \\
&\quad + [R_- \sin(\Omega_- t + \alpha_-) - R_+ \sin(\Omega_+ t + \alpha_+)] \, S_r(t) \, , \\
y &= -\varepsilon \, [R_- \sin(\Omega_- t + \alpha_-) + R_+ \sin(\Omega_+ t + \alpha_+)] \, C_r(t) \\
&\quad + [R_- \cos(\Omega_- t + \alpha_-) - R_+ \cos(\Omega_+ t + \alpha_+)] \, S_r(t) \, , \\
z &= R_z [\cos(\omega_z t + \alpha_z) C_z(t) - \cos(\omega_z t + \alpha_z) S_z(t) \, , \quad (4.10)
\end{aligned}$$

where $\varepsilon = \text{sgn}(\omega_0)$, and

$$\Omega_\pm = \frac{1}{2}(\omega_c \pm \Omega\beta_r'), \quad \omega_z = \frac{1}{2}\Omega\beta_z,$$

$$C_r(t) = \sum_{n=-\infty}^{\infty} c_{2n,r} \cos n\Omega t, \quad S_r(t) = \sum_{n=-\infty}^{\infty} c_{2n,r} \sin n\Omega t,$$

$$C_z(t) = \sum_{n=-\infty}^{\infty} c_{2n,z} \cos n\Omega t, \quad S_z(t) = \sum_{n=-\infty}^{\infty} c_{2n,z} \sin n\Omega t.$$

We obtain an oscillation spectrum with frequencies

$$\omega_{zn} = \left(n + \frac{\beta_z}{2}\right)\Omega, \tag{4.11}$$

and

$$\omega_{rn} = \left(n + \frac{\beta_r'}{2} \pm \frac{g}{2}\right)\Omega, \tag{4.12}$$

with $n = 0, \pm 1, \pm 2, \ldots$. It should be noted that β_r', and hence ω_r, has a nonlinear dependence on the magnetic field.

The Paul and Penning traps are clearly limiting forms of the combined trap for vanishing magnetic field, and vanishing rf-field, respectively. This can be seen when we plot the oscillation frequencies as a function of B and V_0, as shown in Fig. 4.2 [131]. On the left hand side of Fig. 4.2 we start with $B = 0$ and a constant rf-amplitude, representing a Paul trap, with oscillation frequencies ω_r and ω_z. When the magnetic field is increased up to a value B_0, ω_r is split into two components, while ω_z remains unchanged. Starting from the right hand side of the figure we have a constant B-field (B_0) and vanishing rf-amplitude, representing a Penning trap: application of an rf-voltage changes the motional eigenfrequencies. In the center of the figure, for a given value of B and V_0, we have the eigenfrequencies of the combined trap which follow from (4.11) and (4.12). The magnetron frequency is negative for small V_0, because it has a negative total energy associated with it in this region.

4.1.2 Magnetron-free Operation

The magnetron frequency of the Penning trap changes with the application of an rf-voltage to $\omega_m' = \beta_r'\Omega/2 - \omega_c/2$. This expression becomes zero when $\beta_r' = g$, representing an iso-β line in the stability diagram (see Fig. 4.1). A physical picture of the magnetron-free motion can be obtained if we compare the motions of a charged particle in the Penning and Paul traps. The dc electric field of a Penning trap confines the particle in the axial direction but leads to an *outwardly* directed velocity component in the radial plane, whereas the rf-field of a Paul trap leads to confinement forces in all directions and an *inwardly* directed radial velocity component. The presence of the magnetic field causes an $\mathbf{E} \times \mathbf{B}$ drift around the trap axis in opposite directions for

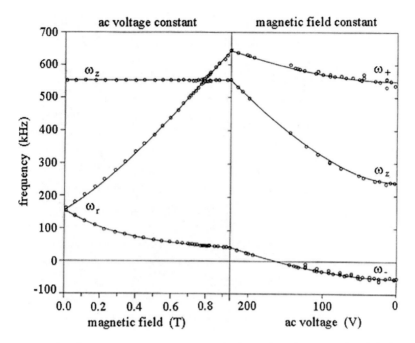

Fig. 4.2. Oscillation frequencies in an ideal combined trap. The *left boundary* represents a Paul trap. A superimposed B-field increases from left to right. The *right boundary* represents a Penning trap. A superimposed rf-field increases from right to left [131]. Copyright (2003) by Taylor & Francis. http://www.tandf.co.uk/journals/tf/09500340.html

the two cases. When the two effects cancel each other, the magnetron motion vanishes. Thus the iso-β line $\beta_r = g$ divides the stability diagram into two regions: one, a more Penning-like, and the other a more Paul-like domain, above and below this line, respectively. This line is shown as dashed line in Fig. 4.1. Yan et al. [132] have performed simulations of ion trajectories in a combined trap showing that, in the magnetron-free mode, the particle may be localized in an off-axis region of the trap, whose spatial dimension becomes smaller with decreasing values of the parameters a_z and q_z.

4.1.3 Quantum Dynamics in Combined Traps

A particle of mass M and electric charge Q confined in a combined trap is described by the quantum Hamiltonian

$$H = \frac{1}{2M}(\boldsymbol{p} - Q\boldsymbol{A})^2 + Q\Phi , \qquad (4.13)$$

where $\boldsymbol{p} = -i\hbar\boldsymbol{\nabla} = (p_x, p_y, p_z)$ is the momentum operator. The magnetic field $\boldsymbol{B} = (0, 0, B_0)$ is static and homogeneous. For the Paul trap $B_0 = 0$. We consider an ideal combined trap with the vector potential $\boldsymbol{A}$ and the electric potential Φ defined by

$$\Phi = \frac{M}{4Q}\omega_z^2(2z^2 - x^2 - y^2) , \quad \boldsymbol{A} = \frac{1}{2}B_0(-y, x, 0) , \qquad (4.14)$$

where the axial frequency ω_z is a function of period $T = 2\pi/\Omega$, with Ω the trap driving frequency. It follows that

$$H = \frac{1}{2M}\boldsymbol{p}^2 + \frac{1}{8}M\omega_1^2 + \frac{1}{2}M\omega_z^2 - \frac{1}{2}\omega_0 L_z , \qquad (4.15)$$

where $\omega_0 = QB_0/M$, $|\omega_0| = \omega_c$, $\omega_1 = (\omega_c^2 - 2\omega_z^2)^{1/2}$, and L_z is the axial angular momentum operator. According to the Heisenberg equations, we have

$$\frac{d^2 x}{dt} - B_z \frac{Q}{M}\frac{dy}{dt} - \frac{1}{2}\omega_z^2 x = 0 ,$$

$$\frac{d^2 y}{dt^2} + B_z \frac{Q}{M}\frac{dx}{dt} - \frac{1}{2}\omega_z^2 y = 0 ,$$

$$\frac{d^2 z}{d\tau^2} + \omega_z^2 z = 0 . \qquad (4.16)$$

We introduce the scaled time τ and the axial rotation angle α defined by

$$\tau = \frac{1}{2}\Omega t , \quad \alpha = \frac{1}{2}\omega_0 t = \frac{\omega_0}{\Omega}\tau , \qquad (4.17)$$

and new coordinate operators

$$u_1 = x\cos\alpha + y\sin\alpha , \quad u_2 = y\cos\alpha - x\sin\alpha , \quad u_3 = z . \qquad (4.18)$$

We find from (4.16) the Hill equations

$$\frac{d^2 u_j}{d\tau^2} + f_j(\tau) u_j = 0 , \quad j = 1, 2, 3 , \qquad (4.19)$$

where

$$f_1 = f_2 = \frac{\omega_1^2}{\Omega^2} , \quad f_3 = 4\frac{\omega_z^2}{\Omega^2} . \qquad (4.20)$$

We introduce the Floquet stable functions w_j such that

$$\frac{d^2 w_j}{d\tau^2} + f_j(\tau) w_j = 0 , \quad w_j^* \frac{dw_j}{d\tau} - w_j \frac{dw_j^*}{d\tau} = 2i\delta_j , \qquad (4.21)$$

$$w_j = \exp(i\beta_j t)(C_j + iS_j) , \qquad (4.22)$$

where C_j and S_j are time-periodic functions of period π. Here $\delta_j > 0$. From (4.21), we find

$$\frac{d^2 C_j}{d\tau^2} - 2\beta_j \frac{dS_j}{d\tau} + (f_j - \beta_j^2)C_j = 0 , \qquad (4.23)$$

$$\frac{d^2 S_j}{d\tau^2} + 2\beta_j \frac{dC_j}{d\tau} + (f_j - \beta_j^2)S_j = 0 ,$$

$$C_j \frac{dS_j}{d\tau} - S_j \frac{dC_j}{d\tau} + \beta_j(C_j^2 + S_j^2) = \delta_j . \qquad (4.24)$$

We now introduce the time-periodic canonical transformations

$$q_j = \frac{1}{\sqrt{\beta_j \delta_j}} \left[\left(\frac{dS_j}{d\tau} + \beta_j C_j\right) u_j - S_j v_j \right] ,$$

$$p_j = \sqrt{\frac{\beta_j}{\delta_j}} \left[\left(-\frac{dC_j}{d\tau} + \beta_j S_j\right) u_j + C v_j \right] , \qquad (4.25)$$

where v_j is the canonical momentum operator conjugate to u_j. Using (4.19) and (4.23)–(4.25), we obtain the Heisenberg equations for a three-dimensional harmonic oscillator of frequencies β_1, β_2, and β_3:

$$\frac{dq_j}{d\tau} = p_j , \quad \frac{dp_j}{d\tau} = -\beta_j^2 q_j . \qquad (4.26)$$

The solutions of the equation (4.23) are

$$q_j(\tau) = \cos(\beta\tau) q_j(0) + \frac{1}{\beta} \sin(\beta\tau) p_j(0) ,$$

$$p_j(\tau) = \cos(\beta\tau) p_j(0) - \beta \sin(\beta\tau) q_j(0) . \qquad (4.27)$$

Thus the quantum Hamiltonian of a charged particle confined in an ideal combined trap is reduced to the time independent quantum Hamiltonian of a three-dimensional harmonic oscillator.

It is convenient to introduce the polar coordinates $x = r\cos\theta$, $y = r\sin\theta$. A solution for the Hamiltonian (4.15) can be written

$$\Psi(r,\theta,z,t) = \frac{1}{\sqrt{2\pi}} \exp(i\theta l) \chi(r,t) \phi(z,t) , \qquad (4.28)$$

where

$$i\hbar \frac{\partial \chi}{\partial t} = \left[-\frac{\hbar^2}{2M}\left(\frac{\partial^2}{\partial r^2} + \frac{1}{r}\frac{\partial}{\partial r} - \frac{l^2}{r^2}\right) + \frac{1}{8}M\omega_1^2 r^2 - \frac{1}{2}\hbar\omega_0 l \right] \chi , \qquad (4.29)$$

$$i\hbar \frac{\partial \phi}{\partial t} = -\frac{\hbar^2}{2M}\frac{\partial^2}{\partial z^2} + \frac{1}{2}M\omega_z^2 z^2 \phi . \qquad (4.30)$$

Using the scaled variables

$$\tilde{r} = \sqrt{\frac{M\Omega}{2\hbar}} r , \quad \tilde{z} = \sqrt{\frac{M\Omega}{2\hbar}} z , \quad \tau = \frac{1}{2}\Omega t , \qquad (4.31)$$

(4.29) and (4.30) become

$$i\frac{\partial \tilde{\chi}}{\partial \tau} = \left[-\frac{1}{2}\left(\frac{\partial^2}{\partial \tilde{r}^2} + \frac{1}{\tilde{r}}\frac{\partial}{\partial \tilde{r}} - \frac{l^2}{\tilde{r}^2} \right) + \frac{1}{2}g(\tau)\tilde{r}^2 - \frac{\omega_0}{\Omega}l \right]\tilde{\chi} , \qquad (4.32)$$

$$i\frac{\partial \tilde{\phi}}{\partial \tau} = \left(-\frac{1}{2}\frac{\partial^2}{\partial \tilde{z}^2} + \frac{1}{2}g'(\tau)\tilde{z}^2 \right)\tilde{\phi} , \qquad (4.33)$$

where

$$\tilde{\chi}(\tilde{r},t) = \left(\frac{M\Omega}{2\hbar}\right)^{1/2}\chi(r,t) , \quad \tilde{\phi}(\tilde{z},t) = \left(\frac{M\Omega}{2\hbar}\right)^{1/4}\phi(z,t) , \qquad (4.34)$$

$$g(\tau) = \frac{\omega_1^2}{2\Omega^2} , \quad g'(\tau) = \frac{2\omega_z^2}{\Omega^2} . \qquad (4.35)$$

It is convenient to introduce the stable complex solutions w_j of the Hill equations

$$\frac{d^2 w}{d\tau^2} + g(\tau)w = 0 , \quad \frac{d^2 w'}{d\tau^2} + g'(\tau)w = 0 , \qquad (4.36)$$

with the Wronskians

$$w^*\frac{dw}{d\tau} - w\frac{dw^*}{d\tau} = 2i\delta , \quad w'^*\frac{dw'}{d\tau} - w'\frac{dw'^*}{d\tau} = 2i\delta' , \qquad (4.37)$$

such that

$$w = \rho\exp(i\gamma) , \quad w' = \rho'\exp(i\gamma') , \qquad (4.38)$$

$$\frac{d\gamma}{d\tau} = \frac{\delta}{\rho^2} , \quad \frac{d\gamma'}{d\tau} = \frac{\delta'}{\rho'^2} , \quad \beta = \frac{\delta}{\pi}\int_0^\pi \frac{1}{\rho^2}d\tau , \quad \beta' = \frac{\delta'}{\pi}\int_0^\pi \frac{1}{\rho'^2}d\tau' . \quad (4.39)$$

Here $\gamma = \arg w$, $\gamma' = \arg w'$, and $\delta, \delta', \rho, \rho' > 0$. Then

$$\tilde{\chi}_{nl}(\tilde{r},\tau) = \left(\frac{\sqrt{\delta}}{\rho}\right)^{1/2}\varphi_{nl}\left(\frac{\sqrt{\delta}}{\rho}\tilde{r}\right)\exp\left[-i(2n+l+1)\gamma\right]\exp\left(\frac{i}{2\rho}\frac{d\rho}{d\tau}\tilde{z}^2\right) , \qquad (4.40)$$

$$\tilde{\phi}_{n'}(\tilde{z},\tau) = \left(\frac{\sqrt{\delta'}}{\rho'}\right)^{1/2}\varphi_n\left(\frac{\sqrt{\delta'}}{\rho'}\tilde{z}\right)\exp\left[-i\left(n'+\frac{1}{2}\right)\gamma'\right]\exp\left(\frac{i}{2\rho'}\frac{d\rho'}{d\tau}\tilde{z}^2\right) , \qquad (4.41)$$

where the Laguerre functions φ_{nl} and the Hermite functions φ_n are given by (D.28) and (D.16), respectively:

$$\varphi_{nl}(\xi) = \sqrt{2\frac{n!}{(n+l)!}}\xi^{l+1/2}L_n^l\left(\xi^2\right)\exp\left(-\frac{1}{2}\xi^2\right), \qquad (4.42)$$

$$\varphi_n(\xi) = \left(\sqrt{\pi}2^n n!\right)^{-1/2}H_n(\xi)\exp\left(-\frac{1}{2}\xi^2\right). \qquad (4.43)$$

We obtain the quasienergy functions

$$\Psi_{n'nl}(r,\theta,z,t) = \left(\frac{2\hbar}{M\Omega}\right)^{3/4}\frac{1}{\sqrt{2\pi}}\exp(i\theta l)\tilde{\chi}_{nl}(\tilde{r},\tau)\tilde{\phi}_{n'}(\tilde{z},\tau), \qquad (4.44)$$

with the quasienergies

$$E_{nl\,n'} = \frac{1}{2}\hbar\Omega\left[(2n+l+1)\beta + \left(n'+\frac{1}{2}\right)\beta'\right]. \qquad (4.45)$$

For the ideal combined trap described in Sect. 4.1.1, β and β' are the characteristic exponents of the Mathieu equation, depending on the trap parameters.

The set of quasienergy functions is complete and orthonormal. Moreover, $L_z\Psi_{n'nl} = \hbar l \Psi_{n'nl}$ where $L_z = -i\hbar\partial/\partial\theta$.

4.2 Cylindrical Traps

It has been general practice to shape trap electrodes to approximate as far as possible the hyperboloids used originally by Paul et al. This was done mainly in order to simplify the theoretical prediction of the stability of entrapment of ions, and their oscillation frequencies, in order to identify them with certainty from the observed resonance spectrum. However, as pointed out in Sect. 1.2 it was early recognized that any rotationally symmetric electrode geometry which produces a saddle point in the equipotential surfaces, would have the potential field in the neighborhood of that point in fact precisely of the Paul form. Thus it was realized that simply using electrodes consisting of a right circular cylinder with two plane end caps, the motion of the confined particles, at least near the center of the trap, would be governed by the same Mathieu equations as for hyperbolic electrodes. The relative ease with which cylindrical electrodes could be fabricated, and particularly the availability of analytical expressions for the microwave field modes in a such a cylindrical cavity, were strong inducements to use this electrode geometry. However to ensure a reasonable trapping volume near the center, cylindrical traps tended to be of large dimensions, requiring higher operating voltages for a given well depth, and reduced electrical coupling between the trapped ions and the outside circuits. However, as will be elaborated below, the use of additional compensation electrodes have largely overcome these drawbacks and made the cylindrical geometry adaptable even for precise frequency measurements.

4.2.1 Electrostatic Field in a Cylindrical Trap

Although simple cylindrical electrodes were used as early as 1959 [133], the first theoretical treatment of the motion of ions in a cylindrical trap was given by Benilan and Audoin [134]. They describe the confinement properties of a cylindrical trap consisting of a right circular cylinder of radius r_0 with flat end plates at $z = \pm z_0$. A numerical integration (4th order Runge–Kutta) of the equations of motion was carried out for two cases of particular interest: $r_0^2 = 2z_0^2$, which conforms to the pure ("symmetric") quadrupole ratio, and therefore expected to approximate the quadrupole field over a larger space, and $r_0^2 = z_0^2$, chosen not only for comparison, but also as a favorable proportion for a microwave cavity of superior quality factor, Q.

As with the other electrode geometries, the problem of finding the potential field responsible for the trapping of ions can be treated, with negligible error, as an electrostatic one, in spite of the fact that high frequency fields may be involved, because the frequencies are so low that retardation effects are entirely negligible. Unlike the case of finite hyperbolically shaped electrodes, the potential field inside a cylindrical trap can be solved analytically using standard techniques for solving electrostatic boundary-value problems. Thus we seek a series expansion in terms of orthogonal harmonic functions satisfying the boundary conditions. One such expansion, as used by Benilan and Audoin [134], which takes into account the rotational symmetry about the z-axis, and the reflection symmetry in the $z = 0$-plane, is the following

$$\Phi(r, z) = \Phi_0 \sum_i A_i J_0(m_i r) \cosh(m_i z) , \qquad (4.46)$$

where $J_0(m_i r)$ is a Bessel function of the first kind of zero order, and A_i and m_i are constants to be determined to fit the boundary conditions. Equivalently, if m_i is replaced by ik_i, an expansion in terms of the harmonic functions $I_0(k_i r) \cos(k_i z)$ is obtained in which $I_0(x)$ is the modified Bessel function. This form is used by Gabrielse et al. [135], and will be pursued later in dealing with the compensation for anharmonicities of the cylindrical geometry.

Using standard methods of imposing the boundary conditions $\Phi = \Phi_0$ on the cylinder $r = r_0$, and $\Phi = 0$ on the end plates at $z = \pm z_0$, one finds

$$A_i = \frac{2\Phi_0}{\lambda_i J_1(\lambda_i) \cosh(\lambda_i z_0/r_0)} , \qquad (4.47)$$

where the λ_i are the roots of the Bessel function $J_0(x)$, and $J_1(x)$ is the Bessel function of the first order. Hence finally:

$$\Phi(r, z) = 2\Phi_0 \sum_i \frac{J_0(\lambda_i r/r_0) \cosh(\lambda_i z/r_0)}{\lambda_i J_1(\lambda_i) \cosh(\lambda_i z_0/r_0)} . \qquad (4.48)$$

From this solution we obtain the electric field components by taking the derivatives with respect to r and z, recalling that $\frac{d}{dx}J_0(x) = -J_1(x)$. Unlike the decoupled equations describing the motion in a pure quadrupole field, the ion motion in this field is governed by nonlinear coupled equations involving transcendental functions. However, as previously shown, in the neighborhood of the origin, which is a saddle point, the field approaches the pure quadrupole form, and therefore, if these transcendental functions are expanded in a power series about the origin, then to first order in r^2 and z^2 the equations will be independent. Thus for $\lambda_i r \ll 1$ and $\lambda_i z \ll 1$, the form of the potential in the neighborhood of the origin is found to be, as expected, that of a pure quadrupole, thus

$$\Phi(r,z) = 2\Phi_0 f(z_0/r_0) - \frac{\Phi_0 g(z_0/r_0)}{2r_0^2}(r^2 - 2z^2), \qquad (4.49)$$

where we have defined, for brevity, the geometric functions $f(z_0/r_0)$ and $g(z_0/r_0)$ as follows:

$$f(z_0/r_0) = \sum_i \frac{1}{\lambda_i J_1(\lambda_i) \cosh(\lambda_1 z_0/r_0)},$$

$$g(z_0/r_0) = \sum_i \frac{\lambda_i}{J_1(\lambda_i) \cosh(\lambda_i z_0/r_0)}. \qquad (4.50)$$

If the cylindrical geometry is used in a Paul trap, these results permit the calculation of the characteristic parameters a and q of the Mathieu equations for motion in the neighborhood of the origin. For the particular case of $z_0/r_0 = 1$ the numerical value of $g(z_0/r_0)$ is 0.712, while for $z_0/r_0 = 1/\sqrt{2}$, the value is 1.103. Perhaps not surprisingly in the case which conforms to the $1/\sqrt{2}$ ratio, characteristic of the pure quadrupole field, the applied potential produces a greater quadrupole component near the origin.

This is confirmed by the detailed numerical analysis carried out by Benilan and Audoin [134] on particle trajectories in Paul traps using such cylindrical electrodes. By numerical integration of the equations of motion for a range of different initial coordinates and velocities over a fixed number of periods (100) of oscillation of the radiofrequency field, they found the calculated trajectories to be similar to those in a pure quadrupole field: The same secular low frequency oscillation giving the appearance of a Lissajous pattern, on which is superposed a high frequency oscillation at the frequency Ω of the field. By varying the parameters corresponding to the a and q of the Mathieu equations, they studied the boundaries beyond which the motion was unstable. The results show a pronounced distortion in the case of $z_0/r_0 = 1$ characterized by a widening of the equivalent a-parameter range for stability.

This says little of course about the nature of the particle trajectories far from the origin, where departures from the pure quadrupole field lead to nonlinear behavior as well as coupling between the axial and radial coordinates.

4.2.2 Inherent Anharmonicity of the Field

Having obtained the analytical solution for the field throughout the interior of a cylindrical trap, we can begin to examine the degree to which this departs from the ideal quadrupole field as we go farther out from the origin. Such a departure would manifest itself, as we recall from the discussion of real Penning and Paul hyperbolic traps, as a nonlinearity in the equations of motion and coupling between them. These have important consequences with regard to the performance of these traps, including amplitude dependent frequency shifts and instabilities. This anharmonicity is quantifiable in terms of the next higher order terms in the power series expansions of the Bessel and hyperbolic cosine functions. Because of the assumed symmetry, the next higher term in the series expansion in r/r_0 and z/z_0 is the 4th, which fortunately involves a coefficient that is a function of the ratio z_0/r_0. A numerical computation, carried out by varying the value of z_0/r_0 as a parameter, shows that this coefficient can in fact be made to vanish; but not as one might expect, at $z_0/r_0 = 1/\sqrt{2}$ characteristic of the hyperbolic case, but rather at $z_0/r_0 \simeq 0.83$. Therefore by constructing the trap electrodes in such a way that allows for the fine adjustment of this ratio, it should be possible to reduce the anharmonicity of the cylindrical trap.

An alternative representation useful in the context of studying the anharmonicities in the trapping field, is the *multipole* expansion, already introduced in the section on imperfect hyperbolic traps, which again assumes rotational symmetry about the z-axis, and reflection symmetry in the $z = 0$-plane. It has the following form:

$$\Phi(\varrho, \theta, \phi) = \Phi_0 \sum_{n(\text{even})} c_n \left(\frac{\varrho}{d}\right)^n P_n(\cos\theta) , \qquad (4.51)$$

where ϱ, θ, ϕ are here *spherical coordinates*, and d is a measure of the trap dimensions, defined as $d^2 = z_0^2 + \frac{1}{2}r_0^2$, for the sake of comparison with the hyperbolic trap. The assumed symmetry limits the index n to even integers, the lowest significant value, of course, being $n = 2$, the quadrupole term. This representation is valid for the spherical domain $\varrho < d$. Experimental techniques to compensate for imperfections in the field, that is, departures from the pure quadrupole, amount to attempting to reduce or cancel the $n = 4$ octupole and higher order multipoles. Following Gabrielse et al. [135], the coefficients c_n of the multipole expansion are found by matching the analytical solution for the cylinder to the multipolar form along the z-axis in the neighborhood of the origin. This is applied to an alternative form of the analytical solution for a cylinder, with symmetric boundary values $\Phi = +\Phi_0/2$ at $z = \pm z_0$ and $\Phi = -\Phi_0/2$ at $r = r_0$, given as

$$\phi = \phi_0 \left[\frac{1}{2} + \sum_i \frac{2(-1)^{i+1} I_0(k_i r)\cos(k_i z)]}{(i + \frac{1}{2})\pi I_0(k_i r_0)}\right] , \qquad (4.52)$$

where $k_i = (i+1/2)\pi/z_0$ to fit the boundary at $z = \pm z_0$. By expanding the cosine function in a power series one finally obtains the following expression:

$$c_n = \frac{(-1)^{\frac{n}{2}}}{(n)!} \frac{\pi^{n-1}}{2^{n-3}} \left(\frac{d}{z_0}\right)^n \sum_{i=0} \frac{(-1)^{i+1}(2i+1)^{n-1}}{I_0(k_i r_0)} . \qquad (4.53)$$

Again we note that the multipole moments c_n are functions of $k_i r_0$, that is $(i+1/2)\pi r_0/z_0$. The series for c_n converges rapidly (but not uniformly) and is readily calculated numerically. It is found that $c_4 = 0$ at $r_0/z_0 = 1.203$ [135], at which value $c_2 = 1.126$ and $c_6 = -0.095$. However, as might be expected, an uncompensated cylindrical trap may typically have anharmonic terms orders of magnitude larger than those of a well constructed hyperbolic trap, some of which have been constructed with c_4 in the 10^{-2}–10^{-3}-range.

4.2.3 Control for Anharmonicity

Mechanical Adjustment

An important advantage of the cylindrical geometry is the ability to compute analytically the magnitudes of the higher order multipole contributions to the field throughout the cylindrical space, and to predict the effect on them of varying the parameters of a simple trap, or of adding correcting electrodes to achieve a more complex structure. The two main areas of application, namely high resolution spectroscopy of ions and fundamental particles, and mass spectrometry, have different objectives with regard to anharmonicities: the former strives to eliminate them as far as possible, while the latter strives to introduce a controlled presence of multipole fields. Exhaustive studies on the reduction of anharmonicities in Penning traps using hyperbolic electrodes for precision measurements on *"geonium"* [34], and other elementary particles, have been extended to the cylindrical geometry [135]. These studies are motivated by the possibility of exploiting the relative ease of fabricating precision miniature cylindrical traps, which yet compete with respect to anharmonicities with the hyperbolic ones.

In the last section the moments c_n were shown to be functions of $k_i r_0 = (i+1/2)\pi r_0/z_0$, and that $c_4 = 0$ at $z_0/r_0 \simeq 0.831$. One possibility therefore for "tuning" out the anharmonicity to 4th order is simply to adjust mechanically the ratio z_0/r_0 to this value. Fortunately, in doing so, the value of c_2, which determines the fundamental oscillation frequency of the particle's motion along the z-axis, is little affected, since, as a function of z_0/r_0, it happens to have a broad maximum where c_4 passes through zero; in fact $\Delta c_2/\Delta c_4 \simeq -0.095$. It must be said, however, that in practice the actual 4th order term may result from imperfections in the construction and alignment of the cylindrical trap, in addition to the contribution inherent to the cylindrical geometry, hence the predicted value of z_0/r_0 for compensation can only be taken as a guide; the final adjustment should be carried out in the course of the actual operation of the trap to reach the optimum compensation. As a practical matter, an important consideration is the sensitivity of c_4/c_2 to the adjustment of z_0/r_0, that is $\partial(c_4/c_2)\partial(z_0/r_0)$, since this sets the required tolerance in adjusting the position of the end caps. The differential ratio is given as 0.81, which means that to achieve the desired value for c_4/c_2 within (say) 10^{-3} it would be necessary to adjust the dimensions to within nearly the same tolerance. This would be difficult in a small trap of only a few millimeters, particularly in the environment of an ultrahigh vacuum system. An alternative approach is to incorporate additional electrodes, whose correcting potential can be precisely controlled.

Electrical Adjustment

The use of correcting ring electrodes in the hyperbolic Penning trap was originally applied by the Dehmelt group with great success in its work on *geonium* and other elementary particles. The idea is simply to insert a *guard ring* between the hyperbolic cylinder and each of the two end caps, the potential on which introduces a precisely variable correction to the field near the center of the trap, resulting in the reduction or elimination of the c_4 term. The axial dimension of the two correcting rings, which must be identical to preserve the symmetry, is an additional available parameter to optimize the compensation. The same compensation method is fortunately applicable to the cylindrical geometry, where an effective method is indispensable. Such an arrangement is shown in Fig. 4.3.

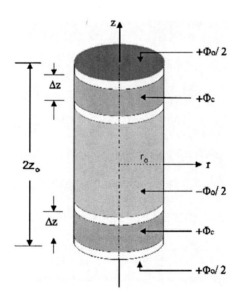

Fig. 4.3. Cylindrical electrode geometry with additional narrow rings for voltage compensation of anharmonicities in the electric field

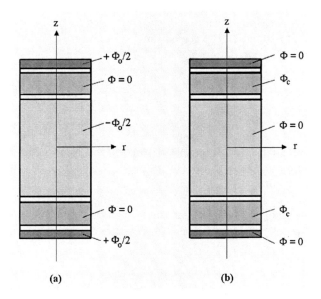

Fig. 4.4. The boundary values resulting from separating the problem of the compensated cylindrical trap into two simpler problems

The electrostatic boundary-value problem may be broken up into two simpler problems by taking advantage of the linearity of Laplace's equation and the uniqueness of harmonic functions satisfying all the given values on the boundary defining the given space. Thus the desired solution may be taken to be the sum of the solutions for the boundary values shown in Fig. 4.4a,b. Writing the field, as before, as a series of harmonic functions, in the following form:

$$\Phi = \frac{\Phi_0}{2} + \sum_i (A_i \Phi_0 + C_i \Phi_c) I_0(k_i r) \cos(k_i z) , \qquad (4.54)$$

the boundary conditions are met if

$$k_i = \left(i + \frac{1}{2}\right) \frac{\pi}{z_0} , \qquad (4.55)$$

and the expansion coefficients A_i and C_i are given by

$$A_i = \frac{(-1)^{i+1} 2 \cos^2(k_i \Delta z/2)}{(i + \frac{1}{2}) \pi I_0(k_i r_0)} , \quad C_i = \frac{(-1)^i 4 \sin^2(k_i \Delta z/2)}{(i + \frac{1}{2}) \pi I_0(k_i r_0)} . \qquad (4.56)$$

The field in the neighborhood of the origin is now expressed in the form of a multipole expansion by defining the coefficients c_n as follows:

$$\Phi = \frac{1}{2} \sum_{n \text{ even}} (\Phi_0 c_n^0 + \Phi_c c_n^c) \left(\frac{\varrho}{d}\right)^n P_n(\cos\theta) , \qquad (4.57)$$

and as before, matching the two representations along the z-axis. Thus by setting $r = 0$ and expanding the cosine function in a power series, and equating equal powers of z, one finds

$$\Phi_0 c_n^0 + \Phi_c c_n^c = \frac{2}{n!} \pi^n \left(\frac{d}{r_0}\right)^n \sum_i \left(i + \frac{1}{2}\right)^n (\Phi_0 A_i + \Phi_c C_i). \qquad (4.58)$$

Here c_n^0 is the coefficient corresponding to applying zero potential to the compensation ring, and c_n^c the coefficient for the compensation field. From this solution a number of figures of merit for the cylindrical trap are readily derived: for example, the quality factor γ of the "orthogonality" of compensation, defined as the degree of independence of the quadrupole field seen by the particles from the adjustment of the compensation field; this is determined by the relative magnitudes of the compensation moments c_2^c, and c_4^c, thus[1]

$$\gamma = c_2^c / c_4^c. \qquad (4.59)$$

For a given value of $\Delta z/z_0$, Gabrielse et al. [135] have shown that it is possible to make $\gamma = 0$ by choosing a particular value of the ratio r_0/z_0, a value which depends weakly on the choice of $\Delta z/z_0$. They point out that this ratio varies around the value $r_0 \simeq 1.16 z_0$, the same value that makes $\gamma = 0$ in the case of hyperbolic electrodes, and also that in the limit of $\Delta z \longrightarrow 0$, the value of r_0/z_0 which makes $\gamma = 0$ is the same as makes $c_4 = 0$ in the case of a simple cylindrical trap without a compensation ring.

In Fig. 4.5 are reproduced plots of c_4^c and c_6^c as a functions of $\Delta z/z_0$ for r_0/z_0 chosen to make $\gamma = 0$. Since the magnitudes of these and higher order moments in the field distribution are orders of magnitude larger than for hyperbolic traps, this implies in practice that the adjustment of Φ_c/Φ_0 is far more critical and must be set at the optimum value with much tighter tolerance. Nevertheless for small oscillation amplitudes only the octupole term is significant and $\Phi_0 c_4^0 + \Phi_c c_4^c$ can be drastically reduced by a practicable adjustment of Φ_c.

4.2.4 Dipole Field in a Cylindrical Trap

We will now consider the electrostatic field distribution in a cylindrical trap produced by the presence of potentials of opposite polarity on the two end caps, resulting in a nonzero electric field intensity at the origin. Again the results will also apply in practice equally well to radiofrequency potentials, since retardation effects are entirely negligible. Such potentials antisymmetric with respect to the $z = 0$-plane, may be produced either by the currents induced by an oscillating trapped ion in an external resistor, connected between the end caps, or are externally applied to the end caps, either in the

---

[1] Gabrielse et al. [135] define coefficients D_k so that $D_k^c = \phi_0 \frac{\partial D_k}{\partial \phi_c}$.

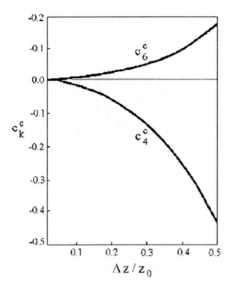

Fig. 4.5. The compensation coefficient $c_4{}^c$ and $c_6{}^c$ for the trap of Fig. 4.3 with z_0/r_0 chosen to make $\gamma = 0$, as a function of $\Delta z/z_0$ [135]

form of a high frequency potential for exciting resonant oscillation in the particles along the z-axis, or in the form of a constant potential to displace the center of oscillation. In all these cases a knowledge of the field distribution in the neighborhood of the origin is needed to predict the degree of coupling between the oscillation of a charged particle in the trap and the external sources connected to the electrodes. In this case then, the boundary conditions are assumed to be as shown in Fig. 4.6. Although the boundary condition $\Phi_A = 0$ for $r = r_0$ can be met directly by an expansion in terms of the harmonic functions $J_0(m_k r)\sin(m_k z)$, we will nevertheless consistently follow Gabrielse et al. and develop the function ϕ', such that $\Phi = \Phi_A(\phi' + z/2z_0)$, as a series in the harmonic functions $I_0(k_n r)\sin(k_n z)$, satisfying the boundary condition $\phi' = 0$ at $z = \pm z_0$ and $\phi' = -z/2z_0$ at $r = r_0$. The result is as follows:

$$\Phi = \Phi_A \left[\frac{z}{2z_0} + \sum_{i=1}^{\infty} \frac{(-1)^i I_0(k_i r)\sin(k_i z)}{i\pi I_0(k_i r_0)} \right] , \tag{4.60}$$

where $k_i = i\pi/z_0$. Again expanding Φ along the z-axis in the neighborhood of the origin, we can match the solution to an expansion in multipoles, thus

$$\Phi = \frac{\Phi_A}{2} \sum_{n \text{ odd}} a_n \left(\frac{z}{z_0}\right)^n P_n(\cos\theta) , \tag{4.61}$$

where the coefficients a_n are given by

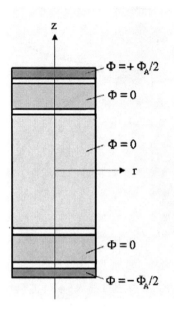

Fig. 4.6. Boundary values for a cylindrical trap with an antisymmetric electric field with respect to reflection in the $z = 0$-plane

$$a_n = \delta_{n,1} + \frac{2\pi^{n-1}}{n!} \sum_{i=1}^{\infty} (-1)^i \frac{i^{n-1}}{I_0(i\pi r_0/z_0)} \, . \tag{4.62}$$

As expected, for $r_0/z_0 \longrightarrow \infty$, corresponding to parallel end plates with a cylinder infinitely far removed, all the coefficients a_n tend to zero except $a_1 = 1$, that is, $\Phi \longrightarrow z/2z_0 \Phi_A$, a uniform field. At the other extreme where $r_0 \longrightarrow 0$, the sum of the series in a_1 converges (nonuniformly) to $-1/2$, and $a_n \longrightarrow 0$ for all n corresponding to total attenuation of the field by the shielding effect of the intervening narrow cylinder. For values of r_0/z_0 of practical interest a_1 falls in the range 0.8–0.9, indicating a slight but often negligible shielding effect of the cylinder.

The presence of an appreciable c_3 in the quadrupole field, due to a practical failure to achieve perfect symmetry about the $z = 0$-plane, leads to a shift in the axial resonance frequency in a Penning trap; which is compounded if a uniform field component is also present, represented by a finite a_1. Thus the effect of a_1 is to shift the center of oscillation by an amount $\Delta z = -\Phi_A a_1/2z_0 m\omega_z^2$, and if c_3 is nonzero, this leads to a change in the frequency of oscillation about this displaced center amounting to

$$\frac{\Delta \omega_z}{\omega_z} \simeq -\frac{3}{4} \frac{a_1 c_3}{c_2^2} \left(\frac{\Phi_A}{\Phi_0}\right)^2 \left(\frac{d}{z_0}\right)^4 \, . \tag{4.63}$$

4.2.5 Open-ended Cylindrical Traps

A central problem in the design of particle traps is to reconcile the opposing requirements, on the one hand of providing a metal enclosure to establish the desired field geometry, and on the other, of providing accessibility to the trapped particles, for probing beams of particles or radiation. Various schemes have in the past been tried, including most notably the use of very fine wires on supporting solid rings, and even conductive coatings of aluminum on quartz for UV beam transparency. In this section we examine an option which promises to be particularly suitable for introducing a beam into the trapping region without degrading the purity of the quadrupole field. This consists of replacing the end caps with open, extended cylinders. Such an arrangement has been used by Byrne and Farago [136] in a polarized electron beam source, by Gabrielse et al. [137] in work on antiprotons, and by Häffner et al. in g-factor experiments on highly charged ions [138].

Following Gabrielse et al. [139] we will summarize an approximate theory of such an open ended cylindrical trap having the added refinement of field compensation rings, by assuming long, but finite, closed end caps, as shown in Fig. 4.7. The long end cylinders, which replace the usual end plates, are assumed to be closed by caps held at zero potential, rather than its actually falling to zero at infinity. This admits the use of discrete Fourier series expansions in the manner already familiar from treatments in earlier sections. That this is a useful approximation is attested by the fact that for $z_e/z_0 > 3$ the predictions come within 1% of the true potential function for an open-ended cylinder based on a Fourier integral solution.

As before the boundary-value problem is separated into two simpler problems by writing $\Phi = \Phi_I + \Phi_{II}$, with nearly identical boundary values as in Fig. 4.4, differing only in having $\Phi = 0$ at $z = \pm(z_0 + z_e)$. The potentials are

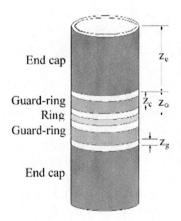

Fig. 4.7. Open-ended cylindrical trap

expressed, as usual, as series of harmonic functions of the form:

$$\Phi_{\mathrm{I}} = \Phi_0 \sum_{i=0}^{\infty} A_i I_0(k_i r) \cos(k_i z) , \qquad (4.64)$$

$$\Phi_{\mathrm{II}} = \Phi_c \sum_{i=0}^{\infty} C_i I_0(k_i r) \cos(k_i z) , \qquad (4.65)$$

where $k_i = (i + \frac{1}{2})\pi/(z_0 + z_e)$, in order to satisfy the assumed boundary condition $\Phi = 0$ at $z = \pm(z_0 + z_e)$. Using the orthogonality of the cosine functions as usual, we find the expansion coefficients to be given by

$$A_i = \frac{(-1)^i - \sin(k_i z_0) - \sin[k_i(z_0 - z_c)]}{(i + \frac{1}{2})\pi I_0(k_i r_0)} ,$$

$$C_i = \frac{2\{\sin(k_i z_0) - \sin[k_i(z_0 - z_c)]\}}{(i + \frac{1}{2})\pi I_0(k_i r_0)} . \qquad (4.66)$$

Again it is convenient, for the purposes of anharmonic analysis in the neighborhood of the origin, to rewrite the potential in the form of a multipole expansion, thus

$$\Phi = \frac{1}{2} \sum_{n(\text{even})} (\Phi_0 c_n^0 + \Phi_c c_n^c) \left(\frac{\varrho}{d}\right)^n P_n(\cos\theta) . \qquad (4.67)$$

The multipole coefficients c_n^0 and c_n^c, giving the contributions arising from the applied trapping potential Φ_0 and the correcting potential Φ_c, are obtained by expanding the cosine function in a power series and equating, as before, the powers of z in the two series along the z-axis. The result is as follows:

$$\Phi_0 c_n^0 + \Phi_c c_n^c = \frac{(-1)^{n/2}}{n!} \frac{\pi^n}{2^{n-3}} \left(\frac{d}{z_0 + z_e}\right)^n \sum_{i=0}^{\infty} (\Phi_0 A_i + \Phi_c C_i)(2i+1)^n . \qquad (4.68)$$

There are three parameters: Φ_c/Φ_0, z_c/z_0, and r_0/z_0, available to adjust in order to correct for the higher order multipoles in the potential, and approach the ideal of a pure quadrupole field extending as far as possible from the center of the trap. It happens that for any given value of z_c/z_0 in a certain range, r_0/z_0 can be chosen so that $c_2^c = 0$, a condition which makes the setting of the quadrupole component of the field, which determines the axial frequency of oscillation in a Penning trap, independent of, or "orthogonal" to, the adjustment of Φ_c. This leaves the parameter z_c/z_0 still to be chosen. Fortunately, according to Gabrielse et al. at $z_c/z_0 \simeq 0.835$, the coefficients c_4^0, c_4^c and c_6^0, c_6^c are such that the homogeneous equations $\Phi_0 c_4^0 + \Phi_c c_4^c = 0$ and $\Phi_0 c_6^0 + \Phi_c c_6^c = 0$ can have a nonzero solution, and therefore not only is it possible to annul the octupole component, but also the dodecapole as well, fulfilling the desired aim.

4.3 Nested Traps

One of the essential differences between the Paul and Penning traps is the fact that particle storage in Paul traps is independent of the sign of the charge of the particles; this is not the case in conventional Penning traps. A modular design consisting of multiple cylindrical Penning traps, however, also offers the possibility of the simultaneous trapping of positive and negative charges. Such a trap is formed of a coaxial set of cylinders to which voltages can be applied in such a way that potential minima in the axial direction are created at different positions along the axis. Clearly, in the regions between these minima the potential has to assume a relative maximum value. Thus while particles having a positive charge are trapped at the potential minima, those having a negative charge will be trapped at the potential maxima appearing between two minima. This arrangement is called a nested trap [140]. If the energy of the positive and negative particles is low enough, they remain trapped in their respective potential energy minima, at different positions in the trap. Assuming a sufficiently uniform magnetic field, this allows the simultaneous measurement of ion cyclotron frequencies of oppositely charged particles. If the energy of one species of particles is raised above the potential barrier that separates its own potential energy minima, the particle will oscillate between the two minima, passing through the position of the other species of the opposite sign. This leads to interaction between the different particles, which can be used, for example, for cooling a hot ion cloud by Coulomb interaction with cold electrons [141]. The technique also has successfully enabled the capture of positrons by antiprotons to form cold antihydrogen atoms in sufficient numbers to carry out studies on this basic form of antimatter [142, 143]. A sketch of the trap used for this experiment is shown in Fig. 4.8.

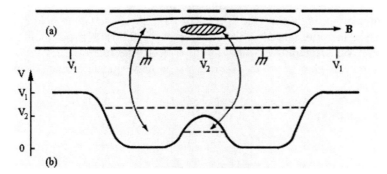

Fig. 4.8. Electrodes (a) and axial potential (b) for a nested pair of Penning traps. The example shows negatively charged particles stored in the center and positive particles in the adjacent potential minima. The *dashed lines* indicate the energies of the particles

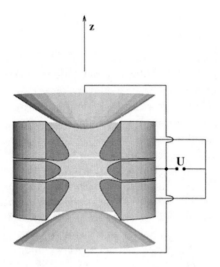

Fig. 4.9. Electrode structure of the rf-octupole trap

4.4 Multipolar Traps

As previously discussed, the field in a real Paul trap, even one with hyperbolically shaped electrodes, will contain in addition to the quadrupole field, higher order multipole components due to holes in truncated electrodes, errors in fabrication and assembly, and for microtraps, surface charges. In fact, in mass spectroscopic applications it has been shown that the presence of certain multipole fields have a beneficial function and may be purposely introduced by varying the electrode spacing. The best ion spectra are not necessarily obtained when the condition $z_0 = r_0/\sqrt{2}$ between the semi-axes of the trap is fulfilled, other ratios may be better suited.

Interest in exploiting the special properties of multipole field traps has led to the introduction of additional electrodes. Thus an rf-octupole trap has five electrodes, including beside the three electrodes of a Paul / Penning trap, two more ring-shaped electrodes, one placed between the ring and each of the end caps (Fig. 4.9).

The potentials applied to these electrodes generate a field whose multipole expansion will have, as lowest order, the octupole term of the form

$$\Phi(\rho, \theta, \phi) = A\rho^4 P_4(\cos\theta) , \qquad (4.69)$$

where P_4 is the 4th order Legendre polynomial,

$$A = \Phi_0/r_0^4 , \quad r_0^4 = (3/8)r_c^4 + (3/7)r_i^4 , \qquad (4.70)$$

where r_c is the center ring radius, and r_i the intermediate ring radius [144]. Converting to cylindrical coordinates, we have

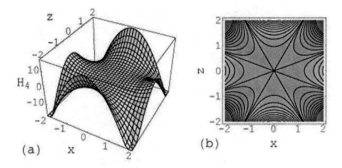

Fig. 4.10. The octupole potential H_4 for $y = 0$, $|x| < 2$ (a) Surface plot (b) Contour plot

$$\Phi = \frac{V_0 \cos \Omega t}{2z_0^4}\left(\frac{3}{8}r^4 - 3r^2 z^2 + z^4\right) . \quad (4.71)$$

The equations of motion of a particle in such an rf-octupole field are clearly intractable nonlinear, coupled, time-dependent differential equations, with no known analytical solution. Nevertheless some important general properties can be anticipated and have been confirmed experimentally. The most significant is the absence of simple periodic orbits, and the relatively large "field-free" volume compared to the quadrupole because of the higher power $(r/r_0)^4$ and $(z/r_0)^4$ dependence of the field amplitude. Furthermore, octupole traps are expected to have stored particles with a narrower energy distribution, superior ion resonance signals and longer lifetimes than the quadrupole [144]. A possible explanation may be that in a pure quadrupole the resonance frequency is independent of amplitude and a relatively small perturbation can stay on resonance even as the amplitude becomes very large and the ion is lost, whereas in the octupole field the ion's oscillation may tend to go off resonance as their motional amplitude builds up.

4.5 Linear Traps

4.5.1 The Ideal Linear Trap

The most striking and transforming development in the evolution of the Paul trap is arguably the demonstration of the possibility of cooling and observing a single ion in the lowest vibrational level near the center of the trap. This has led to extraordinarily high spectral resolution, and the possibility of observing entangled states between cooled ions for quantum computing. But the rf-field is zero only at one point, the trap center, where the particle micromotion at the driving frequency is absent. Over a finite space about the center, which multiple ions will necessarily occupy because of their mutual repulsion,

110   4 Other Traps

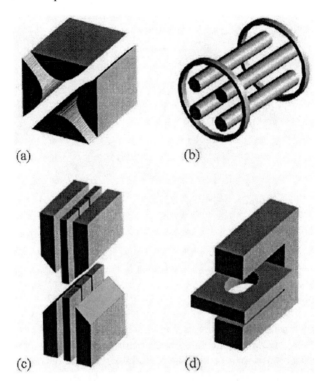

Fig. 4.11. Different realizations of the linear ion trap. (**a**) linear quadrupole trap; (**b**) four rod trap; (**c**) linear end cap trap; (**d**) Paul trap with elongated ring electrode [151]

the kinetic energy associated with the particle micromotion sets a limit to how low a particle temperature can be reached. For such applications as quantum computing where several cold ions must be in close proximity for their wavefunctions to overlap, a string of ions suggests itself, which would be confined by a linear quadrupole field similar to the original Paul mass filter [49]. This field, which provides the radial confinement, clearly has the rf-field zero along the central axis, while axial confinement is ensured by a parabolic static field. Many practical variants of such linear devices have been widely used in different experiments, e.g. [145–152].

The two-dimensional quadrupole field is produced by applying the trapping rf-voltage $V_0 \cos \Omega t$ between diagonally connected pairs of cylinders mounted parallel and equally spaced around the central axis. The particle motion in the plane perpendicular to the trap axis is governed by the same Mathieu type of equation as the cylindrical three-dimensional Paul trap. The confinement along the trap axis is achieved with a weak, approximately parabolic, field created by a dc-voltage U_0 applied to end electrodes, which may take various forms, including rings, cones, or in-line cylinders insulated

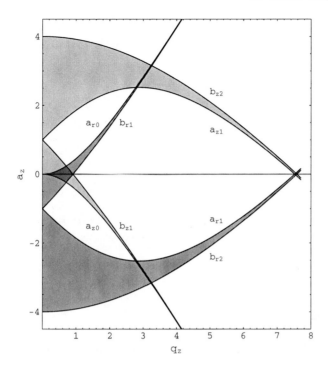

Fig. 4.12. The stability domains for the ideal linear Paul trap: *light grey*: z-direction; *dark grey*: r-direction

from the rf-electrodes. Assuming the length of the trap is very much larger than the spacing of the cylinders, the trapping rf-voltage $V_0 \cos \Omega t$ generates far from the ends, near the trap axis, a potential approximately of the form

$$\Phi(x,y,t) = \frac{x^2 - y^2}{2R^2}(U_0 + V_0 \cos \Omega t) \, , \quad x, y \ll R \, , \quad (4.72)$$

where R is the radial distance from the axis to the quadrupole electrodes.

The equations of motion for a single ion of mass M and charge Q in the x–y-plane are given by

$$\ddot{x} = -\frac{Q}{MR^2}(U_0 + V_0 \cos \Omega t)x \, , \quad \ddot{y} = \frac{Q}{MR^2}(U_0 + V_0 \cos \Omega t)y \, . \quad (4.73)$$

With the substitution

$$a_x = -a_y = \frac{4QU_0}{M\Omega^2 R^2} \, , \quad q_x = -q_y = \frac{2QV_0}{M\Omega^2 R^2} \, , \quad \tau = \frac{\Omega t}{2} \, , \quad (4.74)$$

the equations take the form of the Mathieu equation

$$\frac{d^2 x}{d\tau^2} + (a_x + 2q_x \cos \Omega t)x = 0 \, , \quad \frac{d^2 y}{d\tau^2} + (a_y + 2q_y \cos \Omega t)y = 0 \, . \quad (4.75)$$

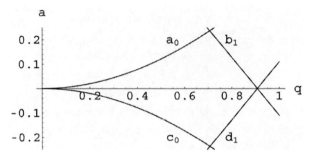

Fig. 4.13. The lowest stability domain for the linear ideal Paul trap

Under conditions where the adiabatic approximation is valid ($|a_x|, |a_y| \ll 1$ and $|q_x|, |q_y| \ll 1$), the solutions to these equations are stable and have the approximate form [85]:

$$x(t) = x_0 \cos(\omega_x t + \varphi_x) \left(1 + \frac{q_x}{2} \cos \Omega t\right) ,$$
$$y(t) = y_0 \cos(\omega_y t + \varphi_y) \left(1 + \frac{q_y}{2} \cos \Omega t\right) , \quad (4.76)$$

where $x_0, y_0, \varphi_x, \varphi_y$ are determined by the initial conditions, and

$$\omega_x = \frac{\Omega}{2}\sqrt{\frac{q_x^2}{2} + a_x} , \quad \omega_y = \frac{\Omega}{2}\sqrt{\frac{q_y^2}{2} + a_y} . \quad (4.77)$$

Hence, the motion of a single ion in the x–y-plane is an harmonic oscillation with frequencies ω_x and ω_y (*secular motion*), having the amplitude modulated at the drive frequency Ω (*micromotion*). Neglecting the micromotion, the ion's secular motion in the x–y-plane of the trap is identical to the motion of a particle in a harmonic pseudopotential Ψ of the form

$$\Psi = \frac{M}{2}(\omega_x^2 x^2 + \omega_y^2 y^2) . \quad (4.78)$$

The depth of the pseudo-potential well in the two directions is then given by

$$\Psi_x = \frac{M}{2}\omega_x^2 R^2 = \frac{Q^2 V_0^2}{4M\Omega^2 R^2} + \frac{QU_0}{2} ,$$
$$\Psi_y = \frac{M}{2}\omega_y^2 R^2 = \frac{Q^2 V_0^2}{4M\Omega^2 R^2} - \frac{QU_0}{2} . \quad (4.79)$$

The axial confining field created by the potential applied to the end electrodes is to a first approximation parabolic near the trap center, whose depth can be approximated by setting $M\omega_z^2 z_0^2/2 = kQU_0$, where z_0 is the distance from the trap center to the end electrodes, ω_z is the observed axial oscillation frequency, and k is a geometric factor ($k \cong 0.025$). This factor accounts phenomenologically for the complicated geometry of conductors involved in determining the field produced by the end electrodes along the trap axis [153]. It follows that in a linear rf-trap the resulting harmonic pseudopotential is anisotropic, given by

$$\Psi = \frac{M}{2}(\omega_x^2 x^2 + \omega_y^2 y^2 + \omega_z^2 z^2) \,. \tag{4.80}$$

In summary, the linear ion trap, which combines both dynamical and static forms of particle trapping, has the advantage that the rf-quadrupole field is zero not just at a single point in space, but along a line, the symmetry axis of the trap. Hence more than one ion can be cooled in this trap to the minimum point of the (pseudo) potential without the mutual repulsion of charges forcing one or more of the particles to occupy positions of higher energy.

4.5.2 Electrostatic Field in a Linear Quadrupole Trap

Suppose the electrode geometry consists of four thin linear segmented conductors set parallel and equally spaced around the z-axis, and at an equal distance r_0 from it, as shown in Figs. 4.14 and 4.15. Let the length of the inner segments to which the quadrupole potentials $\pm V_0$ are applied be $2z_0$ and the outer segments each carrying the potential U_0 have length $z_c - z_0$.

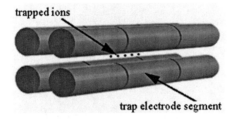

Fig. 4.14. Linear trap geometry with segmented rod electrodes

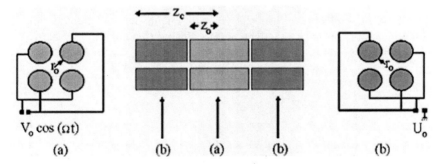

Fig. 4.15. Linear trap power supply connections

4 Other Traps

We will seek an approximate solution to the boundary value problem by assuming the conductors to be narrow strips, circular arcs in cross section, centered on the z-axis, so that the boundary conditions are as follows: for $-z_c < z < -z_0$ and $z_0 < z < z_c$ the potential is U_0 on the cylinder radius r_0 at $0 < \Phi < \Delta\Phi$, $\pi/2 < \Phi < (\pi/2 + \Delta\Phi)$, $\pi < \Phi < (\pi + \Delta\Phi)$, and $3\pi/2 < \Phi < (3\pi/2 + \Delta\Phi)$, where $\Delta\Phi \ll \pi$; while for $-z_0 < z < +z_0$ the potentials on the same circular arcs are assumed to be $+V_0$, $-V_0$, $+V_0$, $-V_0$ respectively. We will seek a solution in the form of a series of harmonic functions of the form:

$$\Phi = \sum_{n,m=0}^{\infty} A_{nm} I_m(n\pi r/z_c) \cos(n\pi z/z_c) \cos(m\phi) , \qquad (4.81)$$

where $I_m(x)$ is the modified Bessel function of the first kind. We write $\Phi = \Phi_1 + \Phi_2$, where Φ_1 is the solution for U_0 alone being present, with $V_0 = 0$, and Φ_2 corresponding to the presence of V_0 with $U_0 = 0$, and assume that $z_0 \ll z_c$ so that there is negligible error incurred in the neighborhood of the origin by assuming periodic boundary conditions satisfied by the two-variable Fourier series expansion with coefficients A_{nm}. Following the usual methods of Fourier series analysis, one finds the following:

$$A_{nm}^{(1)} = \frac{4U_0}{\pi^2 nm} \frac{\sin(n\frac{z_0}{z_c}\pi)}{I_m(\frac{n\pi}{z_c}r_0)} \left[1 + \cos(m\pi) + 2\cos\left(\frac{m\pi}{2}\right)\right] \sin\left(\frac{m\Delta\phi}{2}\right) , \qquad (4.82)$$

$$A_{nm}^{(2)} = \frac{4V_0}{\pi^2 nm} \frac{\sin(n\frac{z_0}{z_c}\pi)}{I_m(\frac{n\pi}{z_c}r_0)} \left[1 + \cos(m\pi) - 2\cos\left(\frac{m\pi}{2}\right)\right] \sin\left(\frac{m\Delta\phi}{2}\right) , \qquad (4.83)$$

where $A_{nm} = A_{nm}^{(1)} + A_{nm}^{(2)}$ for $n, m = 1, 2, 3 \ldots$ and

$$A_{n0} = -\frac{8U_0}{\pi n} \frac{\sin\left(n\frac{z_0}{z_c}\pi\right)}{I_0\left(\frac{n\pi}{z_c}r_0\right)} \frac{\Delta\phi}{\pi}, \quad A_{00} = 4U_0 \left(1 - \frac{z_0}{z_c}\right) \frac{\Delta\phi}{\pi} . \qquad (4.84)$$

Of particular interest are the relative magnitudes of the dc- and rf-components of the field along the z-axis. Noting that for $m \neq 0$, $I_m(0) = 0$, we find that the dc potential along the z-axis is given by

$$\Phi_{r=0}^{dc} = 4U_0 \frac{\Delta\phi}{\pi} \qquad (4.85)$$

$$\times \left\{ \left(1 - \frac{z_0}{z_c}\right) - \frac{2}{\pi} \sum_{n=1}^{\infty} \frac{1}{nI_0\left(\frac{n\pi}{z_c}r_0\right)} \sin\left(n\frac{z_0}{z_c}\pi\right) \cos\left(n\frac{z}{z_c}\pi\right) \right\} ,$$

and the quadrupole high frequency potential is given by

$$\Phi^{rf} = \frac{8V_0}{\pi^2}\sin(\Delta\phi)\left\{\sum_{n=1}^{\infty}\frac{1}{n}\sin\left(\frac{n\pi}{z_c}z_0\right)\frac{I_2\left(\frac{n\pi}{z_c}r\right)}{I_2\left(\frac{n\pi}{z_c}r_0\right)}\cos\left(\frac{n\pi}{z_c}z\right)\right\}\cos(2\phi) \ . \tag{4.86}$$

In the vicinity of the origin the quadrupole field is approximated by

$$\Phi^{rf} \approx \frac{8V_0}{\pi^2}\sin(\Delta\phi)\sum_{n=1}^{\infty}\frac{1}{n}\sin\left(\frac{n\pi}{z_c}z_0\right)\cos\left(\frac{n\pi}{z_c}z\right)\frac{1}{r_0^2}\left(x^2-y^2\right) \ , \tag{4.87}$$

for $r \ll r_0 \ll z_c$.

4.5.3 Electric Field in a Linear Multipole Trap

The theory of the last section lends itself to a straightforward generalization to a linear multipole trap of higher order. Such a trap would consist of an even number of parallel, segmented linear electrodes, equally spaced around a circular cylinder. The application of a potential which alternates in polarity between consecutive middle segments of the linear conductors will produce a multipole field whose radial intensity near the axis remains small over a greater range from the axis than the quadrupole field.

Let there be $2N$ conductors equally spaced around a cylinder of radius r_0 at angular positions $i\pi/N$, where $i = 1, 2, 3 \ldots$ (Fig. 4.16). Again assuming the approximation that the conductors are narrow strips of circular cross sections around the cylinder, and that the inner segments bear the potentials $\pm V_0$ while the outer segments at both ends all have the same potential U_0, we find for the harmonic expansion coefficients $A_{nm}^{(1)}$ and $A_{nm}^{(2)}$ of the potential Φ (see (4.81)) the following expressions:

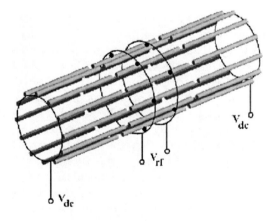

Fig. 4.16. Multipolar linear traps of higher order. Geometry and power supply

$$A^{(1)}_{nm} = \frac{4}{\pi^2} \frac{U_0}{nm} \frac{\sin\left(n\frac{z_0}{z_c}\pi\right)}{I_m\left(n\frac{r_0}{z_c}\pi\right)} \sum_{i=0}^{2N-1} \cos\left(\frac{i}{N}m\pi\right) \sin\left(m\frac{\Delta\phi}{2}\right), \qquad (4.88)$$

$$A^{(2)}_{nm} = \frac{4V_0}{\pi^2 nm} \frac{\sin\left(n\frac{z_0}{z_c}\pi\right)}{I_m\left(n\frac{r_0}{z_c}\pi\right)} \sum_{i=0}^{2N-1} (-1)^i \cos\left(\frac{i}{N}m\pi\right) \sin\left(m\frac{\Delta\phi}{2}\right). \qquad (4.89)$$

One can show by examining the sum of the geometric series whose general term is $\exp(im\pi/N)$ that the coefficients $A^{(1)}_{nm}$ assume nonzero values only for $m = 2pN$, while the coefficients $A^{(2)}_{nm}$ are nonzero for $m = (2q-1)N$, where p, q are integers. It follows that in the neighborhood of the center of the trap, where the terms involving $A^{(2)}_{nm}$ dominate, the first nonzero term will be A_{nN} and therefore the lowest order Bessel function in the series, which determines the radial dependence of the field will be $I_N(n\frac{r}{z_c}\pi)$. For $r \ll r_0 \ll z_c$ therefore, the radial dependence is expected to be $\sim (r/r_0)^N$, rising very slowly with r, from zero on the axis. That is, the high frequency potential for a given z near the center of the trap is given by

$$\Phi = V_0 \left(\frac{r}{r_0}\right)^N \cos(N\phi) \cos(\Omega t), \qquad (4.90)$$

from which follows the time-independent pseudo-potential, or effective potential Φ_{eff} as

$$\Phi_{\text{eff}} = \frac{QV_0^2 N^2}{4M\Omega^2 r_0^2} \left(\frac{r}{r_0}\right)^{2N-2}. \qquad (4.91)$$

For large values of N, the potential remains nearly constant in a region around the axis, and changes rapidly only in the vicinity of the electrodes. A comparison of the potential of the quadrupole trap and a 22-pole trap is shown in Fig. 4.17, in which it is seen that the latter has a greatly extended region of weak high frequency field intensity. This is of considerable practical significance since the trapped particles would gain very little energy in collisions with neutral background molecules in such a low field region. Radio-frequency heating takes place only during the short time that the ions spend near the electrodes. Consequently they are thermalized to the temperature of the ambient gas, with a corresponding energy distribution very similar to the thermal one of the neutrals. A numerical calculation of the trajectories, as shown in Fig. 4.18, confirms the fact that the ions follow nearly a straight line most of the time, and the micromotion amplitude becomes significant only near the turning points. More details on the behavior of charged particles in multipole traps can be found in [155].

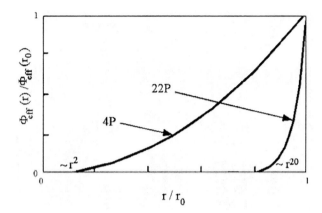

Fig. 4.17. Effective potential for a quadrupole (4P), and a 22-pole as a function of the distance from the center line [154]

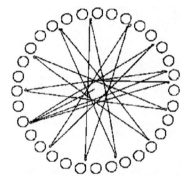

Fig. 4.18. Radial trajectory of an ion in a 32-pole trap. In the electrode vicinity the rf-field influence can be seen [154]

4.6 Ring Traps

A variant of the linear Paul trap is the ring trap. Here the four electrodes to which an rf-potential is applied are in the form circles equally spaced on the surface of a toroid, typically several centimeters in diameter. The quadrupole potential near the central circle of the arrangement provides confinement in the radial direction, but no confining potential exists along the circular axis; stored ions can freely move along this direction. In practice, small stopping potentials may exist arising from surface charges on the trap electrodes, and stored ions will assume a position fixed in space when their energy is small. A cloud of stored particles will be distributed in such a way that the average distance between the ions is constant. On the other hand, additional time varying potentials applied in the axial direction between different segments on the electrodes may be used to accelerate the stored ions. As such, the

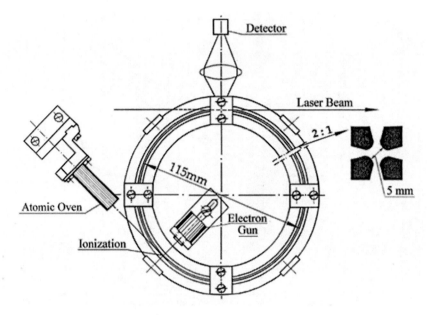

Fig. 4.19. Quadrupole storage ring. The *cross section* of the electrode configuration is shown on the *insert* on the righthand side of the figure [159, 160]

device can serve as a small storage ring which can be used to investigate properties of stored charged particles at medium to low energies.

The principle and first experimental realizations of quadrupole storage rings are described by Drees and Paul [156] and by Church [157]. An example of the ring trap dimensions and operation parameter are given in [158].

More recently Walther and coworkers have used a ring trap (Fig. 4.19) to investigate transitions between the gaseous, liquid and crystalline phases of an ion cloud under the influence of laser cooling forces [159]. Schätz et al. used a similar ring trap ("PALLAS") to accelerate ion clouds and to investigate the stability of accelerated ion crystals [161].

4.7 Planar Paul Traps

Due to the difficulty of achieving the requisite precision in the micromachining of hyperboloidal surfaces for a Paul micro-trap, alternative geometries producing the essential saddle point in the potential, have been developed. Of these, the planar geometry is ideally suited to precise fabrication on a micrometer scale. Hence planar quadrupole traps involving one or more conducting rings, or hole(s) in one or more conducting sheets (fig. 4.20) have come into common use.

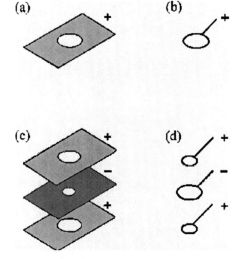

Fig. 4.20. Schematic of planar Paul traps made from conducting rings or holes in conducting sheets [162]. (a) One-hole trap. (b) One-ring trap. (c) Three-hole trap. (d) Three-ring trap. Copyright (2003) by the American Physical Society

In comparison with the conventional Paul trap, the planar traps permit the trapped ions to be optically accessible through a larger solid angle, thereby improving the optics for ion detection and laser-ion interaction, an advantage of particular importance in a micro-trap.

The field in the planar trap approximates the Paul trap near the trap center where it is harmonic; however, away from that point the field becomes increasingly anharmonic. But as discussed in connection with the cylindrical traps and others, the anharmonic terms can be compensated by the adjustment of potentials applied to electrodes incorporated for that purpose.

Of the planar geometries, the one commonly referred to as the Paul–Straubel trap [158, 163] consists essentially of a small ring carrying the high frequency potential, placed symmetrically between grounded conducting sheets parallel to the plane of the ring. The boundary value problem for the potential field around such a ring of finite thickness, within open boundaries, is a difficult one having "mixed boundary conditions". Computerized iterative methods[2], such as the "Boundary Element", and the "Finite Difference: Relaxation" methods are highly developed for the numerical solution of such complex electrostatic field problems, mostly in the context of electron lens design (see for example [164]). However, an approximate analytical solution is possible, and may be useful for an overall understanding of the relative importance of the physical parameters of the system. Suppose that a thin ring, radius R, carrying the high frequency voltage, is placed in the plane

---

[2] For example SIMION 7.0, and EStat 4.0

$z = L/2$ symmetrically between two grounded, plane conducting sheets at $z = 0$ and $z = L$, normal to the z-axis, which passes through the center of the ring. We assume initially that we have a circle of charge rather than a conducting ring at a specified potential, and solve the potential field problem using the Dirichlet Green function [165] for the space between the two conducting planes. The potential function for $r < R$, referred to the plane of the ring, $z = L/2$, and satisfying the boundary conditions, can be written

$$\Phi(r) = A \sum_{n \text{ odd}} \cos\left(\frac{n\pi}{L}\varsigma\right) I_0\left(\frac{n\pi}{L}r\right) K_0\left(\frac{n\pi}{L}R\right) , \qquad (4.92)$$

where $z = L/2 + \varsigma$, and A is a constant proportional to the charge on the ring. We can now determine an approximate value for A in terms of the voltage applied to the ring, by using the fact that at points sufficiently close to the circle of charge, the equipotential surfaces approximate toroids surrounding the line of charge, with an approximately circular meridian cross section. It remains only to compute the potential at the torus coinciding with the surface of the given actual ring, and set that potential equal to the applied voltage.

For large separations of the grounded planes, that is $R/L \ll 1$, the potential field approaches that of an isolated ring of charge in unbounded free space, to which the following multipole expansion applies in spherical coordinates ϱ, θ:

$$\Phi = B \sum_{n \text{ even}}^{\infty} \left(\frac{\varrho^n}{R^{n+1}}\right) P_n(0) P_n(\cos\theta) , \qquad (4.93)$$

where $\varrho < R$ and P_n is the Legendre function. Again B should be obtainable using the same approximate procedure as described above. The extent of anharmonicities in the field produced by an isolated ring ($L/R \geq 3$) is shown, as usual, by a series expansion in powers of the cylindrical coordinates r, z. Thus for $r/R \ll 1$ and $z/R \ll 1$, we have

$$\Phi = B\left[1 - \frac{1}{4R^2}(2z^2 - r^2) + \frac{9}{8R^4}\left(\frac{1}{3}z^4 - r^2z^2 + \frac{1}{8}r^4\right) + \ldots\right] . \qquad (4.94)$$

To find the effect of the presence of the grounded plane electrodes, we expand the general solution in powers of the cylindrical coordinates in the neighborhood of the point $(0, L/2)$, to obtain

$$\Phi \approx A\left[c_0 - \frac{c_2}{R^2}(2z^2 - r^2) + \frac{c_4}{R^4}\left(\frac{1}{3}z^4 - r^2z^2 + \frac{1}{8}r^4\right) + \ldots\right] , \qquad (4.95)$$

where

$$c_0 = \sum_{n \text{ odd}}^{\infty} \frac{4k}{\pi} K_0(kn) , \quad c_2 = \sum_{n \text{ odd}}^{\infty} \frac{k^3}{\pi} n^2 K_0(kn) , \quad c_4 = \sum_{n \text{ odd}}^{\infty} \frac{1}{2}\frac{k^5}{\pi} n^4 K_0(kn) ,$$

and $k = \pi R/L$. As expected, for $R/L \ll 1$, the values of c_2 and c_4 approach the isolated ring values, but of greater interest is the fact that the ratio c_4/c_2 exhibits a minimum as a function of L/R, occurring at $L/R \approx 1.7$. As already mentioned, more effective manipulation of the higher order anharmonicities can be achieved by appropriate choice of the diameter and positions of additional compensation plates. In this way the anharmonicity up to the octupole term may be readily suppressed.

The trapping of particles at the saddle point in the potential field of circular holes in plane conductors, goes back to a proposal in 1977 [166] involving a large array of micro-traps to increase the effective number of ions contributing to a resonance signal. Currently the motivation is in the opposite direction: to trap and observe *individual* charged particles in a single micro-trap. Again an analytical solution of the boundary value problem is tractable only for the limiting case of unbounded parallel plane electrodes being placed sufficiently far from the trap that the field near them is uniform. Assume we have a circular hole of radius a centered at the origin in a plane conducting sheet taken to be the x–y-plane, with two parallel grounded conducting planes at $z = \pm L/2$, with $L \gg a$. Let the voltage applied to the center conductor be Φ_0. It can be shown that a solution for the potential field for $r < a$ can be written in the form [165]

$$\Phi = \Phi_0 \left(1 - \frac{2|z|}{L}\right) - \frac{4\Phi_0 a}{\pi L}\left[\sqrt{\frac{R-\lambda}{2}} - \frac{|z|}{a}\arctan\sqrt{\frac{2}{R+\lambda}}\right], \quad (4.96)$$

where

$$\lambda = \frac{1}{a^2}(z^2 + r^2 - a^2), \quad R = \sqrt{\lambda^2 + 4\frac{z^2}{a^2}}. \quad (4.97)$$

In the neighborhood of the center of the trap, for $r/a \ll 1$ and $z/a \ll 1$, we have the expansion up to the octupole term

$$\Phi \approx \Phi_0 \left\{ 1 - \frac{4a}{\pi L}\left[1 + \frac{1}{2a^2}(2z^2 - r^2) - \frac{1}{a^4}\left(\frac{1}{3}z^4 - r^2 z^2 + \frac{1}{8}r^4\right)\right]\right\}. \quad (4.98)$$

As with the ring trap, the anharmonic terms can be reduced by appropriate choice of the diameter and position of the two parallel plane electrodes, and the voltage applied to them.

Planar micro-traps have also been fabricated in multiple arrays that can be used to confidently compare data from different traps within the same array, thereby avoiding many factors that could not be controlled between individually fabricated traps. In Fig. 4.21 a double ring-and-fork Paul trap with two rings is shown [167], that allows one also to investigate, in a controlled way, the dependence of the trapping properties on the size of the trap. By using the planar geometry, stable ion trapping has been obtained. For example, in experiments made with a three-hole micro-trap with a center

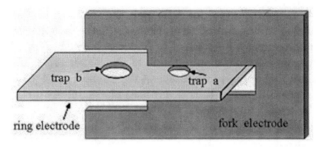

Fig. 4.21. Schematic of the double planar Paul trap. The distance between the two rings is 1.7 mm, and between the "end caps" (fork electrode) is 0.35 mm for the trap a and 0.79 mm for the trap b [167]

hole radius of 80 μm, laser-cooled Ba$^+$ ions have been trapped, both individually and as a dense cloud, in which the two-ion distance is compressed to 1 μm [162].

The possibility of merging the ion trap and integrated-circuit technologies in the form of a single chip, containing the trap, diode lasers, and associated electronics, was long ago foreseen [162]. Currently lithographically fabricated rf-traps are available, allowing new experiments in quantum optics (Fig. 4.22) [168].

The idealized four-rod linear trap geometry is approximated using a wafer stack. It has laser-machined slots and evaporated gold traces of 0.5 μm thickness, deposited through a shadow mask onto the alumina substrate. The wafers are held together with two screws (one on each end of the wafer stack) passing through the trap axis (Fig. 4.22). Stable ion confinement in such a

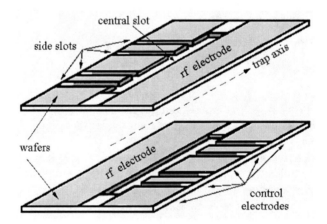

Fig. 4.22. Planar photolithographic double wafer-stack trap (drawing not to scale). Wafers are made from gold coating bare alumina [168]. The central slot width and wafer distance are of the order of 400 μm

trap is obtained for a peak amplitude of the applied rf-voltage of about 500 V at 230 MHz [169].

Such traps are particularly appealing in the case of the very important applications to atomic clocks using a multiple ion trap array to improve the accuracy over a one-ion clock, and quantum computation with trapped ions.

4.8 Electrostatic Traps

We recall that according to Earnshaw's theorem, it is not possible to maintain a charged particle in three-dimensional stable equilibrium in free space by electrostatic forces alone. A purely electrostatic ion trap therefore seems to contradict this theorem; however, the theorem precludes only static equilibrium of the ions, whereas dynamical confinement may be possible in the reference frame of fast moving ions.

Zajfman and coworkers [170] have developed a device (Fig. 4.23) which is capable of confining ions having kinetic energies in the keV range, for finite times using only electrostatic fields. It is based on the analogy with optical resonators: ions are reflected between two electrostatic "mirrors" when their energy is smaller than the potential applied to the electrodes. The mirrors consist of a set of five electrodes, to which different potentials are applied to focus an ion beam in accordance with the stability criterion for radial

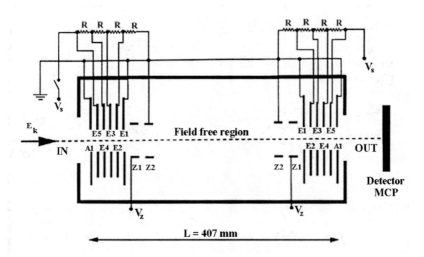

Fig. 4.23. Schematic view of the ion trap. The ion beam is injected from the *left*, when the entrance electrodes are grounded. The three electrodes $E5$, $Z1$, and $Z2$ form an asymmetric Einzel lens, which is used for focusing the ions. Neutral particles escaping the trap are monitored by a microchannel plate detector downstream. The drawing is not to scale [170]. Copyright (2003) by the American Physical Society

confinement in the "cavity": $L/4 < f < \infty$, where L is the distance between the mirrors, and the focal length f is given by $f = R/2$, where R is the radius of curvature of the potential at the electrodes.

The space between the mirrors is approximately field-free. Experimentally, a device of about 40 cm in length and a few centimeters inner diameter has been demonstrated to achieve stable confinement of N_2^+ ions over a range of a few hundred volts around a total trapping voltage of 4.2 keV. The ion lifetime is mainly limited by neutralisation due to charge transfer collisions with background particles, and is typically of the order of 1 s at residual pressures on the order of 10^{-8} Pa.

The energy of the stored ions, amounting to several keV, is much larger than is typical of conventional Paul or Penning traps. Thus the device represents an intermediate step between low energy ion traps and storage rings where particles having MeV energies are confined. As such, it may have applications in various areas of atomic and molecular physics. The fact that the beam has a well defined direction in a field free region makes it attractive for impact studies with charged or neutral particles, or photons at intermediate energies. The first applications of the technique have been carried out on photodissotiation of molecular ions.

4.9 Kingdon Trap

The Kingdon trap consists of a straight wire of small diameter placed along the axis of a hollow cylinder, with isolated end cap electrodes at both ends. An electrostatic potential difference Φ_0 is applied between the wire and the cylinder. Since this is a purely electrostatic device, no potential minimum in free space can, of course, be produced in which ions could be stored. If, however, the ions have a finite angular momentum about the central wire, the electric field can provide a restoring force in the radial plane, leading to dynamical stability of the ion trajectories. Confinement in the axial direction is provided by a static potential applied to the end caps. This configuration was first studied by Kingdon [171] for the neutralisation of (negative) space-charge, originating from thermionic emission from a central wire cathode, by trapped positive ions, between the wire cathode and the cylindrical anode.

The potential field is logarithmic near the central wire, but is expected to have a complicated distribution throughout the trap, determined by the potentials on the wire, the cylinder and the end caps. The geometry of the boundary surfaces makes it difficult to derive an analytical solution useful as a basis for a general discussion of the motion of the ions. The simplest to analyze are the circular orbits close to the midplane; they are the most stable, since they remain farthest from the electrodes, where the potential is given by [172]

$$\Phi = \Phi_0 \left[\frac{\ln(r/a)}{\ln(b/a)} \right] \;, \quad r = \sqrt{(x^2 + y^2)} \;, \tag{4.99}$$

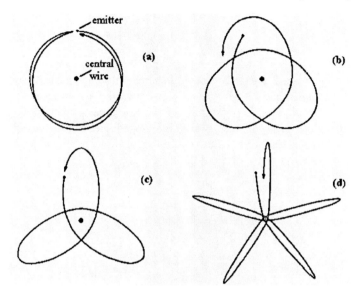

Fig. 4.24. Examples of orbits of particles in a logarithmic potential for different trapping conditions (see [173]). Copyright (2003) by the American Physical Society

Φ_0 is the potential difference between the outer cylinder and the central wire, and a and b are the radii of the inner and outer electrodes, respectively. It is assumed that the length of the device is much larger than the radius of the inner electrode.

The equations of motion for these orbits are [172, 173]

$$M\frac{d^2r}{dt^2} - \frac{L^2}{Mr^3} = -\frac{dU}{dr} , \quad M\frac{d}{dt}\left(r^2\frac{d\theta}{dt}\right) = 0 , \quad M\frac{d^2z}{dt^2} = -\frac{dU}{dz} ,$$
(4.100)

where $U = q\Phi$.

The angular momentum L and the total energy E of the ions are constants of the motion; from this one finds a lower limit on the radius r:

$$r > L\left[2M(E - U)\right]^{-1/2} ,$$
(4.101)

or, equivalently, an upper bound on the angular momentum for orbits intersecting the central wire:

$$L < a(2ME)^{1/2} .$$
(4.102)

The solution of the equations of motions can, for small deviations from the central plane, be described by two uncoupled harmonic oscillations with frequencies

$$\omega_r^2 = 3\omega^2 + \frac{1}{M}\frac{d^2V}{dr^2} , \quad \omega_z^2 = \frac{1}{M}\frac{d^2V}{dz^2} ,$$
(4.103)

where ω is defined by
$$\omega_r^2 + \omega_z^2 = 2\omega^2 \ . \tag{4.104}$$

For particular values of the ion energy and angular momentum, Hooverman [173] has calculated ion trajectories in the midplane of the Kingdon trap (Fig. 4.24). The device is simple to construct and to operate; it has been used in atomic physics, for example, for lifetime measurements of metastable states of highly charged ions [174]. The storage times of ions is limited by collisions with background gas molecules; several seconds have been obtained at pressures on the order of 10^{-8} Pa.

Part II

Trap Techniques

The practical realization of charged particle traps and the exploitation of their unique experimental potential have depended on the development of a number of inseparable techniques. The most basic are the production and measurement of ultrahigh vacuum, made possible by the titanium ion pump, and the production of intensely bright and finely tunable laser radiation.

The first experiments were characterized by the fact that the density of trapped particles was limited by Coulomb repulsion to about $10^6\,\mathrm{cm}^{-3}$, for a depth of the potential minimum of the order of several eV. This corresponds to a neutral gas pressure of about 10^{-8} Pa and is a small number compared to typical densities in neutral atom spectroscopy. In order to deal with such low particle numbers, sensitive detection methods had to be developed. The relatively diffuse clouds of trapped ions meant initially that only electronic detection was feasible, using resonant rf excitation, as originally devised by Paul and his students. This essentially involved the detection of the change in the quality factor of an external resonant LC circuit. Before laser sources became available, detection using resonance fluorescence induced by light from conventional lamps was extremely difficult. Now however, the observation of laser induced fluorescence when the ion under investigation has a suitable energy level scheme is commonplace. Also the electronic method evolved in the direction of detecting the currents induced in the trap electrodes by the ion's oscillation, using sensitive narrow band detectors. Both techniques have advanced to a state where it is possible to monitor continuously a single stored particle, thus achieving the ultimate possible sensitivity. This is not only of academic interest, but allows the investigation of species that are not available in large quantities such as rare radioactive isotopes or antiparticles. Moreover precise measurements of single particle properties avoid possible systematic uncertainties, imposed by interaction with simultaneously trapped ions. Thus recent experiments of extremely high precision rely on single particle operation of traps.

The many varied applications of particle traps have led to special techniques to load the trap. The most common is performed by ionization of neutral atoms or molecules inside the trap. In many cases, however, the ions are created outside and then injected to the trap. This requires in general some kind of damping mechanism since otherwise the particles would leave the potential minimum. Great effort has been devoted to the specific case of mass spectrometry using the Paul trap. Charged molecules are created at high background pressure by electrospray ionisation. A liquid containing the anion of interest is pumped through a metal capillary which has an open end with a sharply pointed tip which is attached to a voltage supply. The liquid becomes charged and enters the apparatus through a number of apertures which allow differential pumping to the high vacuum region at the trap position. This technique will not be discussed in this chapter since many review articles and books are available from the mass spectrometry community [175–177].

One of the basic ingredients for successful experiments using charged particle traps is the operation at ultra-high vacuum, since collisions with neutral background molecules might limit the storage time. They also will lead to shifts of the energy levels and thus be the source of systematic uncertainties in spectroscopy. Titanium ion getter pumps or molecular pumps provide easily base pressures of the order of 10^{-8} Pa, after cleaning and bake-out of the vacuum vessel. Standard ultra-high vacuum practices must be followed requiring the use of heat resistive materials with low vapour pressure. Stainless-steel and oxygen-free copper are the main materials used when building a trap apparatus.

5 Loading of Traps

5.1 Ion Creation Inside Trap

In the static conservative fields of a Penning trap, ions created within the space defined by the electrodes of the trap will remain trapped, provided their energy is smaller than the depth of the effective trapping potential, as determined by the applied field strengths. Such ions are generally created from a background gas or atomic beam passing through the trap, either by photoionization or electron impact (Fig. 5.1).

Electron impact ionisation usually is not specific to one element and investigation of a particular ion species requires the ejection of the unwanted species from the trap. This can be achieved by excitation of one of their (mass dependent) oscillation frequencies to such amplitudes that the ion touches an electrode and is lost. More specific is resonant ionisation by narrow band lasers: by proper selection of laser frequencies, a specific isotope or any desired mixture of isotopes can be loaded into the trap [178, 179].

In the Paul trap the requirement of low energy is not sufficient: at the instant of its creation even with zero kinetic energy the ion may experience a high amplitude of the radio-frequency trapping field and may be ejected out of the trap. Thus the probability that the ion remains confined depends on the phase of the rf-field. Fischer has calculated curves of equal maximum

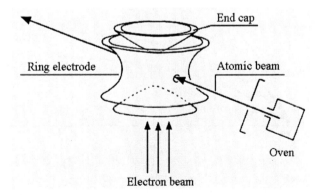

Fig. 5.1. Atomic beam ionization by electrons for ion creation inside the trap

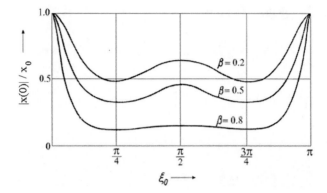

Fig. 5.2. Curves of equal maximal ion oscillation amplitude as function of the ion initial position $x(0)$ (in units of the trap size x_0) and trapping field phase ξ_0 for zero ion initial velocity [56]

amplitude of the oscillation, expressed by the ratio of the initial position $|x(0)|$ to the trap size x_0, as a function of the rf-phase for different values of the stability parameter β [56]. The result is shown in Fig. 5.2. If the initial position in the diagram is below the curve, then the maximum amplitude is smaller than the trap size and the ion remains confined. If the time of creation is statistically distributed over all phases, the ratio of the area below the curve to the total area gives the overall confinement probability.

The increase of the stored ion number during the creation process can be well described by a function of the form $A[1 - \exp(-t/\tau)]$ (Fig. 5.3), where

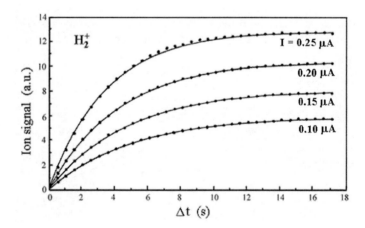

Fig. 5.3. Signal from trapped H_2^+ ions created by electroionisation from background gas as function of the electron pulse length Δt for different electron emission currents I. The background gas pressure was 10^{-7} Pa

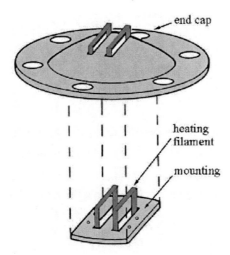

Fig. 5.4. Lower end cap with retractable heating filament mounting

the characteristic time constant τ depends on the creation rate, and the equilibrium ion number A is given by the ratio of the creation to loss rate.

A convenient way to load the trap with ions is to produce them at the border of the trap. A filament can be placed in a slot in an end cap electrode in such a way that it is approximately continuous with the end cap surface (Fig. 5.4). Upon heating the filament ions are released if the ionisation potential of the atoms is smaller than the work function of the filament metal. At sufficiently high temperatures, electrons are also released which may ionize any neutral atoms evaporated from the filament. The efficiency of this method has been estimated for filaments loaded with neutral atoms injected from a mass separator at 60 keV energy, where it was found that a load of 10^{14} atoms could be used about 100 times to fill a trap with 10^5 ions each [180].

5.2 Ion Injection from Outside the Trap

An ion injected into a Penning trap from an outside source under high vacuum condition may be captured by proper switching of the potentials on the trap electrodes: When an ion approaches the first end cap along the magnetic field lines, as it passes through that end cap its potential is set to zero, while the second end cap is held at some retarding potential. If the axial energy of the ion is smaller than the retarding potential it will be reflected; but before the ion leaves the trap through the first end cap, its potential is raised and prevents the ion from escaping (Fig. 5.5).

This simple method requires that the arrival time of the ion at the trap be known; this can be achieved by pulsing the ion source. Moreover the

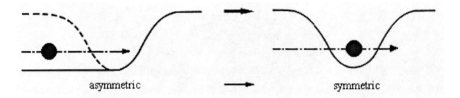

Fig. 5.5. Simple model describing the ion capture in a Penning trap

switching has to be performed in a time shorter than the transit time of the ions through the trap. For ion kinetic energies of a few 100 eV and a trap size of one centimeter this time is of the order of 100 ns. Successful capture of ions from a pulsed source is routinely used at the ISOLDE-facility, CERN, where efficiencies up to 50% at ion energies of 40 keV are achieved [181, 182]. Injection into a Paul trap under similar conditions is more difficult, since the time-varying trapping potential typically in the MHz frequency range cannot be switched from zero to full amplitude in a time of the order of less than a microsecond. The ion longitudinal kinetic energy, however, may be transferred into transverse components by the inhomogeneous electric trapping field and thus the ions may be confined for some finite time. Schuessler and coworkers [183] have made extensive simulations and phase studies and have found that ions injected at low energy, during a short interval when the ac trapping field has zero amplitude, may remain in the trap for some finite time. Moore et al. [184] have achieved trapping efficiencies up to 0.2% from a dc beam by manipulation of the phase space volume.

The situation is different if the ions undergo some kind of dissipative force while they pass through the trap. This damping is most easily obtained by collisions with a light buffer gas. The density of the buffer gas has to be at least of such a value that the mean free path of the ions between collisions is less than the size of the trap. Coutandin et al. [185] have shown that the trap is filled up to its maximum capacity in a short time at pressures around 10^{-3} Pa in a 1 cm size Paul trap when ions are injected along the trap axis with a few keV kinetic energy (Fig. 5.6).

Injection into a Penning trap using buffer gas collisions for friction requires some care: the ion's motion becomes unstable since collisions lead to an increase of the magnetron radius. This can be overcome if the trap ring electrode is split into four segments and an additional radio-frequency field is applied between adjacent parts to create a quadrupole rf-field in the radial plane. At the sum of the perturbed cyclotron frequency ω_+ and the magnetron frequency ω_- this field couples the two oscillations. The damping of the cyclotron motion by collisions with the background atoms, overcomes the increase of the magnetron radius and as a result the ions aggregate near the trap's center [186]. Since $\omega_+ + \omega_- = \omega_c$, the free ion cyclotron frequency, it is possible to stabilize a particular isotope or even isobar in the trap by proper

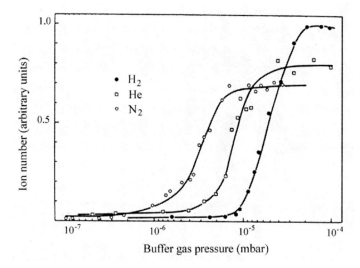

Fig. 5.6. Trapped Ba$^+$ ion number vs. buffer gas pressure for He, H$_2$, and N$_2$ when injected at a few keV kinetic energy into a Paul trap. The maximum ion number for H$_2$ is arbitraryly set to 1 [185]

choice of the rf-frequency [36] while ions of unwanted mass do not remain in the trap. The technique of mode coupling by an radio-frequency field is also of importance for buffer gas cooling of ions in a Penning trap and will be outlined in more detail in the part on ion cooling.

5.3 Positron Loading

The loading of positrons into Penning traps is of particular interest for experiments bearing on such fundamental questions as the ratios of the mass and magnetic moment of the positron to those of the electron [187], and the formation of antihydrogen [143, 188]. Special techniques are required because radioactive sources are involved, and the possibility exists of annihilation with background gas molecules.

Positron sources are usually β^+ emitters, such as ^{22}Na or ^{64}Cu. A fraction, typically 10^{-3} to 10^{-4}, of the high energy positrons leaving the source can be slowed down by scattering from a moderator surface, such as tungsten or solid neon, which they leave at much reduced energies [189], and can then be captured in a trap.

The first successful confinement, in ultra-high vacuum, of single positrons in a Penning trap was reported by Dehmelt et al. in 1978 [190]. Positrons from a 1 mCi radioactive source emerge from the surface of a moderating foil with energies of about 50 keV, and spiral along the lines of a 5 T magnetic field

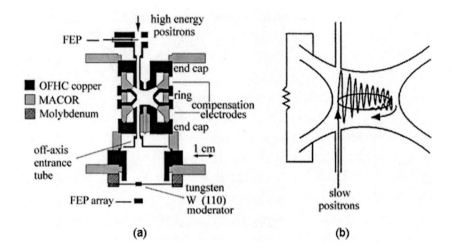

Fig. 5.7. (a) Penning trap and moderator for trapping of positrons. (b) Exaggerated view of the trajectory of a slow positron which enters the trap, and makes axial (*vertical*) oscillations of decreasing amplitude (due to the electrical damping) as the positron circles in a magnetron orbit. Small cyclotron orbits are not visible [193]. Copyright (2003) by the American Physical Society

with a longitudinal velocity that is retarded by a -3000 V bias placed on the foil with respect to the trap. Those particles that have enough longitudinal energy to overcome the potential barrier enter the trap through an opening off-axis by a few millimeters. The energy loss by synchrotron radiation in the magnetic field, and by capacitive coupling to an axial LC damping circuit, is sufficient to ensure that on completing a magnetron orbit, the positrons do not hit the electrode at the injection point.

Accumulation of larger numbers of positrons in a Penning trap has been achieved using a refinement of the same technique (Fig. 5.7). Conti et al. [191] have found that the accumulation rate of positrons in a cylindrical Pening trap was increased when the operating conditions were chosen in such a way that the cyclotron frequency was an integer multiple of the axial frequency. This leads to a coupling of the two motions, and energy is transfered from the axial into the cyclotron mode, from which it is dissipated by synchrotron radiation.

A method based on the field ionisation of positronium atoms in high Rydberg states, created on the surface of a moderator, has been described by Gabrielse et al. [192–194], starting with positrons from a 2.5 mCi ^{22}Na source outside the trap and using tungsten transmission and reflection moderators at both end of a 14 cm long cylindrical trap, held at cryogenic temperatures. When a moderated positron leaves the transmission moderator and is followed by a secondary electron the axial spacing between them can be reduced by biasing the transmission moderator potential which accelerates one species

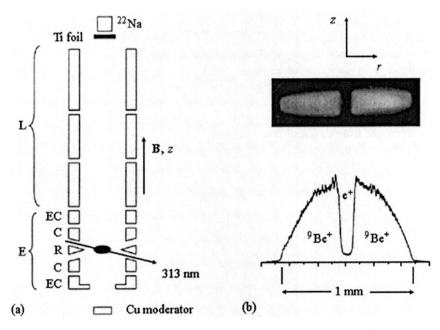

Fig. 5.8. (a) Schematic diagram of the load (L) and experimental (E) cylindrical Penning traps (drawn to scale); EC, end cap; C, compensation; R, ring of the experimental trap. (b) Two-species $^9\text{Be}^+$–e^+ plasma: side-view camera image (*top*) and radial variation of fluorescence signal integrated over z (*bottom*) [195]

and reduces the velocity of the other. If their Coulomb attraction energy exceeds their kinetic energy in the center-of-mass frame they are bound in a high Rydberg state of positronium. The positronium atom is ionized in the electric trapping field, and if the kinetic energy of the positron is sufficiently low it will be captured. Using this technique a total number of more than 10^6 positrons could be accumulated in 15 hours.

Loading of positrons into a trap can also be achieved by sympathetic cooling with simultaneously trapped cold particles. Jelenković et al. [195] have slowed down positrons from a 2 mCi ^{22}Na source by a Cu moderator, then injected them into a cylindrical Penning trap, in which a cloud of Be$^+$ ions had been laser cooled to low temperatures, to a density of 3×10^9 cm^{-3}. Coulomb collisions serve to exchange kinetic energy between the particles, thereby cooling the positrons and allowing a few thousands of them to be stored for several days in ultra-high vacuum. The density of the positron cloud was about the same as that of the Be$^+$ ions, however the positrons formed a column along the trap axis, while the Be$^+$ cloud extended away from the trap axis by centrifugal forces (Fig. 5.8).

The largest number of trapped positrons has been achieved by a method developed by Surko et al. [196, 197]. In this method, positrons injected into a trap at low energies lose their energy by collisions with N_2 background molecules. Starting with 6×10^6 positrons per second from a 90 mCi ^{22}Na source, moderated by a solid Ne surface, the equilibrium trapped positron number at a background pressure of 10^{-3} Pa is about 3×10^8 obtained in a collection time of 8 min. Rapid decrease of the pressure within 1 min to 5×10^{-8} Pa leaves the positrons in the trap for further use in experiments. The first formation of cold antihydrogen atoms by merging positrons and antiprotons from different traps as observed at CERN [143] was mainly based on this method of achieving a comparatevely high positron density.

6 Trapped Charged Particle Detection

6.1 Destructive Detection

6.1.1 Nonresonant Ejection

Ions confined in a trap may leave the trap either by lowering the potential of one end cap electrode or by application of a voltage pulse of high amplitude. They can be counted by suitable detectors such as ion multiplier tubes or channel plate detectors. A possible experimental arrangement is shown in Fig. 6.1.

By different arrival times at the detector, ions of different charge to mass ratio can be distinguished. Figure 6.2 shows an example where simultaneously stored H^+, H_2^+, H_3^+ ions ejected from a Paul trap arrive at time intervals of several µs at a detector placed 5 cm above the end cap electrode. In a Paul trap the amplitude of the detector pulse, however, depends on the phase of the leading edge of the ejection pulse with respect to the phase of the rf-trapping voltage. Figure 6.3 gives an experimental example for H_2^+ ions.

In principle, the arrival time of ions contains also information on the trapped ion energy. Vedel et al. [198, 199] have attempted to determine the energy distribution of trapped N^+ and N_2^+ ions by deconvolution of the time distribution of the detection pulse. The inhomogeneous rf-trapping field, how-

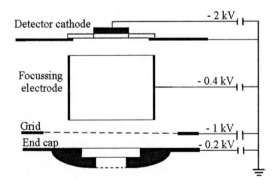

Fig. 6.1. Arrangement for ion detection by extraction from the trap

140   6 Trapped Charged Particle Detection

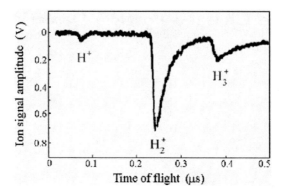

Fig. 6.2. Multiplier output after pulsed ion extraction. The trapping parameters are choosen to allow simultaneous trapping of H^+, H_2^+, H_3^+ [200]

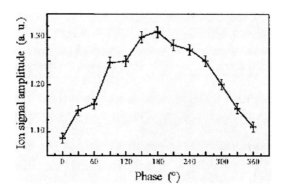

Fig. 6.3. Ion signal amplitude as function of the extraction pulse phase with respect to the rf-trapping field [201]

ever, makes it difficult to obtain reliable results because it changes the ion energy during the passage through the trap.

6.1.2 Resonant Ejection

When several ion species are simultaneously trapped, a particle of specific Q/M ratio can be selectively ejected from the trap and counted by a detector using a weak rf-field resonant with the axial oscillation frequency of the ion species of interest. This so-called "tickle" field may be applied in a dipolar or quadrupolar mode between the electrodes of the trap.

Figure 6.4 shows a timing sequence employed by Vedel et al. [203], where specific ions are detected during the tickle period and the remaining ones are detected after the trap's drive voltage is switched off.

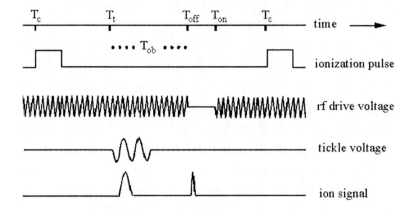

Fig. 6.4. Timing sequence in the "tickle" detection method [202]. Copyright (2003) by the American Physical Society

6.2 Nondestructive Detection

6.2.1 Electronic Detection

The mass dependent oscillation frequencies of ions in a trap can be used for detection without ion loss. A resonant circuit consisting of an inductance L and a capacitance C is connected to the trap (which has its own capacitance) and weakly excited at its resonance frequency ω_{LC}. The ion axial oscillation frequency ω_z can be swept through resonance with the LC circuit by varying the electric trapping field. At the point of resonance, energy is transferred from the circuit to the ions leading to a damping of the circuit, and a decrease in the voltage across the circuit. Modulation of the trap voltage around the operating point at which resonance occurs (see Fig. 6.5) and detection of the voltage across the circuit leads to a repeated voltage drop whose amplitude is

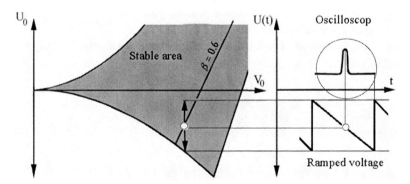

Fig. 6.5. Principle of electronic detection [56]

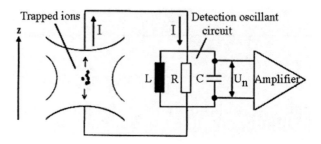

Fig. 6.6. Principle of bolometric ion detection

proportional to the number of trapped ions. When different ions are simultaneously confined, signals appear at different values of the modulated trapping voltage. The sensitivity of the method depends on the quality factor Q of the resonance circuit. With the moderate values of the order of 50, about 1000 trapped ions lead to an observable signal.

6.2.2 Bolometric Detection

In the bolometric detection of trapped ions, first proposed and realised by Dehmelt et al. [204–206] the ions are kept in resonance with a tuned LC circuit connected to the trap electrodes in a way similar to what was discussed in the previous paragraph (Fig. 6.6).

An ion of charge Q oscillating between the end cap electrodes of a trap induces a current in the external circuit given by

$$I = \Gamma \frac{Q\dot{z}}{2z_0} , \qquad (6.1)$$

where $2z_0$ is the separation of the end caps, and Γ is a correction factor, which accounts for the approximation of the trap electrodes by parallel plates of infinite dimensions. For hyperbolic electrodes $\Gamma \simeq 0.75$. The electromagnetic energy associated with this current will be dissipated as thermal energy in the parallel resonance resistance R of the LC circuit. The increased temperature T of that resistance results in an increased RMS noise voltage V_n per bandwidth $\delta\nu$:

$$V_n = \sqrt{4k_B T R \delta\nu} , \qquad (6.2)$$

with k_B the Boltzmann constant. When the oscillation frequencies of different simultaneously stored ions are brought into resonance with the LC circuit by sweeping the trapping voltage, different charge-to-mass ratios can be distinguished (Fig. 6.7).

If the ions are kept continuously in resonance their oscillation energy is dissipated as thermal energy in the detection circuit and their motion is damped. Thus, from the conservation of energy we have

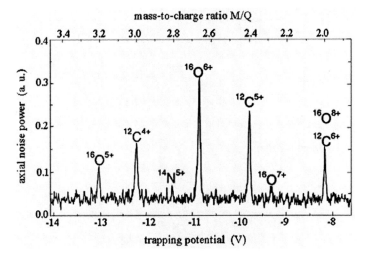

Fig. 6.7. Mass spectrum of ions in a Penning trap measured by induced noise voltage [207]

$$\frac{d\bar{E}}{dt} = -R\bar{I}^2 = -\frac{R\Gamma^2 Q^2}{4z_0^2 M}\bar{E}, \tag{6.3}$$

from which follows immediately the exponential decay of the ion energy

$$E(t) = E_0 e^{-t/\tau}, \tag{6.4}$$

and the time constant τ given by

$$\tau = \frac{(2z_0)^2}{\Gamma^2 R}\frac{M}{Q^2}. \tag{6.5}$$

When the ions reach thermal equilibrium with the circuit, excess noise can no longer be detected; nevertheless, the presence of ions can be detected by a spectral analysis of the circuit noise. The voltage that an ion induces with its remaining oscillation amplitude adds to the thermal noise of the circuit, however with opposite phase. As a result, the total noise voltage at the ion oscillation frequency is reduced, and the spectral distribution of the noise shows a minimum at this frequency if ions are present in the trap. This is shown in Fig. 6.8.

Formally, the spectral distribution of the noise voltage can be derived by simulating the oscillation of the ion by an equivalent resonant circuit. Consider the axial equation of motion of an ion under the action of the trapping field and a potential $V(t)$ between the end caps. Neglecting the correction factor Γ we have

$$\ddot{z} - \frac{Q}{2z_0 M}V(t) + \omega_z^2 z = 0. \tag{6.6}$$

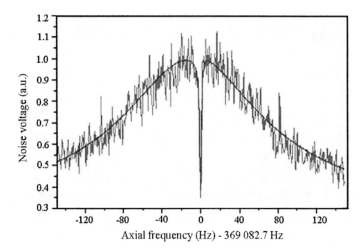

Fig. 6.8. Noise spectrum of an LC detection circuit attached to a Penning trap in the presence of a single stored C^{5+} ion

Noting that the axial ion motion represents the flow of a current between the same electrodes of magnitude $I = Q\dot{z}/(2z_0)$, we can rewrite the equation of motion in the form suggestive of circuit theory, thus

$$V(t) = M\frac{(2z_0)^2}{Q^2}\dot{I} + M\omega_z^2 \frac{(2z_0)^2}{Q^2} \int I dt \,. \tag{6.7}$$

This is an equation representing the voltage across the series resonant circuit with inductance l_i and capacitance c_i given by

$$l_i = M\frac{(2z_0)^2}{Q^2} \,, \tag{6.8}$$

and

$$c_i = \frac{Q^2}{M\omega_z^2(2z_0)^2} \,. \tag{6.9}$$

To account for the finite spectral width of the ion oscillation we can introduce additionally a quality factor Q_z for the ion equivalent circuit

$$Q_z = \frac{\omega_z}{\Delta\omega_z} = \frac{\omega_z l_i}{r} \,, \tag{6.10}$$

where r is the equivalent resonance resistance of the ion.

It is interesting to note the equivalent values l_i and c_i, for example of an electron oscillating in a trap with $z_0 = 1\,\text{cm}$ at a frequency of $10\,\text{MHz}$: we obtain $l_i \simeq 10^5\,\text{Hy}$, $c_i \simeq 1/10^{19}\,\text{F}$. For N stored particles we have to replace l_i by l_i/N and c_i by $c_i N$.

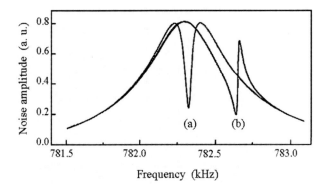

Fig. 6.9. Theoretical noise amplitude of LC detection circuit in the presence of a trapped ion when (**a**) the resonance frequency ω_{LC} is equal to the ion oscillation frequency ω_c, or (**b**) when one of these frequencies is detuned

The equivalent ion circuit is connected across the external tank circuit. The noise voltage measured on it is given by (6.2) if we replace the resistance R by the real part of the impedance $Z(\omega)$ of the complete circuit. The frequency dependence of $Z(\omega)$ can be calculated using standard circuit theory, thus

$$Z^{-1} = Z_{LC}^{-1} + Z_{lc}^{-1} \ . \tag{6.11}$$

Using normalized frequencies $\Omega = \omega/\omega_{LC}$, $\delta = \omega/\omega_z$ and Q_{LC} as the quality factor of the tank circuit, we obtain

$$Z_{LC}^{-1} = \frac{1}{R}\left[1 + iQ_{LC}\left(\Omega - \frac{1}{\Omega}\right)\right] ,$$

$$Z_{lc}^{-1} = \frac{1}{r}\frac{1}{1 + Q_z^2\left(1 - \frac{1}{\delta^2}\right)}\left[1 + iQ_z\left(\frac{1}{\delta} - \delta\right)\right] . \tag{6.12}$$

If we plot the noise voltage in the presence of a trapped ion as a function of the frequency ω, we obtain a minimum in the spectral density having a Lorentzian line shape when the ion and circuit resonance frequencies coincide (Fig. 6.9a). A detuning of either frequency leads to a dispersive type of line shape (Fig. 6.9b). The theory agrees well with the experimental observations.

6.2.3 Fourier Transform Detection

The Fourier transform of the voltage induced by an oscillating ion in the ring electrode of a Penning trap can be used in a similar way to that described previously for ion detection, if the amplifier shown in Fig. 6.6 is replaced by a fast Fourier transform. At room temperature a cloud of stored ions is excited by an electric dipole field to perform coherent cyclotron motions. The ring is split into several segments and the induced time-varying voltage

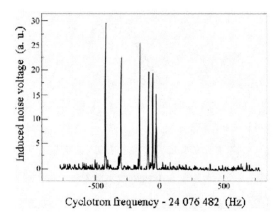

Fig. 6.10. Induced noise voltage in a Penning trap with six ^{12}C^{5+} ions [208]

difference between adjacent segments can be amplified, digitized, and stored in a transient recorder. At room temperature the minimum number of singly charged ions which can be detected is of the order of 100. Higher sensitivity is obtained for highly charged ions because of the larger induced voltage. Further increase of the detection sensitivity is obtained by cooling the trap and the resistance between the trap segments. Figure 6.10 shows a signal of 6 C^{5+} ions confined in a Penning trap with a slightly inhomogeneous magnetic field using a superconducting LC detection circuit. They can be individually distinguished since their magnetron radii are different and consequently the average value of the magnetic field strength along their path. This method of ion detection, introduced by Marshall and Grosshans [209] is extensively used in analytical mass spectrometry because of the high resolving power over a broad mass range [210]. Details and variants of the technique are described in a number of review articles [211].

6.2.4 Optical Detection

A very efficient way to detect the presence of ions in the trap is to monitor their laser induced fluorescence. This method is, of course, restricted to ions which have an energy level scheme that allows excitation by available lasers. It is based on the fact that the lifetime of an excited ionic energy level is of the order of 10^{-7} s when it decays by electric dipole radiation. Repetitive excitation of the same ion by a laser at saturation intensity then leads to a fluorescence count rate of 10^7 photons per second. If a photon detection system is assumed in which the acceptance solid angle is 10%, a photomultiplier detection efficiency of 10%, and filter and transmission losses of 90%, a fraction on the order of 10^{-3} of the photons can be detected leading to an easily observable signal. A prerequisite for this technique is that the trapped ion moves permanently inside the laser beam. This requires cooling of the ion motion to such a degree that the amplitude is smaller than the diameter of the laser beam. Ion cooling methods will be extensively discussed in Part IV of this book. Therefore we restrict ourselves here to very basic considerations only.

The method is most effective when the ion under consideration has a large (electric dipole) transition probability between the electronic ground state and an upper state, as in alkali-like configurations. It becomes particularly favourable when the excited state decays back directly to the ground state. Such two-level systems occur in Be^+ and Mg^+, and consequently these ions are preferred subjects when optical detection of single stored particles is the objective. For other ions of alkali like structure such as Ca^+, Sr^+, or Ba^+ it become slightly more complicated since the excited state may decay into a long lived low lying metastable state, which prevents fast return of the ion into its ground state. In that case an additional laser is required to pump the ion out of the metastable state. Signals of the expected strength have been obtained in the laboratory for all these ions. They allow even the visual observation of a single stored ion, as demonstrated in a pioneering experiment at the University of Heidelberg on Ba^+ [212].

Part III

Nonclassical States of Trapped Ions

The achievement of nearly total isolation from the environment of single charged particles in harmonic trapping fields (first reported around 1980 [212]) and their efficient cooling to the lowest quantum energy levels, has provided the experimental means of generating and studying persistent nonclassical states in a simple physical system [213–215]. Trapped ion techniques have the ability to precisely localize a single ion, to prepare it in a pure quantum state, and to detect its state by the electron shelving technique, making them therefore the ideal tool for the fundamental investigation of nonclassical states. These states have for the most part been the quantum motion of particles in an harmonic oscillator potential, strongly coupled to their internal quantum states. They include the Fock states (also known as number state vectors) and superpositions of such states, squeezed states, nonlinear coherent states as the "Schrödinger cat" states, and many others. A great number of different families of nonclassical states belonging to the Schrödinger equation for the harmonic oscillator can be obtained by choosing different algebras and basic states. One of the first examples was related to the $SU(1,1)$ algebra [216] leading to the "$SU(1,1)$ squeezed states", as coherent states calculated in terms of the infinitesimal generators of the group [217–219]. In the last decade, "dark states" [220] are frequently mentioned as nonlinear coherent states [221]. These are certain superpositions of the atomic eigenstates, whose typical common feature is the existence of some sharp dips in the fluorescence spectra, due to the destructive quantum interference of transition amplitudes between the different energy levels involved [222, 223]. All these states can be considered as squeezed states, since the variances of different canonically conjugated variables can assume values which are less than the ground state variances.

Certainly, the main problem is to create such states in the laboratory and to verify (by detection, "recognition" or reconstruction), that the desired state was indeed obtained. The possibility of applications of trapped ions to quantum computing (as quantum registers for quantum memory) [226, 227], and high resolution spectroscopy [228] fundamentally rely on the ability to prepare nonclassical states of (large) trapped ion strings, cooled to temperatures at which the nonclassical properties of the center-of-mass motion become manifest, requiring a quantum mechanical description.

Following various theoretical schemes to create nonclassical states of motion, in 1996 the NIST group first prepared many of these states using Be^+ trapped ions in a Paul trap [224, 225]. Fock states were created at the University of Innsbruck in 1999 on Ca^+ ions [150]. Most of these states were prepared in the Lamb–Dicke regime, because the starting point in all these preparation processes was the ground state of motion, reached in this regime.

In what follows a brief account is given of the salient points in the theory and engineering of such states; for a fuller account of the engineering and reconstruction of different special nonclassical states for single particles in a Paul trap the reader is referred to articles that have appeared on the subject in recent years ([84] and references therein).

7 Quantum States of Motion

7.1 Fock States

A trapped charged particle can be considered in the quadrupole approximation an harmonic oscillator. Its axial motional states can be express in the basis of eigenstates of the oscillator number operator defined by the relation $N = a^\dagger a$, where a and $a^\dagger$ are the creation and annihilation operators, respectively. Then

$$[a^\dagger, a] = -\mathbf{1}, \quad [N, a^\dagger] = a^\dagger, \quad [N, a] = -a, \qquad (7.1)$$

where $\mathbf{1}$ is the unit operator. The standard orthonormal basis of the Fock space $\mathcal{F}$ consists of the number state vectors

$$|n> = \frac{1}{\sqrt{n!}}(a^\dagger)^n |0>, \qquad (7.2)$$

where n is a nonnegative integer and $|0>$ is the vacuum state vector defined by

$$a|0> = 0, \quad <0|0> = 1. \qquad (7.3)$$

By (7.1)–(7.3), we have

$$a|n> = \sqrt{n}|n-1>, \quad a^\dagger|n> = \sqrt{n+1}|n+1>, \quad N|n> = n|n>. \qquad (7.4)$$

In the Heisenberg-picture N is independent of time, and its eigenstates given by (7.2) are known as *Fock states* or *number states*. These number states vectors satisfy the orthogonality and completeness conditions

$$<m|n> = \delta_{mn}, \quad \sum_{n=0}^{\infty} |n><n| = \mathbf{1}. \qquad (7.5)$$

In the Schrödinger-picture N depends on time, but its eigenstates can be defined analogously as

$$N(t)|n, t> = n|n, t>. \qquad (7.6)$$

These states are not energy eigenstates of the system, since the kinetic energy of the trapped charged particle (having a periodical micromotion) is not a constant; the quantum number n can be connected to the ion energy averaged over a motion period. Any motional state can be expressed as a superposition of the number states

$$|\Psi(t)\rangle = \sum_{n=0}^{\infty} c_n |n, t> , \qquad (7.7)$$

where c_n are some coefficients.

7.2 Oscillator Coherent States

In trying to find quantum-mechanical states which follow the classical motion of a particle in a given potential, Schrödinger constructed in 1926 a Gaussian wave function from a suitable superposition of the stationary wave functions of the harmonic oscillator, by using the generating function of the Hermite polynomials (D.19) [123]. Glauber named the states obtained by Schrödinger *coherent states*, having shown that they can describe a coherent laser field.

The coherent states for the harmonic oscillator are expressed by the vectors $|\alpha\rangle$, where α is a complex parameter, in three equivalent ways:

1. The coherent states minimize Heisenberg's position-momentum uncertainty relation provided that the ground state is a coherent state.
2. The coherent states are eigenstates of the annihilation operator a

$$a|\alpha\rangle = \alpha|\alpha\rangle . \qquad (7.8)$$

3. The coherent states are given by the displacement operator [229–232] acting on the normalized ground state $|0>$

$$|\alpha\rangle = D(\alpha)|0\rangle , \qquad (7.9)$$

where the displacement operator is defined by

$$D(\alpha) = \exp\left(\alpha a^\dagger - \alpha^* a\right) . \qquad (7.10)$$

The creation and annihilation operators satisfy the commutation relation $[a, a^\dagger] = \mathbf{1}$, where $\mathbf{1}$ is the identity operator. The number operator is defined by $N = a^\dagger a$; then $N|n\rangle = n|n\rangle$, where the number or Fock states are defined by

$$|n\rangle = \frac{1}{\sqrt{n!}} (a^\dagger)^n |0\rangle , \qquad (7.11)$$

$$a|0\rangle = 0 , \quad \langle 0|0\rangle = 1 . \qquad (7.12)$$

By using the Weyl operator identity, the displacement operator can be written as [233]:

$$D(\alpha) = \exp\left(-\frac{|\alpha|^2}{2}\right) \exp(\alpha a^\dagger) \exp(-\alpha^* a) . \qquad (7.13)$$

From (7.9)–(7.13), we obtain

$$|\alpha\rangle = \exp\left(-\frac{1}{2}|\alpha|^2\right) \sum_{n=0}^{\infty} \frac{\alpha^n}{\sqrt{n!}} |n\rangle . \tag{7.14}$$

The probability distribution of the coherent states among number states is Poissonian:

$$P_n = |\langle \alpha|n\rangle|^2 = \frac{|\alpha|^{2n}}{n!} \exp\left(-|\alpha|^2\right) . \tag{7.15}$$

The coherent states, as any other motional states, can be expressed as a superposition of the number state vectors. Coherent states are well localized in position and momentum, and approximate most closely the classical motion of a particle.

7.2.1 The Ideal Penning Trap

Consider the Schrödinger equation for the harmonic oscillator of mass M and frequency ω

$$i\hbar \frac{\partial \phi}{\partial t} = H_0 \phi , \quad H_0 = -\frac{\hbar^2}{2M}\frac{\partial^2}{\partial z^2} + \frac{M\omega^2}{2}z^2 . \tag{7.16}$$

The equation (7.16) describes the axial motion of frequency $\omega/(2\pi)$ of a charged particle in an ideal Penning trap. According to (D.16), the solution of the stationary Schrödinger equation (7.16) is given by

$$H_0 \phi_n = E_n \phi_n , \quad E_n = \left(n + \frac{1}{2}\right)\hbar\omega , \tag{7.17}$$

$$\phi_n(z) = \left(\sqrt{\pi} 2^n n! d\right)^{-1/2} H_n\left(\frac{z}{d}\right) \exp\left(-\frac{z^2}{2d^2}\right) , \tag{7.18}$$

where n is a nonnegative integer, H_n is the Hermite polynomial, and

$$d = \left(\frac{\hbar}{M\omega}\right)^{1/2} . \tag{7.19}$$

The generating function ϕ^α of the orthonormalized functions ϕ_n is the Gaussian wave function [234]

$$\phi^\alpha(z) = \left(\sqrt{\pi} d\right)^{-1/2} \exp\left[-\frac{z^2}{2d^2} + \sqrt{2}\frac{\alpha z}{d} - \frac{1}{2}\left(\alpha^2 + |\alpha|^2\right)\right] , \tag{7.20}$$

which can be written as

$$\phi^\alpha(z) = \sum_{n=0}^{\infty} \frac{\alpha^n}{\sqrt{n!}} \exp\left(-\frac{1}{2}|\alpha|^2\right) \phi_n(z) , \tag{7.21}$$

by using the generating function of the Hermite functions (D.19). In the Dirac notation, the functions ϕ^α and ϕ_n are written as $|\alpha\rangle$ and $|n\rangle$, respectively, hence the expansion (7.21) can be rewritten as (7.14).

The coherent states (7.20) can be rewritten as [235]

$$\phi^\alpha(z) = [2\pi(\Delta z)^2]^{-1/4} \exp\left[-\frac{(z-\langle z\rangle)^2}{4(\Delta z)^2} + \frac{i}{\hbar}\langle p_z\rangle\left(z - \frac{\langle z\rangle}{2}\right)\right], \quad (7.22)$$

$$\Delta z = \frac{d}{\sqrt{2}}, \quad \langle z\rangle = \frac{d}{\sqrt{2}}(\alpha + \alpha^*), \quad \langle p_z\rangle = \frac{\hbar}{\sqrt{2}d}(\alpha + \alpha^*), \quad (7.23)$$

where $(\Delta z)^2$ is the position variance; $\langle z\rangle$ and $\langle p_z\rangle$ are the expectation values of the operators z and $p_z = -i\hbar\partial/\partial z$. From (7.22) it can be seen that the coherent wave packets are Gaussians, with widths equal to the ground-state Gaussian.

The time evolution of the coherent states is given by

$$\phi^\alpha(z,t) = e^{-i\omega t/2}\phi^{\alpha(t)}(z), \quad (7.24)$$

where $\alpha(t) = \exp(-i\omega t)\alpha$.

The solutions of the Heisenberg equation of motion are [235]

$$z(t) = z\cos\omega t + \frac{p_z}{M\omega}\sin\omega t,$$
$$p_z(t) = p_z\cos\omega t - M\omega z\sin\omega t. \quad (7.25)$$

The center of the coherent state packet follows the classical motion in time and does not change its shape [235]:

$$\langle\phi^\alpha(z,t)|z|\phi^\alpha(z,t)\rangle = d(\alpha_1\cos\omega t - \alpha_2\sin\omega t),$$
$$\langle\phi^\alpha(z,t)|p|\phi^\alpha(z,t)\rangle = \frac{\hbar}{d}(\alpha_2\cos\omega t + \alpha_1\sin\omega t), \quad (7.26)$$

$$\Delta z = \frac{d}{\sqrt{2}}, \quad \Delta p_z = \frac{\hbar}{\sqrt{2}d}, \quad (7.27)$$

where $\alpha_1 = \text{Re}(\alpha)$ and $\alpha_2 = \text{Im}(\alpha)$. Moreover, the minimum uncertainty property $\Delta z \Delta p_z = \hbar/2$ is preserved in time.

7.2.2 The Harmonic Paul Trap

The axial quantum dynamics of a charged particle of mass M confined in a harmonic Paul trap is described by the Schrödinger equation

$$i\hbar\frac{\partial\psi}{\partial t} = -\frac{\hbar^2}{2M}\frac{\partial^2\psi}{\partial z^2} + \frac{M}{2}g(t)z^2\psi, \quad (7.28)$$

where g is a time-periodic function of period $T = 2\pi/\Omega$.

Gaussian Solutions

In order to describe the quasienergy spectrum of (7.28), recall the derivation in Sect. 2.7. Here again we consider a stable complex solution w of the classical Hill equation

$$\frac{d^2 w}{dt^2} + g(t)w = 0 , \tag{7.29}$$

with the initial conditions

$$w(0) = 1 , \quad \frac{dw(0)}{dt} = i\omega . \tag{7.30}$$

It is convenient to introduce the variables

$$c = \left(\frac{\hbar}{M\omega}\right)^{1/2} |w| , \quad \gamma = \omega \int_0^t \frac{1}{|w(t')|^2} dt' , \tag{7.31}$$

with $w = |w|e^{-i\gamma}$.

The solutions of definite quasienergy of (7.28) are given following (D.46) as

$$\psi_n(z,t) = (\sqrt{\pi} 2^n n! c)^{-1/2} \exp\left[-i\left(n + \frac{1}{2}\right)\gamma\right]$$
$$\times H_n\left(\frac{z}{c}\right) \exp\left[-\left(1 - \frac{i}{c\omega}\frac{dc}{dt}\right)\frac{z^2}{2c^2}\right] . \tag{7.32}$$

The Gaussian solutions of (7.28) are given following (D.47) as

$$\psi^\alpha(z,t) = (\sqrt{\pi}c)^{-1/2} \exp\left[-\frac{1}{2}\left(\alpha^2 e^{-2i\gamma} + |\alpha|^2 + i\gamma\right)\right]$$
$$\times \exp\left[-\left(1 - \frac{i}{c\omega}\frac{dc}{dt}\right)\frac{z^2}{2c^2} + \sqrt{2}\alpha e^{-i\gamma}\frac{z}{c}\right] . \tag{7.33}$$

They are the coherent states for a charged particle trapped in a harmonic Paul trap.

The Gaussians (7.33) give a generating function for (7.32) as (D.52)

$$\psi^\alpha(z,t) = \sum_{n=0}^\infty \frac{\alpha^n}{\sqrt{n!}} e^{-|\alpha|^2/2} \psi_n(z,t) . \tag{7.34}$$

Constants of the Motion

The coherent states for a charged particle in a harmonic Paul trap can be constructed in complete analogy with the Glauber coherent states on the basis of the constants of the motion.

7 Quantum States of Motion

In [236,237] suitable time dependent invariants, linear in position and momentum operators, have been used to rederive the solution of (7.28). Consider the time-invariant operator

$$A = \frac{i}{\sqrt{2\omega d}}\left(\frac{w}{M}p_z - \frac{dw}{dt}z\right), \qquad (7.35)$$

for which the boson commutation relation $[A, A^\dagger] = 1$ is fulfilled, and $dA/dt = 0$. Then A is a constant of the motion, which can be used to write the coherent states, as

$$\psi^\alpha = \exp\left(\alpha A - \alpha^* A^\dagger\right)\psi_0.$$

The general solution of the Schrödinger equation (7.28) can be expressed in terms of the eigenstates of the quadratic invariants up to suitable time-dependent phase factors.

Consider the operator [88]:

$$I = \frac{1}{2}\left[\frac{z^2}{\rho^2} + \left(\rho p_z - \frac{d\rho}{dt}Mz\right)^2\right], \qquad (7.36)$$

where $\rho = (M\omega)^{-1/2}|w|$ is a solution of the Ermakov equation [238]

$$\frac{d^2\rho}{dt^2} + g(t)\rho = \frac{1}{M^2\rho^3}. \qquad (7.37)$$

The quadratic invariant I of Lewis and Riesenfeld [88] is the quantum counterpart of the classical Ermakov–Lewis invariant [238].

According to (7.36) and (7.37), the operator I is a quantum constant of the motion for the Hamiltonian (7.28). Define the operator [88]

$$B = \frac{1}{\sqrt{2\hbar}}\left[\frac{1}{\rho}z + i\left(\rho p_z - \frac{d\rho}{dt}Mz\right)\right], \qquad (7.38)$$

which satisfies the boson commutation relation $[B, B^\dagger] = 1$.

Then I can be rewritten as

$$I = \frac{\hbar}{2}\left(2B^\dagger B + 1\right). \qquad (7.39)$$

The coherent states of a single ion trapped in a harmonic Paul trap are given by

$$\psi^\alpha = \exp\left(\alpha B - \alpha^* B^\dagger\right)\psi_0. \qquad (7.40)$$

7.3 Squeezed States

An analytical description of the ion dynamics in a nonlinear electromagnetic trap can be obtained in terms of the displaced sqeezed states [239], frequently (and in the following) simply called *squeezed states*.

A common method to obtain squeezed states $|\alpha, z\rangle$ is to apply both the squeeze $S(z)$ and displacement $D(\alpha)$ operators (ordering is arbitrary) onto the ground state. The squeezed states are defined by

$$|\alpha, z\rangle = D(\alpha)S(z)|0\rangle ,\qquad(7.41)$$

with the squeeze operator given by

$$S(z) = \exp\left[\frac{z}{2}\left(a^\dagger\right)^2 - \frac{z^*}{2}a^2\right] ,\qquad(7.42)$$

where z is a complex number. The squeezed states are eigenstates of a linear combination of a and $a^\dagger$:

$$\left[(\cosh r)a - (e^{i\varphi}\sinh r)a^\dagger\right]|\alpha, z\rangle = \left[(\cosh r)\alpha - (e^{i\varphi}\sinh r)\alpha^*\right]|\alpha, z\rangle ,\qquad(7.43)$$

where $z = re^{i\varphi}$ and $r > 0$.

With time, the centers of the squeezed wave packets follow the classical motion, but they do not retain their shapes. The squeezed position wave packet becomes wider for half an oscillation period before it contracts back to the original width after a full period. The momentum wave packet contracts and expands accordingly, so that at any time the uncertainty is minimal [130, 240].

The position variances of the sqeezed states is reduced at certain times by $\Delta x = \Delta x_0/\beta$, where Δx_0 is the variance of the ground state [84]. The width of a particular Gaussian oscillates as

$$[\Delta z(t)]^2 = \left[(\cosh r)^2 + (\sinh r)^2 + 2(\cosh r)(\sinh r)\cos(2t - \varphi)\right] ,\qquad(7.44)$$

$$[\Delta p(t)]^2 = \left[(\cosh r)^2 + (\sinh r)^2 - 2(\cosh r)(\sinh r)\cos(2t - \varphi)\right] ,\qquad(7.45)$$

$$4[\Delta z(t)]^2[\Delta p(t)]^2 = 1 + \frac{1}{4}\left(s^2 - \frac{1}{s^2}\right)^2 \sin^2(2t - \varphi) ,\quad s = e^r .\qquad(7.46)$$

The expectation values of quantum observables with respect to the states yield a classical picture and allow one to write the uncertainty relation for a Paul trap. A quasi-classical description for the collective center-of-mass motion of the stored ion system has been obtained, and the stability properties of the trapped ions discussed [239] for the common trap anharmonicities observed [241].

The expectation values of the operators z and $p = -i\hbar\partial/\partial z$ are

$$\bar{z} = \langle\psi^\alpha|z|\psi^\alpha\rangle = \frac{1}{\sqrt{2}}(\alpha w^* + \alpha^* w),$$

$$\bar{p} = \langle\psi^\alpha|p|\psi^\alpha\rangle = \frac{1}{\sqrt{2}}\left(\alpha\frac{dw^*}{d\tau} + \alpha^*\frac{dw}{d\tau}\right). \tag{7.47}$$

The position variance σ_{zz}, the momentum variance σ_{pp}, the covariance σ_{zp}, and the correlation coefficient r_{zp} of the operators z and p are given by

$$\sigma_{zz} = \langle z^2\rangle - \langle z\rangle^2 = c^2, \quad \sigma_{pp} = \langle p^2\rangle - \langle p\rangle^2 = \frac{1}{2c^2} + \frac{1}{2c}\left(\frac{dc}{dt}\right)^2,$$

$$\sigma_{zp} = \frac{1}{2}\langle zp + pz\rangle - \langle z\rangle\langle p\rangle = \frac{c}{2}\frac{dc}{dt}, \tag{7.48}$$

$$r_{zp} = \frac{\sigma_{zp}}{\sqrt{\sigma_{zz}\sigma_{pp}}}\left[1 + \frac{1}{c^2}\left(\frac{dc}{dt}\right)^{-2}\right]^{-1/2}, \tag{7.49}$$

where $\langle\,\rangle$ means the expectation value in ψ^α.

The Schrödinger uncertainty relation for these states [98] can be written in the form

$$\sigma_{zz}\sigma_{pp} - \sigma_{zp}^2 \geq \frac{1}{4}. \tag{7.50}$$

This inequality implies the Heisenberg uncertainty relation

$$\sigma_{zz}\sigma_{pp} \geq \frac{1}{4}. \tag{7.51}$$

The variances, covariances and correlation coefficients are T-periodic functions of t. In terms of these the Gaussian solution ψ^α can be written as

$$\psi^\alpha(q,t) = (2\pi\sigma_{qq})^{-1/2}\exp\left(-\frac{i}{2}\gamma\right)$$

$$\times \exp\left[-\frac{(1-i\sigma_{pq})}{4\sigma_{qq}}(q-\bar{q})^2 + i\bar{p}\left(q - \frac{\bar{q}}{2}\right)\right]. \tag{7.52}$$

For ψ^α the Schrödinger uncertainty relation is satisfied with the equality sign.

Walls and Zoller [242] extended the definition of squeezed states for arbitrary Hermitian operators A and B. Hillery [243] defined second-order squeezing by assuming

$$A = a^2 + a^{\dagger 2}, \quad B = \frac{a^2 - a^{\dagger 2}}{2i}, \tag{7.53}$$

and others have introduced higher order squeezing [244].

Cirac et al. first proposed in 1993 a scheme for preparing squeezed states of motion in an ion trap based on the multichromatic excitation of a trapped ion [245]. Later, Vogel et al. using two waves with beat frequency equal to twice the trap frequency have produced a dark resonance in the ion fluorescence, identified with a squeezed state [221]. Squeezed states of trapped charged particles with $\beta = 40 \pm 10$ were prepared by the NIST group [224].

8 Coherent States for Dynamical Groups

8.1 Trap Symmetries

Group theory can be used to construct coherent states of charged particles confined in electromagnetic traps [246, 247]. The group of linear canonical transformations for a dynamical system with n degrees of freedom is the symplectic group $Sp(2n, \mathbf{R})$. Thus, a single charged particle confined in a quadrupole electromagnetic trap can be explicitly described by coherent states of a subgroup of the symplectic group $Sp(6, \mathbf{R})$.

We consider a particle of mass M and charge Q confined in a static homogeneous magnetic field aligned with respect to the z-axis and an electric field derived from the harmonic potential

$$\Phi = \sum_{i,j=1}^{3} A_{ij} x_i x_j \ . \tag{8.1}$$

Here A_{ij} may be time-periodic functions of period T or constant coefficients, with $A_{11} + A_{22} + A_{33} = 0$. The quantum Hamiltonian (4.13) can be written as

$$H = \frac{1}{2M} \sum_{j=1}^{3} p_j^2 + Q\Phi + \frac{1}{8} M\omega_c^2 (x_1^2 + x_2^2) - \frac{1}{2}\omega_c (x_1 p_2 - x_2 p_1) \ , \tag{8.2}$$

where ω_c is the cyclotron frequency and $p_{\alpha j} = -i\hbar \partial/\partial x_{\alpha j}$, $1 \leq j \leq 3$. Since the Hamiltonian H is a second-order polynomial in x_j and p_j, then the dynamical group of (8.2) is a subgroup of $Sp(6, \mathbf{R})$.

The symplectic coherent states describing the quantum motion of a single trapped charged particle can be written as

$$|w\rangle = \exp[iX(w)] |0\rangle \ , \tag{8.3}$$

where $X(w)$ is a second-order polynomial in x_j and p_j, $1 \leq j \leq 3$. Here w is a point in the classical phase space and $|0\rangle$ is a special fixed state.

The dynamical groups for some quadrupole traps are given in the Table 8.1. Here $A_{ij} = A_j \delta_{ij}$; S_j is the group of axial linear canonical transformations in the (x_j, p_j) phase space, $1 \leq j \leq 3$; S_{rl} is the group of radial linear canonical transformations in the (x_1, p_1, x_2, p_2) phase space of fixed x_3-axis angular momentum $\hbar l$; $SO(2)$ is the group of rotations around the x_3-axis.

Table 8.1. Examples of trap symmetries

quadrupole trap	symmetry	dynamical group
combined trap	$A_1 = A_2$	$S_{rl} \otimes S_3 \otimes SO(2)$
linear Paul trap	$A_1 = -A_2$, $\omega_c = 0$	$S_1 \otimes S_2$
elliptical Paul trap	$A_1 \neq A_2 \neq A_3$, $\omega_c = 0$	$S_1 \otimes S_2 \otimes S_3$

The collective center-of-mass quantum dynamics for a system of N identical charged particles confined in a quadrupole electromagnetic trap can be described by a quantum Hamiltonian similar to (8.2). Some collective dynamical $Sp(2, \mathbf{R})$ groups for quadrupole electromagnetic traps with cylindrical symmetry are presented in [97]. The state vectors $|\psi\rangle = \exp(i\varphi)|\mathbf{w}\rangle$ are solutions of the time-dependent Schrödinger equation for the Hamiltonian provided w and the geometric phase φ satisfy the classical equations of motion. The discrete quasienergy spectra are given by appropriate symplectic coherent states parametrized by the stable solutions of the corresponding classical equations of motion.

8.2 Quasienergy States for Combined Traps

8.2.1 A Single Trapped Charged Particle

Quantum Hamiltonian

We consider the following quantum Hamiltonian for a charged particle of mass M and electric charge Q confined in a quadrupole electromagnetic trap with cylindrical symmetry:

$$H = \frac{1}{2M}\left(-i\hbar\nabla - \frac{1}{2}Q\mathbf{B} \times \mathbf{r}\right)^2 + QA\left(x^2 + y^2 - 2z^2\right) . \tag{8.4}$$

Here $\mathbf{r} = (x, y, z)$ is the position operator and the constant axial magnetic field is given by $\mathbf{B} = (0, 0, B_0)$. The function A is either time-periodic of period $T = 2\pi/\Omega$ for dynamical traps, or stationary for the Penning traps.

The Hamiltonian (8.4) commutes with the z-axis angular momentum operator

$$L_z = i\hbar\left(y\frac{\partial}{\partial x} - x\frac{\partial}{\partial y}\right) . \tag{8.5}$$

We restrict the discussion to the space of eigenvectors of L_z with a fixed eigenvalue $\hbar l$, where l is a nonnegative integer.

We are looking for the solutions of the time-dependent oscillator equations obtained from the Schrödinger equation by the separation of the axial and radial motion. The Hamiltonian for the charged particle is written as

$$H = H^{(a)} + H^{(r)} , \tag{8.6}$$

8.2 Quasienergy States for Combined Traps

where the axial Hamiltonian $H^{(\mathrm{a})}$ and the radial Hamiltonian $H^{(\mathrm{r})}$ are given by

$$H^{(\mathrm{a})} = -\frac{\hbar^2}{2M}\frac{\partial^2}{\partial r^2} + \frac{M}{2}\lambda_a z^2 , \qquad (8.7)$$

$$H^{(\mathrm{r})} = -\frac{\hbar^2}{2M}\left(\frac{\partial^2}{\partial r^2} + \frac{1}{r}\frac{\partial}{\partial r} - \frac{l^2}{r^2}\right) + \frac{M\lambda_r}{2}r^2 - \frac{\hbar\omega_c}{2}l , \qquad (8.8)$$

with $r = \sqrt{x^2 + y^2}$ and

$$\lambda_a = -4\frac{Q}{M}A(t) , \quad A(t) = (r_0^2 + 2z_0^2)^{-1}(U_0 + V_0 \cos\Omega t) , \qquad (8.9)$$

$$\lambda_r = \frac{1}{4}(\omega_c^2 - 2\lambda_a) , \quad \omega_c = \frac{Q}{M}B_0 . \qquad (8.10)$$

Here U_0 and V_0 are the static and the time-varying voltages applied to the trap electrodes of semi-axes r_0 and z_0.

Following Gheorghe and Vedel [94], the solution of the Schrödinger equation for the charged particle can be written as

$$\Psi(\boldsymbol{r}) = \frac{1}{\sqrt{r}}\exp\left[\mathrm{i}l\left(\theta + \frac{\omega_c}{2}t\right)\right]\Psi^{(\mathrm{r})}(r)\Psi^{(\mathrm{a})}(z) , \qquad (8.11)$$

where

$$\mathrm{i}\hbar\frac{\partial\Psi^{(\mathrm{a})}}{\partial t} = H^{(\mathrm{a})}\Psi^{(\mathrm{a})} , \quad \mathrm{i}\hbar\frac{\partial\Psi^{(\mathrm{r})}}{\partial t} = H^{(\mathrm{r})}\Psi^{(\mathrm{r})} . \qquad (8.12)$$

Axial Motion

We define the operators

$$K_0^{(\mathrm{a})} = \frac{1}{4}\left(-\frac{\partial^2}{\partial z^2} + z^2\right) , \quad K_1^{(\mathrm{a})} = \frac{1}{4}\left(\frac{\partial^2}{\partial z^2} + z^2\right) . \qquad (8.13)$$

The raising and lowering operators $K_+^{(\mathrm{a})}$, $K_-^{(\mathrm{a})}$, defined as usual (see Appendix D.5.1) and $K_2^{(\mathrm{a})} = \mathrm{i}[K_1^{(\mathrm{a})}, K_0^{(\mathrm{a})}]$ satisfy the commutation relations (D.68) for the $Sp(2,\mathbf{R})$ axial algebra. The Bargmann index has the values $k_a = 1/4, 3/4$.

We write the axial Hamiltonian $H^{(\mathrm{a})}$ as a linear combination of the infinitesimal generators $K_0^{(\mathrm{a})}$ and $K_1^{(\mathrm{a})}$ of the axial symplectic group:

$$H^{(\mathrm{a})} = \alpha_a K_0^{(\mathrm{a})} + \beta_a K_1^{(\mathrm{a})} , \qquad (8.14)$$

where

$$\alpha_a = M\lambda_a + \frac{\hbar^2}{M} , \quad \beta_a = M\lambda_a - \frac{\hbar^2}{M} . \qquad (8.15)$$

Radial Motion

We introduce the operators

$$K_0^{(\mathrm{r})} = \frac{1}{4}\left[-\frac{\partial^2}{\partial r^2} + r^2 + \left(l^2 - \frac{1}{4}\right)\frac{1}{r^2}\right],$$

$$K_1^{(\mathrm{r})} = \frac{1}{4}\left[\frac{\partial^2}{\partial r^2} + r^2 - \left(l^2 - \frac{1}{4}\right)\frac{1}{r^2}\right]. \quad (8.16)$$

The operators $K_+^{(\mathrm{r})}$, $K_-^{(\mathrm{r})}$, and $K_2^{(\mathrm{r})} = \mathrm{i}[K_1^{(\mathrm{r})}, K_0^{(\mathrm{r})}]$ satisfy the commutation relations (D.68) for the $Sp(2,\mathbf{R})$ radial algebra. The Bargmann index has the values $k_{\mathrm{a}} = (l+1)/2$.

The radial Hamiltonian $H^{(\mathrm{r})}$ can be written as a linear combination of the infinitesimal generators $K_0^{(\mathrm{r})}$ and $K_1^{(\mathrm{r})}$ of the radial symplectic group:

$$H^{(\mathrm{r})} = \alpha_{\mathrm{r}} K_0^{(\mathrm{r})} + \beta_{\mathrm{r}} K_1^{(\mathrm{r})}, \quad (8.17)$$

where

$$\alpha_{\mathrm{r}} = M\lambda_{\mathrm{r}} + \frac{\hbar^2}{M}, \quad \beta_{\mathrm{r}} = M\lambda_{\mathrm{r}} - \frac{\hbar^2}{M}. \quad (8.18)$$

The quasienergy vectors for H can be written as

$$\Psi_{k_{\mathrm{a}} m_{\mathrm{a}} k_{\mathrm{r}} m_{\mathrm{r}} l} = \frac{1}{\sqrt{r}} \exp\left[\mathrm{i}l\left(\theta + \frac{\omega_c}{2}t\right)\right] \psi_{k_{\mathrm{a}} m_{\mathrm{a}}}(z_{\mathrm{a}}) \psi_{k_{\mathrm{r}} m_{\mathrm{r}}}(z_{\mathrm{r}}), \quad (8.19)$$

where m_{a} and m_{r} are nonnegative integers. The state vectors $\psi_{k_{\mathrm{a}} m_{\mathrm{a}}}(z_{\mathrm{a}})$ and $\psi_{k_{\mathrm{r}} m_{\mathrm{r}}}(z_{\mathrm{r}})$ are given by (D.83)–(D.85) for $a = \alpha_{\mathrm{a}}$, $b = \alpha_{\mathrm{a}}$, $z = z_{\mathrm{a}}$, and $a = \alpha_{\mathrm{r}}$, $b = \alpha_{\mathrm{r}}$, $z = z_{\mathrm{r}}$, respectively.

According to (D.86), the quasienergies corresponding to $\Psi_{k_{\mathrm{a}} m_{\mathrm{a}} k_{\mathrm{r}} m_{\mathrm{r}} l}$ can be written as

$$E_{k_{\mathrm{a}} m_{\mathrm{a}} k_{\mathrm{r}} m_{\mathrm{r}} l} = 2\hbar\left[\mu_{\mathrm{a}}(k_{\mathrm{a}} + m_{\mathrm{a}}) + \mu_{\mathrm{r}}(k_{\mathrm{r}} + m_{\mathrm{r}}) - \frac{\omega_c l}{4}\right], \quad (8.20)$$

where μ_{a} and μ_{r} are the Floquet exponents for the axial and radial solutions of (D.85):

$$\frac{\mathrm{d}^2 w_j}{\mathrm{d}t^2} + \lambda_j w_j = 0, \quad (8.21)$$

$$w_j^* \frac{\mathrm{d}w_j}{\mathrm{d}\tau} - w_j \frac{\mathrm{d}w_j^*}{\mathrm{d}\tau} = 2\mathrm{i}, \quad (8.22)$$

where $j = \mathrm{a}, \mathrm{r}$. According to (D.79), the total phase $\varphi_j = \varphi_{jd} + \varphi_{jg}$ is given by

$$\frac{\mathrm{d}\varphi_j}{\mathrm{d}t} = -\frac{1}{2|w_j|^2}, \quad (8.23)$$

where the dynamical phase φ_{jd} and the geometrical phase φ_{jg} (also known as Berry phase [248]) are given by

$$\frac{d\varphi_{jd}}{dt} = -\frac{1}{4}\left[\left(\frac{d|w_j|}{dt}\right)^2 - |w_j|\frac{d^2|w_j|}{dt^2} + \frac{2}{|w_j|^2}\right], \quad (8.24)$$

$$\frac{d\varphi_{jg}}{dt} = \frac{1}{4}\left[\left(\frac{d|w_j|}{dt}\right)^2 - |w_j|\frac{d^2|w_j|}{dt^2}\right]. \quad (8.25)$$

8.2.2 Quantum Multiparticle States

Center-of-Mass Motion

Following Gheorghe and Werth [97], we write the quantum Hamiltonian for a system of N identical charged particles of mass M and electric charge Q confined in an ideal combined trap as

$$H = \sum_{\alpha=1}^{N} H_\alpha + V, \quad (8.26)$$

where H_α is the Hamiltonian for the particle α written in the form (8.4)

$$H_\alpha = -\frac{\hbar^2}{2M}\sum_{j=1}^{3}\frac{\partial^2}{\partial x_{\alpha j}^2} + \frac{M\lambda_{\mathrm{r}}}{2}\left(x_{\alpha 1}^2 + x_{\alpha 2}^2\right) + \frac{M\lambda_{\mathrm{a}}}{2}x_{\alpha 3}^2 - \frac{\omega_c}{2}L_{\alpha 3}, \quad (8.27)$$

and V is the interaction potential between particles. Here $x_{\alpha 1}$, $x_{\alpha 2}$, and $x_{\alpha 3}$ are the coordinates of the particle α. The axial angular momentum operator of the particle α is defined by

$$L_{\alpha 3} = i\hbar\left(x_{\alpha 2}\frac{\partial}{\partial x_{\alpha 1}} - x_{\alpha 1}\frac{\partial}{\partial x_{\alpha 2}}\right). \quad (8.28)$$

The trap parameters ω_c, λ_{r}, and λ_{a} are given by (8.10).

We now introduce the following translation-invariant coordinates $y_{\alpha j}$ and translation-invariant differential operators $D_{\alpha j}$:

$$y_{\alpha j} = x_{\alpha j} - x_j, \quad X_j = \frac{1}{N}\sum_{\alpha=1}^{N}x_{\alpha j}, \quad (8.29)$$

$$D_{\alpha j} = \frac{\partial}{\partial x_{\alpha j}} - D_j, \quad D_j = \frac{1}{N}\sum_{\alpha=1}^{N}\frac{\partial}{\partial x_{\alpha j}}, \quad (8.30)$$

where $\alpha = 1,\ldots,N$ and $j = 1,2,3$. From (8.29) and (8.30) we find

$$\sum_{\alpha=1}^{N} y_{\alpha j} = 0, \quad \sum_{\alpha=1}^{N} D_{\alpha j} = 0, \quad D_{\beta k}(y_{\alpha j}) = \delta_{kj}\left(\delta_{\alpha\beta} - \frac{1}{N}\right), \quad (8.31)$$

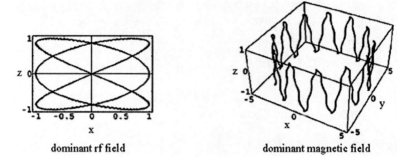

Fig. 8.1. Periodic trajectories of the wave packet center in a combined trap

$$\sum_{\alpha=1}^{N} x_{\alpha j}^2 = NX_j^2 + \sum_{\alpha=1}^{N} y_{\alpha j}^2 , \quad \sum_{\alpha=1}^{N} \frac{\partial^2}{\partial x_{\alpha j}^2} = ND_j^2 + \sum_{\alpha=1}^{N} D_{\alpha j}^2 , \quad (8.32)$$

$$\sum_{\alpha=1}^{N} L_{\alpha 3} = L_{\text{cm} 3} + \tilde{L}_3 , \quad L_{\text{cm} 3} = X_1 P_2 - X_2 P_1 ,$$

$$\tilde{L}_3 = i\hbar \sum_{\alpha=1}^{N} (y_{\alpha 2} D_{\alpha 1} - y_{\alpha 1} D_{\alpha 2}) , \quad (8.33)$$

where $P_j = -i\hbar ND_j$, $L_{\text{cm} 3}$ is the center-of-mass angular momentum and $\tilde{L}_3$ is the intrinsic angular momentum of the N-particle system.

Using (8.29)–(8.33) we can rewrite (8.26) as

$$H = H_{\text{cm}} + \tilde{H} + V , \quad (8.34)$$

where the center-of-mass Hamiltonian is given by

$$H_{\text{cm}} = \frac{1}{2M'} (P_1^2 + P_2^2 + P_3^2) + \frac{M'}{2} \left[\lambda_{\text{r}} \left(X_1^2 + X_2^2 \right) + \lambda_{\text{a}} X_3^2 \right] - \frac{\omega_{\text{c}}}{2} L_3 , \quad (8.35)$$

with $M' = NM$. The intrinsic Hamiltonian is defined by

$$\tilde{H} = \sum_{\alpha=1}^{N} \left[-\frac{\hbar^2}{2M} \sum_{j=1}^{3} D_{\alpha j}^2 + \frac{1}{2} M \lambda_{\text{r}} \left(y_{\alpha 1}^2 + y_{\alpha 2}^2 \right) + \frac{1}{2} M \lambda_{\text{a}} y_{\alpha 3}^2 \right] - \frac{\omega_{\text{c}}}{2} \tilde{L}_3 . \quad (8.36)$$

Collective Intrinsic Models

We consider the model Hamiltonian [97]

$$H_{\text{c}} = \tilde{H} + \frac{2\hbar^2}{M} \left(W^{(\text{a})} + W^{(\text{r})} \right) , \quad (8.37)$$

where $W^{(\text{a})}$ is a function of the axial coordinates $y_{\alpha 3}$, and the $W^{(\text{r})}$ is a function of the radial coordinates $y_{\alpha 1}$ and $y_{\alpha 2}$ such that $[\tilde{L}_3, W^{(\text{r})}] = 0$. Moreover, we suppose that $W^{(\text{a})}$ and $W^{(\text{r})}$ are homogeneous functions of degree (-2). Then the Euler theorem gives

$$\sum_{\alpha=1}^{N} y_{\alpha 3} D_{\alpha 3}(W^{(\text{a})}) = -2W^{(\text{a})} \;, \quad \sum_{\alpha=1}^{N}\sum_{j=1}^{2} y_{\alpha j} D_{\alpha j}(W^{(\text{r})}) = -2W^{(\text{r})} \;. \quad (8.38)$$

A particular axial potential is considered for the one-dimensional N-body exactly soluble Calogero dynamical system with quadratic and inversely quadratic pair potentials [249]:

$$W^{(\text{a})} = \sum_{1 \leq \alpha < \beta \leq N} \frac{g^2}{(y_{\alpha j} - y_{\beta j})^2} \;. \quad (8.39)$$

We introduce the following operators:

$$\tilde{K}_0^{(\text{a})} = \frac{1}{4}\sum_{\alpha=1}^{N}(y_{\alpha 3}^2 - D_{\alpha 3}^2) + W^{(\text{a})} \;,$$

$$\tilde{K}_1^{(\text{a})} = \frac{1}{4}\sum_{\alpha=1}^{N}(y_{\alpha 3}^2 + D_{\alpha 3}^2) + W^{(\text{a})} \;, \quad (8.40)$$

$$\tilde{K}_0^{(\text{r})} = \frac{1}{4}\sum_{\alpha=1}^{N}\sum_{j=1}^{2}(y_{\alpha j}^2 - D_{\alpha j}^2) + W^{(\text{r})} \;,$$

$$\tilde{K}_1^{(\text{r})} = \frac{1}{4}\sum_{\alpha=1}^{N}\sum_{j=1}^{2}(y_{\alpha j}^2 + D_{\alpha j}^2) + W^{(\text{r})} \;. \quad (8.41)$$

According to (8.42), the operators (8.40) and (8.41) satisfy the commutation relations for the axial and radial $Sp(2,\mathbf{R})$ algebras. Note that the axial operators (8.40) commute with the radial operators (8.41). Moreover, the angular momentum $\tilde{L}_3$ commutes with any operator from (8.40) and (8.41). The Hamiltonian $\tilde{H}$ is restricted to the space of $\tilde{L}_3$ with a fixed eigenvalue $\hbar l$, where l is nonnegative integer. The axial Bargmann index k_a and radial Bargmann index k_r are given by

$$k_\text{a} = \frac{1}{4}(N-1) + \frac{1}{2}n_\text{a} \;, \quad k_\text{r} = \frac{1}{2}(N-1) + n_\text{r} \;, \quad (8.42)$$

where n_a and n_r are nonnegative integers.

Using (8.36), (8.40) and (8.41), Gheorghe and Werth have written the Hamiltonian (8.37) as

$$H_\text{c} = H_\text{c}^{(\text{a})} + H_\text{c}^{(\text{r})} - \frac{\hbar\omega_\text{c}}{2}l \;, \quad (8.43)$$

$$H_\text{c}^{(\text{a})} = \alpha_\text{a} K_0^{(\text{a})} + \beta_\text{a} K_1^{(\text{a})} \;, \quad H_\text{c}^{(\text{r})} = \alpha_\text{r} K_0^{(\text{r})} + \beta_\text{r} K_1^{(\text{r})} \;, \quad (8.44)$$

where α_a, β_a, α_r, and β_r are given by (8.15) and (8.18).

Interacting Potentials

Previously the assumption $V = W_a + W_r$ was made, such that $[W_a, W_r] = 0$ and $[L'_3, W_r] = 0$.

For the N pair-wise interacting particles, W_a is the *Calogero potential* for the exactly soluble one-dimensional N-body problem with quadratic and inversely quadratic pair potentials givrn by (8.39).

The model remains valid also if a multiparticle interaction potential of the form

$$W_a = \text{const.} \sum_{\alpha < \beta} \frac{1}{\sqrt{y_{13}^2 + y_{23}^2 + \ldots + y_{N3}^2}} \cdot \frac{1}{|y_{\alpha 3} - y_{\beta 3}|} \qquad (8.45)$$

is considered. It follows that W_a models locally the axial *Coulomb potential*. Around the equilibrium positions of the particles, the square root varies slowly and it can be developed in a series.

A particular form of W_r is

$$W_r = \sum_{1 \leq \alpha < \beta \leq N} g^2 \left[\sum_{j=1}^{2} (y_{\alpha j} - y_{\beta j})^2 \right]^{-1/2}. \qquad (8.46)$$

9 State Engineering and Reconstruction

9.1 Trapped Ion-Laser Interaction

9.1.1 Atom-Field Hamiltonians

The atom–light interaction can be ideally studied by trapping a single charged atomic particle well isolated from the environment, put it into a well-determined state of motion, and then direct laser field radiation onto it in a controlled manner. In present day ion trap experiments, the interaction of the ions with laser radiation plays a central role, e.g. for preparation of ultracold ions and of the nonclassical motional states of the trapped ions [225,250,251]. Due to this interaction, the internal levels of trapped ions can be coupled to each other and to the external motional degrees of freedom of the ions. Different couplings are possible: coupling with exchange of multiple motional quanta (like multiphoton transitions in quantum optics), coupling in which both the internal state of the atomic particle and its motion undergo simultaneous transitions, and also exchange of motional quanta at integer multiples of the rf-driving field or combinations of integer multiples of the driving field and the secular motion (micromotion sidebands) [84].

The interaction of atomic system with quantized electromagnetic fields has been extensively studied [252–260]; for all that, it is yet a largely unsolved problem in quantum optics. The full theoretical treatment of the laser–ion interaction is a nontrivial problem, as it is highly nonlinear. Gheorghe and Collins have developed asymptotic solutions to the Hamiltonian for large photon numbers and large angular momenta, describing an atomic assembly interacting with a single-mode quantized electromagnetic field through dipole coupling [261]. They have been used in the formulation of a multiphoton description of the Autler–Townes effect.

We describe the atom–field interaction by the Hamiltonian H as

$$H = H_0 + H_{\text{int}}, \quad H_0 = H_{\text{at}} + H_{\text{b}}, \quad H_{\text{at}} = \omega J_0, \quad H_{\text{b}} = \sum_{k=1}^{r} \omega_k a_k^\dagger a_k,$$

$$H_{\text{int}} = \sum_{k=1}^{r} \left(\lambda_k a_k J_+ + \lambda_k^* a_k^\dagger J_- + \mu_k a_k J_- + \mu_k^* a_k^\dagger J_+ \right), \quad (9.1)$$

where ω_k and ω are the field and atomic frequencies, respectively. The multicomponent case has been published by Gheorghe [262].

The coupling parameters λ_k and μ_k can be time-dependent functions. Here a_k and $a_k^\dagger$ are the annihilation and creation boson operators for the mode k and $J_\pm = J_1 \pm iJ_2$, where $\boldsymbol{J} = (J_1, J_2, J_3)$ is the collective pseudo-spin vector operator for N identical two-level atoms.

9.1.2 Two-Level Approximation

The various theoretical treatments of the trapped ion-laser field interaction are usually based on models in which the ion is simplified to be of two levels, the trap potential is quantized as a harmonic oscillator, and the laser field assumed to have the classical forms of standing or travelling waves. Moreover, the Lamb–Dicke regime is frequently assumed, in which the ion is supposed to be confined within a region much smaller than the laser wavelength [263, 264]. Many treatments also assume a weak laser-coupling. The resulting Hamiltonians are of the Jaynes–Cummings type [265–267].

The total Hamiltonian describing the system ion-laser can be written as

$$H = H_m + H_e + H_i , \qquad (9.2)$$

where H_m is the motional Hamiltonian along one trap axis, H_e describes the internal electronic level structure of the ion and H_i is the Hamiltonian of the interaction of the applied light fields with the ion.

The Hamiltonian of the motion H_m describes a harmonic oscillator. Following [84], we consider that the time-dependent motion of the ion bound in the trap is harmonic in all three dimensions, and the problem is separable into three one-dimensional problems. The operators x and p are expressed in the Heisenberg picture in terms of a special solution w of the Mathieu equation (7.29), subject to the initial conditions (7.30) and of the known oscillator operators a and $a^\dagger$ as

$$x(t) = \sqrt{\frac{\hbar}{2M\omega}} \left[w^* a + w a^\dagger \right] , \quad p(t) = \sqrt{\frac{\hbar M}{2\nu}} \left[\frac{dw^*}{dt} a + \frac{dw}{dt} a^\dagger \right] . \qquad (9.3)$$

The internal electronic structure of the ion is approximated by a two-level system with levels $|\downarrow\rangle$ and $\langle\uparrow|$ of energies $\hbar\omega_\downarrow$ and $\hbar\omega_\uparrow$, respectively. Using the Pauli matrices (3.161) the two-level Hamiltonian H_e can be expressed as

$$H_e = \frac{1}{2}\hbar \left[\omega \sigma_z + (\omega_\downarrow + \omega_\uparrow) \sigma_0 \right] , \qquad (9.4)$$

where $\omega = \omega_\uparrow - \omega_\downarrow$ is positive.

The electromagnetic field is treated as a classical plane-wave field of the form

$$\boldsymbol{E}(\boldsymbol{x}, t) = 2\boldsymbol{E}_0 \cos\left[(\omega t - kx) \right] , \qquad (9.5)$$

where $\boldsymbol{x} = (x, y, z)$ is the position vector, $\boldsymbol{k} = (k, 0, 0)$ is the wave vector, and E_0 is the real field amplitude.

Consider $\omega - \omega_0 \ll \omega_0$. Then the rotating-wave approximation for the interaction Hamiltonian $U^\dagger H_{\rm i} U$, where $U = \exp[-{\rm i}t\hbar^{-1}(H_{\rm m} + H_{\rm e})]$, can be written as

$$H_{\rm int} = \frac{\hbar\Omega_R}{2} e^{{\rm i}(\phi - \delta t)} \sigma_+ \exp\left[{\rm i}\eta\left(w^* a + w a^\dagger\right)\right] + {\rm H.c.} , \tag{9.6}$$

where the Lamb–Dicke parameter is defined by $\eta = k x_0$ and the coordinate operator in the Heisenberg representation is given by

$$x = x_0 \left(w^* a + w a^\dagger\right) , \quad x_0 = \sqrt{\frac{\hbar}{2M\nu}} . \tag{9.7}$$

Here

$$w = {\rm Ce}(a, q, \tau) + {\rm iSe}(a, q, \tau) = e^{{\rm i}\nu t} \sum_{n=-\infty}^{\infty} c_{2n} e^{{\rm i}n\Omega t} , \tag{9.8}$$

is a solution of the Mathieu equation

$$\frac{{\rm d}^2 w}{{\rm d}t^2} + [a - 2q\cos(\Omega t)] w = 0 , \tag{9.9}$$

with

$$w^* \frac{{\rm d}w}{{\rm d}\tau} - w \frac{{\rm d}w^*}{{\rm d}\tau} = 2{\rm i}\nu . \tag{9.10}$$

The interaction Hamiltonian (9.6) simplifies to

$$H_{\rm int} = \frac{\hbar\Omega_R}{2} e^{{\rm i}(\phi - \delta t)} \sigma_+ \exp\left[{\rm i}\eta\left(e^{-{\rm i}\nu t} a + e^{{\rm i}\nu t} a^\dagger\right)\right] + {\rm H.c.} \tag{9.11}$$

Lamb–Dicke Regime

In the Lamb–Dicke regime [263, 264], the extension of the ion wavefunction is much smaller than $1/k$. Accordingly, by an expansion retaining the lowest order in η, the Lamb–Dicke Hamiltonian can be written as [84]

$$H_{\rm LD}(t) = \frac{1}{2}\hbar\Omega_0 \left[1 + {\rm i}\eta\left(e^{-{\rm i}\nu t} a + e^{{\rm i}\nu t} a^\dagger\right)\right] \exp\left[{\rm i}(\phi - \delta t)\right] \sigma_+ + {\rm H.c.} \tag{9.12}$$

The first resonance for $\delta = 0$ is called the *carrier-resonance* and has the Hamiltonian

$$H_{\rm car}(t) = \frac{1}{2}\hbar\Omega_0 \left(e^{{\rm i}\phi}\sigma_+ + e^{-{\rm i}\phi}\sigma_-\right) , \tag{9.13}$$

giving rise to transitions $|n\rangle|\downarrow\rangle \longleftrightarrow |n\rangle|\uparrow\rangle$ with Rabi-frequency Ω_0. These transitions do not affect the motional state.

The resonant part for $\delta = -\nu$ is called the *first red sideband* and has the Hamiltonian

$$H_{\rm rsb}(t) = \frac{1}{2}\hbar\Omega_0 \left(e^{i\phi}a\sigma_+ + e^{-i\phi}a^\dagger\sigma_-\right), \tag{9.14}$$

giving rise to transitions $|n\rangle|\downarrow\rangle \longleftrightarrow |n-1\rangle|\uparrow\rangle$ with Rabi-frequency $\Omega_{n,n-1} = \eta\sqrt{n}\Omega_0$ that entangle the motional state with the internal state of the ion. This Hamiltonian is also responsible for the remarkable similarity of investigations done in cavity QED [268–270].

The resonant part for $\delta = \nu$ is called the *first blue sideband* and has the Hamiltonian

$$H_{\rm bsb}(t) = \frac{1}{2}\hbar\Omega_0 \left(e^{i\phi}a^\dagger\sigma_+ + e^{-i\phi}a\sigma_-\right), \tag{9.15}$$

giving rise to transitions of the type $|n\rangle|\downarrow\rangle \longleftrightarrow |n+1\rangle|\uparrow\rangle$ with Rabi-frequency $\Omega_{n,n-1} = \eta\sqrt{n+1}\Omega_0$ that entangle the motional state with the internal state of the ion.

Resolved Sidebands

Without dissipative terms the time evolution of the general state

$$|\Psi(t)\rangle = \sum_{n=0}^{\infty} c_{\uparrow n}(t)|\uparrow\rangle|n\rangle + \sum_{n=0}^{\infty} c_{\downarrow n}(t)|\downarrow\rangle|n\rangle \tag{9.16}$$

is governed by the Schrödinger equation $(i\hbar\partial/\partial t - H_{\rm int}(t))|\Psi(t)\rangle = 0$, which is equivalent to the equations [84]

$$\frac{dc_{\downarrow n}(t)}{dt} = -\frac{1}{2}i^{(1-|n-n'|)}e^{i(\delta't-\phi)}\Omega_{n',n}c_{\uparrow n'}(t), \tag{9.17}$$

$$\frac{dc_{\uparrow n'}(t)}{dt} = -\frac{1}{2}i^{(1+|n-n'|)}e^{-i(\delta't-\phi)}\Omega_{n',n}c_{\downarrow n}(t). \tag{9.18}$$

Using the method of Laplace transforms, we obtain

$$\begin{bmatrix} c_{\uparrow n'}(t) \\ c_{\downarrow n}(t) \end{bmatrix} = T_{nn'}(t) \begin{bmatrix} c_{\uparrow n'}(0) \\ c_{\downarrow n}(0) \end{bmatrix}, \tag{9.19}$$

where

$$T_{nn'}(t) = \cos\left(\frac{1}{2}\omega_{nn'}t\right)\begin{pmatrix} e^{-i\delta't/2} & 0 \\ 0 & e^{i\delta't/2} \end{pmatrix}$$
$$+ \frac{i}{\omega_{nn'}}\sin\left(\frac{1}{2}\omega_{nn'}t\right)\begin{pmatrix} \delta'e^{-i\delta't/2} & -\lambda_{n'n} \\ -\lambda^*_{n'n} & -\delta'e^{i\delta't/2} \end{pmatrix}, \tag{9.20}$$

$$\omega_{nn'} = \sqrt{\delta'^2 + \Omega^2_{nn'}}, \quad \lambda_{n'n} = i^{|n-n'|}e^{-i(\phi-\delta't/2)}\Omega_{n'n}. \tag{9.21}$$

The solution describes a generalized form of sinusoidal Rabi-flopping between the states $|\downarrow\rangle|n\rangle$ and $|\uparrow\rangle|n'\rangle$ and is essential for the quantum state preparation and analysis experiments.

Unresolved Sidebands

A convenient picture of the dynamics is provided by the master-equation formalism [130,267]. The exact derivation will not be outlined here; the reader is reffered to the references [271, 272] for details of the theory.

Detection of Motional State Populations

By resonantly driving an ion in the initial state

$$|\Psi(0)\rangle = \sum_{n=0}^{\infty} c_{\downarrow n}(0) |\downarrow\rangle |n\rangle \qquad (9.22)$$

on the blue sideband one obtains the signal

$$P_{\downarrow}(t) = \frac{1}{2}\left[1 + \frac{1}{2}\sum_{n=0}^{\infty} P_n \cos\left(\Omega_{n,n+1} t\right)\right] . \qquad (9.23)$$

Here $P_n = |c_{\downarrow n}(0)|^2$ is the probability of finding the ion in the Fock state $|n\rangle$. In the Lamb–Dicke regime, the $\Omega_{n,n+1}$ are distinct frequencies, and the occupation of all number states P_n can be found by Fourier-transforming the signal.

9.2 State Creation

9.2.1 Number States

These express the motional quantum state of a trapped ion in the basis of the eigenstates of the reference oscillator number operator. If the oscillator is independent of time, they are just the familiar number or Fock states [84]. Number states up to $N = 16$ have so far been created by a simple technique involving multiple π pulses.

At NIST [273] the ion is initially cooled to the $|g, 0\rangle$ number state. Higher n number states are created by applying a sequence of resonant π pulses of laser radiation on the blue sideband, red sideband, or carrier (Fig. 9.1).

For constructed number states $n = 0$ and $n = 1$ at Innsbruck, a single ^{40}Ca$^+$ ion was cooled to the ground state on the quadrupole transition $S_{1/2} \to D_{5/2}$, leaving the ion in the $n = 0$ number state; then the number state $n = 1$ was created by applying a π pulse on the blue sideband and incoherently repumping the excited electronic state $|e\rangle$ (Fig. 9.2). Rabi oscillations for these two states are presented for comparison in reference [274].

174   9 State Engineering and Reconstruction

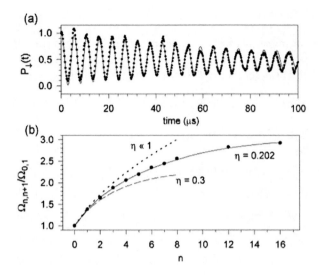

Fig. 9.1. Number states in the NIST experiment. The detected signal is the probability $P_\downarrow(t)$ to find the ion in a particular internal states of the ground state of the $^9\text{Be}^+$, noted $|\downarrow\rangle$. (a) $P_\downarrow(t)$ for the initial state $|\downarrow\rangle|0\rangle$. (b) Observed ratios of the Rabi frequencies $\Omega_{n,n+1}/\Omega_{0,1}$. The *lines* represent calculated values for different values of the Lamb–Dicke parameter η. The measured value of η was 0.202 ± 0.005 [224]

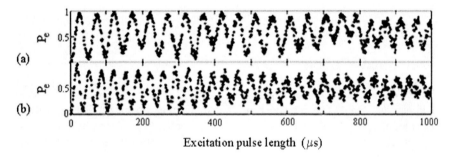

Fig. 9.2. Rabi oscillations on the blue sideband. (a) The Fock state $|0\rangle$. (b) The Fock state $|1\rangle$ [274]

9.2.2 Coherent States

Coherent states have been created so far at NIST from the state $|n=0\rangle$ by spatially uniform oscillatory field [232], or by a moving standing-wave [275].

In the first method, a sinusoidally varying potential at the trap oscillation frequency, with varying amplitudes was applied to one of the trap compensation electrodes for about $10\,\mu\text{s}$. It changes the quantum state of the oscillator only if the drive is resonant with its motion, in which case it generates an approximately spatially homogeneous force on the ion, associated with an

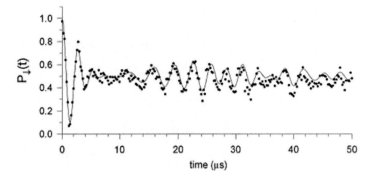

Fig. 9.3. $P_\downarrow(t)$ for a coherent state with $\bar{n} = 3.1 \pm 0.1$ [224]

interaction potential depending on time. In the rotating-wave approximation the nonstationary terms in this interaction potential are neglected, making it independent of time. It can then be integrated, yielding the evolution operator $U(t) = D(\Omega_d t)$, where Ω_d depends on the sinusoidally varying potential applied to the trap and on the resonance frequencies of the system. The drive coherently displaces an initial motional state $|\Psi\rangle$ to $|\Psi'\rangle = D(\Omega_d t)|\Psi\rangle$. The coherent displacement is proportional to the time the driving field is applied and its amplitude, displaying the general behavior of a coherent state.

In the moving standing-wave approach, two laser beams with a frequency difference of $\Delta\omega$ and detuned by Δ_{drive} from the $S_{1/2} \to P_{1/2}$ transition in Be$^+$ were used. The detuning was much bigger than the linewidth Γ of the $P_{1/2}$ state, so this state was populated with only an extremely small probabilty.

The difference frequency term leads to an intensity that varies in space and time, with a corresponding ac Stark shift of the level that can resonantly drive the motion, if $\Delta\omega = \nu$, where ν is the ion's resonant frequency. The dipole force on the ion is proportional to the spatial derivative of this ac Stark shift, having exactly the same form, as discussed in the context of the previous method of creation of the coherent states at NIST. The result is a coherent displacement of the motional wave function in the Lamb–Dicke regime. In Fig. 9.3 is shown a coherent state created at NIST. A coherent state has a definite phase relationship between the Fock state components; the phase coherence of the created states has been experimentally demonstrated [224].

9.2.3 Squeezed States

Squeezed states of motion can be created by a parametric drive at 2ν [213, 276, 277], by a combination of standing- and traveling-wave laser fields [214], or by a nonadiabatic change in the trap spring constant [213, 276, 277].

Squeezed states were created at NIST by cooling the ion to the ground state (the Lamb–Dicke parameter of 0.202) and then irradiating it with two

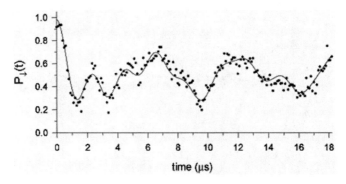

Fig. 9.4. Rabi oscillations for a squeezed state with $\bar{n} \approx 7.1$ [224]

Raman beams that differed in frequency by 2ν, driving Raman transitions between the even-n levels within the same hyperfine state [278]. As previously, the interaction can be thought of as a parametric drive induced by an optical dipole force modulated at 2ν. The squeezed state prepared in this way at NIST is presented in Fig. 9.4 [224]. A squeezed state is characterized by a squeeze parameter β, defined as the factor by which the variance of the squeezed quadrature is decreased. It grows exponentially with the driving time. The squeezed state presented in Fig. 9.4 had a squeeze parameter of $\beta = 40 \pm 10$. For this squeezing parameter, the average motional quantum number was $n \simeq 7.1$ [84].

Since 1990 various features of squeezed states including phase properties and photon statistics have been studied [279, 280]. "Cranked oscillator states" (or sheared states) [281], "multimode squeezed states" [282] have also been considered. The special case of Gaussian states (in one and many dimensions) sometimes called "mixed squeezed states" or "squeezed thermal states" have been discussed [283]. "Graybody states" [284] and "squashed states" [285] have been introduced recently. In the case of an arbitrary (inhomogeneous) electromagnetic field or an arbitrary potential, the quasiclassical Gaussian packets centered on the classical trajectories (frequently called "trajectory-coherent states") have been studied in reference [286].

9.2.4 Arbitrary States

Arbitrary states of motion in a QED cavity were introduced in 1996 by Law and Eberly [287]. Later this method was extended to create superpositions of internal and arbitrary motional states of a trapped ion [288], using hyperfine states of a single trapped Be$^+$ ion [289].

For an atomic system with four energy eigenstates $|0\rangle$, $|1\rangle$, $|2\rangle$, $|3\rangle$ arbitrary states of motion are written as $c_0|0\rangle + c_1|1\rangle + c_2|2\rangle + c_3|3\rangle$, with complex, normalized coefficients c_i, $i = 0, 1, 2, 3$ [290]. The four states are assumed to be nondegenerate levels with different energy separations. Coherent transi-

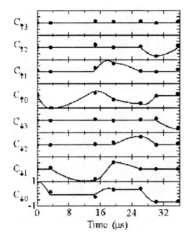

Fig. 9.5. Initial to final state Hilbert space trajectory. The real amplitudes $c_{\downarrow n}$ and $c_{\uparrow n}$ are shown for the sequence used to generate $|\Psi_{03}\rangle$. The *solid lines* are the theoretical prediction, while the *solid circles* are measured probabilities [291]

tions $|0\rangle \leftrightarrow |i\rangle$ could be driven by applied radiation. For a single $^9\text{Be}^+$ ion trapped in a linear Paul trap, the states contributing to the creation of arbitrary states of motion are the hyperfine states $F = 1$, $m_F = -1$, $F = 2$, $m_F = -2$ of the $2S_{1/2}$ electronic ground state, denoted $|\uparrow\rangle$ and $|\downarrow\rangle$.

Coherent Raman transitions generated by a pair of laser beams well detuned from the $^2S_{1/2}$–$^2P_{1/2}$ electronic transitions couple the spin states and motional levels [224]. The probabilities $|c_{\uparrow i}|^2$ and $|c_{\downarrow i}|^2$ were measured after each step in generating arbitrary states of motion and compared with theoretical predictions [290]. As shown in Fig. 9.5, the Hilbert space trajectories from the initial to final state show that at least temporarily a probability appears in the $|\downarrow\rangle|n = 0, 1, 2, 3\rangle$ and $|\uparrow\rangle|n = 0, 1, 2\rangle$ states.

9.2.5 Thermal States

Thermal states (Fig. 9.6) or more appropriately "Thermal Distribution" are defined for a single ion in thermal equilibrium with an external reservoir at temperature T, assuming that the number operator N is measured many times and the ion reequilibrates after each measurement. Then one can extract a temperature from the average result $\bar{n}$ according to [84]

$$T = \frac{\hbar \nu}{k_B \ln\left(\frac{\bar{n}+1}{\bar{n}}\right)}, \qquad (9.24)$$

where k_B is the Boltzmann constant.

If an ensemble of ions is considered, it is appropriate to characterize the state by a density matrix with zero off-diagonal elements corresponding to

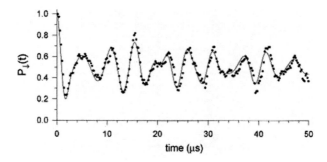

Fig. 9.6. $P_\downarrow(t)$ for a thermal state with $\bar{n} = 1.3 \pm 0.1$ [292]

maximum entropy. Consequently the thermal distribution cannot be written in the form

$$\Psi = \sum_0^\infty c_n |n, t\rangle \,, \tag{9.25}$$

which corresponds to a density matrix with nonzero off-diagonal elements for $T > 0$. By normalizing the trace of the states weighted by the Boltzmann factors, the density matrix may be written [84]

$$\rho_{th} = \frac{1}{\bar{n}+1} \sum_{n=0}^\infty \left(\frac{\bar{n}}{\bar{n}+1}\right)^n |n\rangle\langle n| \,, \tag{9.26}$$

with level population probability

$$P_n = \frac{\bar{n}^n}{(\bar{n}+1)^{n+1}} \,. \tag{9.27}$$

9.2.6 Schrödinger-Cat States

Schrödinger-cat states (or "spin superposition states") contradict our conception of reality, because in our experience of the macroscopic world, objects are either in one state or another, but never in a superposition of several states at the same time. Such an apparent contradiction arises whenever the state of macroscopic objects is correlated with that of microscopic objects, where quantum effects are evident. It is highlighted by the the Schrödinger's "thought experiment" [45], in which a cat (a macroscopic object) is placed inside a closed box together with a vial of a lethal substance and a radioactive atom (a quantum object) prepared in the metastable state. The radioactive atom has a probability of 50% of decaying in one hour. If it decays, the emitted particle releases the lethal substance and the cat dies; if it does not decay, the lethal substance is not released and the cat remains alive. So, the cat's life is determined by a single quantum event: whether or not an unstable atom decays.

The paradox appears because in this experiment the atom, a microscopic object, must be described by quantum mechanics, according to which the state of the atom and cat can be "collapsed" to one state by making a measurement – by opening the box. Until then, after one hour the atom is in a strange superposition of two states of having decayed and not decayed. By the intermediary of the lethal substance in this "thought experiment" the state of the cat of being alive or dead is correlated with the state of the atom. But quantum mechanics, as a universal and complete theory, must describe the whole system. Consequently the cat (macroscopic object) must also be in a superposition of being dead and alive, contradicting our experience of reality; we know that this is not possible.

A Schrödinger cat-like situation [45] was created at NIST [224], by putting an atom in a superposition of two electronic levels, each of which is correlated with a different state of the atom's motion. The microscopic electronic state plays the role of the radioactive atom, whereas the coherent state of motion plays the role of the cat's viability.

A Schrödinger-cat state is a superposition of two coherent states $|\alpha\rangle$ and $|\beta\rangle$ with a separation in phase space $|\alpha - \beta|$ much larger than the variance of each coherent state. It allow one to distinguish the positions of the two coherent contributions by a position measurement, because their maximum spatial separation is much larger than the single-component wave-packet extension. In this sense, the Schrödinger-cat state is "macroscopic"; the widely separated coherent states replace the classical notions of "dead" and "alive" in Schrödinger's original experiment.

If the familiy of the Schrödinger-cat states is written in a more convenient form

$$|\alpha, \varphi\rangle = \left[2 \left(1 + \cos\varphi\, e^{-2|\alpha|^2} \right) \right]^{-1/2} \left(|\alpha\rangle + e^{i\varphi} |-\alpha\rangle \right) , \qquad (9.28)$$

where φ is a phase, it is easy to distinguish its special cases: *even* states (for $\varphi = 0$), *odd* states (for $\varphi = \pi$), and *Yurke–Stoler* states (for $\varphi = \pi/2$). The probabilty densities for different kinds of cat states as function of position and time are shown in Fig. 9.7 and in Fig. 9.8.

Creation of Schrödinger-Cat States

A Schrödinger-cat state for a single trapped ion was engineered by the NIST group by using the "walking standing light wave" technique [224]. A combined state of the motional and internal states ($|e\rangle$ and $|g\rangle$) of the form $|\downarrow\rangle|\alpha\rangle + |\uparrow\rangle|\alpha \exp^{i\Phi}\rangle$ was created in which the two coherent motional components had the same amplitude but different phases (Fig. 9.9).

The phase signal of a created Schrödinger-cat state for different magnitudes of the coherent displacement was measured [225]. Coherent-state amplitudes as high as $\alpha \simeq 2.97(6)$ were measured, corresponding to an average

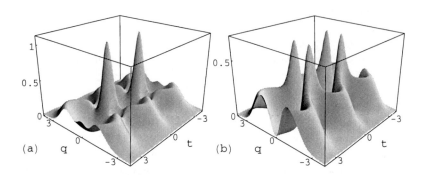

Fig. 9.7. Schrödinger-cat coherent state probability densities as functions of position and time for Re $\alpha = \sqrt{2}$ and Im $\alpha = 0$. (a) The even state probability density $P_+(q,t)$. (b) The odd state probability density $P_-(q,t)$

of $\bar{n} \simeq 9$ vibrational quanta in the state of motion, indicating a maximum spatial separation of 83(3) nm, which was significantly larger than the single wave-packet size characterized by $x_0 = 7.1(1)$ nm. The individual wave packets were thus clearly spatially separated and also separated in phase space. As the separation of the cat state is made larger, the decay from superposition to statistical mixture accelerates.

The key to cat production is to adjust the polarization of the laser beams so that only the $^2S_{1/2}(F = 1, m_F = 1)$ state interacts with the laser, while the $^2S_{1/2}(F = 2, m_F = 2)$ state does not. Thus, the laser only creates a coherent state out of the part of the wavefunction in the upper electronic state. The sequence of laser pulses used is represented in the Fig. 9.10. The two

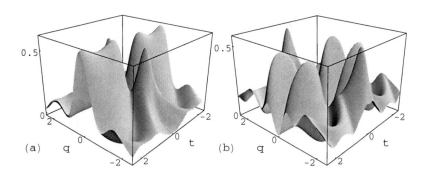

Fig. 9.8. Schrödinger-cat squeezed state probability densities as functions of position and time for Re $\alpha = \sqrt{2}$, Im $\alpha = 0$ and $s = 3/2$, where s is defined by (7.43) and (7.46). (a) The even state probability density $P_+(q,t,s)$. (b) The odd state probability density $P_-(q,t,s)$

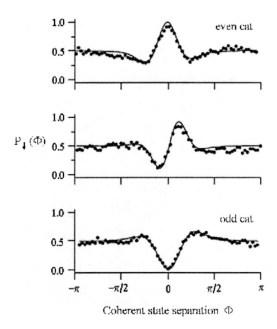

Fig. 9.9. Measured interference signal $P_\downarrow(\phi)$ for three values of δ ($\alpha \simeq 1.5$) $\delta = 1.03\pi$ (approximate even cat state), $\delta = 1.48\pi$ (approximate Yurke–Stoler cat state), and the $\delta = 0.06\pi$ (approximate odd cat state exhibiting destructive interference) [225]

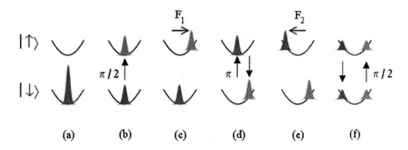

Fig. 9.10. Evolution of the position-space wavepacket entangled with the internal states $|\downarrow\rangle$ and $|\uparrow\rangle$, during creation of a Schrödinger-cat state with $\alpha = 3$ and $\phi = \pi$. (a) The initial wavepacket corresponds to the quantum ground state of motion following laser-cooling. (b) The wavepacket is split following a $\pi/2$-pulse on the carrier. (c) The $|\uparrow\rangle$ wavepacket is excited to a coherent state by the force $\boldsymbol{F}$. (d) The $|\uparrow\rangle$ and $|\downarrow\rangle$ wavepackets are exchanged following a π-pulse on the carrier. (e) The $|\uparrow\rangle$ wavepacket is excited to a coherent state by the force $-\boldsymbol{F}$. The state corresponds most closely to Schrödinger's-cat. (f) The $|\uparrow\rangle$ and $|\downarrow\rangle$ wavepackets are combined following a $\pi/2$-pulse on the carrier [293]

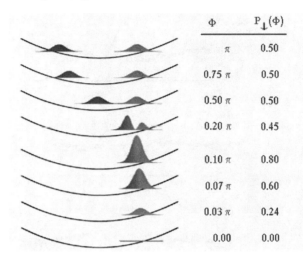

Fig. 9.11. Evolution of the position-space wavepacket superposition correlated with the $|\downarrow\rangle$ internal state as the phase separation ϕ of the two coherent states is varied, for $\alpha = 3$ and $\delta = 0$. The expected signal $P_\downarrow(\phi)$ is the integrated area under wavepackets [293]

independent coherent states were excited by a pair of Raman laser beams, both polarized σ_+, generating a sequence of laser pulses. The two coherent motional components of the superposition had the same amplitude but different phases. Each of these states is correlated with an internal state of the trapped ion, designated $|\uparrow\rangle$ or $|\downarrow\rangle$. The resulting Schrödinger-cat state is a superposition of these states, written for a relative phase $\Phi = \pi$ as [84]:

$$|\Psi\rangle = \frac{1}{\sqrt{2}}\left(|\alpha\rangle|\downarrow\rangle + |-\alpha\rangle|\uparrow\rangle\right) . \qquad (9.29)$$

With the notation introduced in the section on displaced states, the Schrödinger-cat states corresponding to displaced squeezed states can be written as

$$|n, \alpha, z\rangle_\pm = D(\alpha)_\pm U(z)|n\rangle , \qquad (9.30)$$

where

$$D_\pm(\alpha) = \left[2\left(1 \pm e^{-2|\alpha|^2}\right)\right]^{-1/2} [D(\alpha) \pm D(-\alpha)] . \qquad (9.31)$$

The process to create the Schrödinger-cat states at NIST starts in the ground motional and electronic states, as shown in Fig. 9.10. By using different laser pulses, a microscopic superposition within the atom was transformed into a large superposition of atomic motion [224].

In Fig. 9.11 is presented the evolution of the wave-packet superposition in a Schrödinger-cat as function of the phase separation Φ between the two coherent components of the Schrödinger-cat.

9.3 State Reconstruction

9.3.1 Wigner Functions

Several methods to reconstruct quantum states are known [294], ranging from quantum tomography [295], quantum state endoscopy [296], to Wigner function determination from outcome probabilities [297].

Using reconstructed density matrices and Wigner functions of arbitrary motional quantum states, the transition from quantum to classical behavior can be studied [298–302]. The Wigner function [299] describes quantum phenomena as characterized by probability amplitudes, using the classical-like concept of a phase space distribution function. The properties of the Wigner function can be used to obtain and visualize states of motion of a harmonic oscillator in any time-dependent quadratic potential.

For an arbitrary $\psi(q,t)$ the Wigner function is defined by

$$\psi(q,p,t) = \pi^{-1} \int_{-\infty}^{\infty} \psi^*(q+y,t)\psi(q-y,t) e^{2ipy} dy \ . \tag{9.32}$$

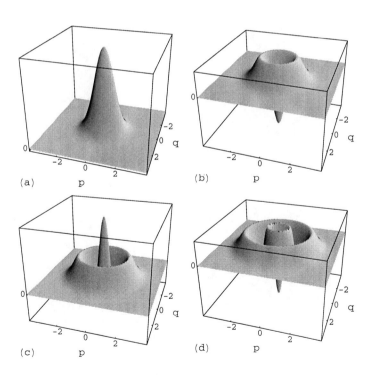

Fig. 9.12. Axial Wigner functions in a Penning trap. (a) $W_0(q,p,\tau)$. (b) $W_1(q,p,\tau)$. (c) $W_2(q,p,\tau)$. (d) $W_3(q,p,\tau)$

184   9 State Engineering and Reconstruction

If ψ is a ground state vector, (9.32) can be written as

$$W_0(q,p,t) = \pi^{-1}\exp\left[-E(q,p,t)\right], \qquad (9.33)$$

where

$$E(x,p,t) = |w|^2 p^2 - \left(w\frac{dw^*}{dt} + w^*\frac{dw}{dt}\right)px + \left|\frac{dw}{dt}\right|^2 x^2. \qquad (9.34)$$

If ψ is a n-number state vector, the corresponding Wigner function can be written as

$$W_n(x,p,t) = \pi^{-1}(-1)^n L_n\left[2E(q,p,t)\right]\exp\left[-E(q,p,t)\right], \qquad (9.35)$$

where L_n denotes the n-th Laguerre polynomial.

Some axial Wigner functions W_n, $0 \le n \le 3$ for a Penning trap are shown in Fig. 9.12. Squeezing and rotations of the axial Wigner functions W_0 and W_2 for different trapping conditions of a Paul trap are represented in Fig. 9.13 and Fig. 9.14.

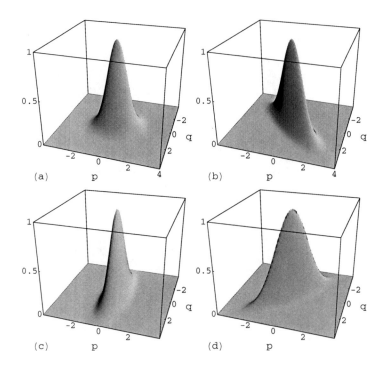

Fig. 9.13. Squeezing and rotations of the axial Wigner function $W_0(q,p,\tau)$ in a Paul trap with $a_z = 0.01$ and $q_z = 0.5$. (a) $\tau = 0$. (b) $\tau = \pi/3$. (c) $\tau = \pi/2$. (d) $\tau = 2\pi/3$

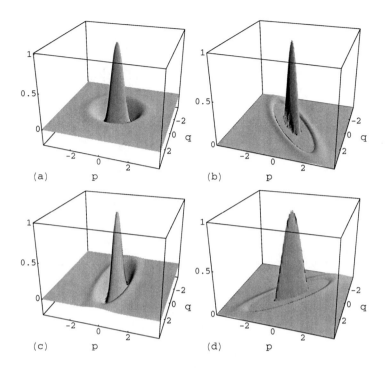

Fig. 9.14. Squeezing and rotations of the axial Wigner function $W_2(q,p,\tau)$ in a Paul trap with $a_z = 0.01$ and $q_z = 0.5$. (**a**) $\tau = 0$. (**b**) $\tau = \pi/3$. (**c**) $\tau = \pi/2$. (**d**) $\tau = 2\pi/3$

9.3.2 Experimental State Reconstruction

The density matrix and the Wigner function provide a complete description of a quantum state of motion. Consequently, methods for reconstructing the density matrix and the Wigner function of a quantum state of motion of a harmonically bound atomic particle have been experimentally realized [303]. These methods depend on controllably displacing the state in phase space, applying radiation to drive the first blue sideband for time t, and then measuring the averaged, normalized, signal $P_\downarrow$ given by

$$P_\downarrow(t,\alpha) = \frac{1}{2}\left[1 + \sum_{k=0}^{\infty} Q_k(\alpha)\cos(\Omega_{k,k+1}t)e^{-\gamma_k t}\right]. \quad (9.36)$$

Here the complex number α represents the amplitude and phase of the displacement, and $Q(\alpha)$ is the occupation probability of the vibrational state $|k\rangle$ for the displaced state. The number state populations $Q_k(\alpha)$ are related to the density matrix ρ by

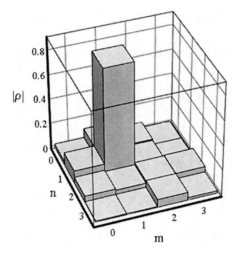

Fig. 9.15. Reconstructed number state density matrix amplitudes ρ_{nm} for an approximate $|1\rangle$ number state [303]

$$Q_k(\alpha) = \langle \alpha, k| \rho |\alpha, k\rangle = \sum_{m,n=0}^{\infty} D(\alpha)^*_{mk} D(\alpha)_{nk} \rho_{mn} \ . \tag{9.37}$$

To separate the contributions of different matrix elements ρ_{mn}, the coherent displacements can be chosen to lie on a circle in phase space. If the $Q_k(\alpha)$ coefficients are derived for a series of values of α lying in a circle,

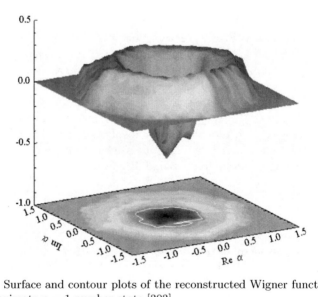

Fig. 9.16. Surface and contour plots of the reconstructed Wigner function $W_1(\alpha)$ of an approximate $n = 1$ number state [303]

9.3 State Reconstruction   187

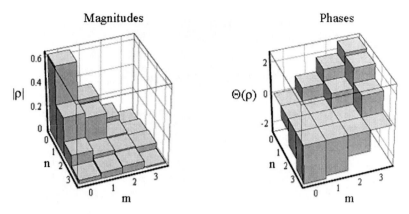

Fig. 9.17. Reconstructed amplitudes ρ_{nm} and phases $\Theta(\rho_{nm})$ of a coherent state [303]

$$\alpha_p = r_p \exp\left(i\pi \frac{p}{N}\right), \qquad (9.38)$$

where $r_p > 0$ and $p = -N, \ldots, N-1$, then the density matrix elements ρ_{nm} can be determined for values of n and m up to $N-1$.

Figure 9.15 shows the reconstructed density matrix amplitudes for an approximate $n = 1$ state.

Figure 9.17 shows the reconstructed density matrix for a coherent state with amplitude 0.67.

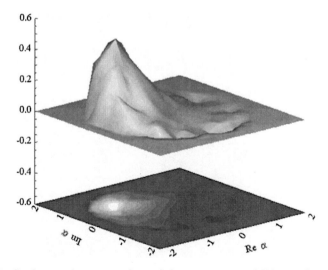

Fig. 9.18. Surface and contour plots of the reconstructed Wigner function $W(\alpha)$ of a coherent state [303]

The Wigner function for a given value of the complex parameter α can be determined from the sum

$$W(\alpha) = \frac{2}{\pi} \sum_{n=0}^{\infty} (-1)^n Q_n(\alpha) \ . \tag{9.39}$$

Figure 9.16 shows the reconstructed Wigner function for an approximate $n = 1$ state. The fact that it is negative in a region around the origin highlights the fact that it is a nonclassical state.

Figure 9.18 shows the reconstructed Wigner function for a coherent state with amplitude 1.5. It is positive, which is not surprising, since the coherent state is the quantum state that most closely approximates a classical state.

Part IV

Cooling of Trapped Charged Particles

When ions first come under the action of the trapping fields, whether they are formed internally by some ionization process, or injected from outside, they typically have mean energy falling in the electron-volt range. Prior to the development of efficient ways of cooling them, they filled traps of what would now be considered large dimensions with a tenuous distribution, making it nearly impossible to observe light scattering from them. And so initially, ion traps were primarily limited (with some notable exceptions) to applications in mass spectrometry, where the properties of mass-selective ion accumulation, and resonance detection made them attractive for residual gas analyzers and vacuum leak detectors. It was the achievement of ways of cooling the ions to temperatures approaching absolute zero in miniaturized traps, that transformed the technique into one of exquisite power and broad application. It has made possible the ultimate level of precision in magnetic resonance and mass spectrometry in elementary particles and ions, the manipulation of ion motional and electronic quantum states, and the study of chaos-order phase transitions in cold ion clouds.

Among the ways in which ion cooling makes all that possible are: first, it dramatically lengthens particle observation time, and second, it concentrates the particle distribution in coordinate and velocity space toward zero. This is of profound benefit to high resolution spectroscopy in eliminating the first order Doppler broadening through the Dicke effect, but more uniquely, in minimizing the *second-order* Doppler shift. Also, the concentration of the ion distribution in space reduces the anharmonicities in the trapping fields as seen by the ions, and, more significantly, increases the brightness of the ions as a source, improving ion optics for mass spectrometry, but critically raising the signal-to-noise ratio of optical fluorescence signals, to the point where *individual* ions are observable.

The use of the term "cooling" implies lowering the particle temperature, a concept that applies to a state of thermal equilibrium; however, in the Paul and Penning ion traps, no such simple thermal equilibrium generally obtains. We must distinguish between a part of the ion motion that remains coherent with the trapping fields that are (apart from electrical noise) determinate, from the random thermal part that results, for example, from collisions with other particles. Thus in the Paul trap the high frequency micromotion is determined at all times by the field, and remains independent of any randomizing collisions, which affect only the secular motion. Also in the Penning trap the magnetron rotation remains coherent, determined by the ratio of electric to magnetic fields. Therefore in what follows "cooling" simply means reduction in the kinetic energy associated with any or all modes of motion of the ions.

There are a number of ways that have been developed by which the kinetic energy of trapped ions can be reduced. Among them are collisions with a light buffer gas at a lower temperature; resistive cooling through currents induced in an external resistance, with possible negative feedback; radiative cyclotron / synchrotron cooling in a magnetic field, useful for light particles such as electrons and positrons; and the method that stands above all the rest, made possible by the advent of the laser: laser cooling. As a monochromatic excitation source for resonance fluorescence, the laser is so powerfully adapted to ion field suspension in vacuum that it has transformed the latter from one championed by a few as one of future promise, into one which has burgeoned into an extraordinarily fruitful field with applications in several fundamental areas. In cases where the quantum level structure does not permit resonance scattering at a laser wavelength that can be synthesized in practice, it may be possible to use what has come to be called *sympathetic* cooling. In this, particles that cannot be cooled directly are cooled through thermal contact with other types of particles for which an efficient cooling method exists.

There have been a number of review articles on the laser cooling of atoms and ions, among them [272, 304, 305], which describe the mechanism and different methods to reduce the particle's temperature. More specifically for the case of trapped ions we refer to [306, 307].

10 Trapped Ion Temperature

The initial kinetic energy distribution of the different random modes of motion of a cloud of trapped ions is usually far from thermal equilibrium, and depends on the ions' initial positions in the trap, and for the Paul trap, the initial rf-phase. Subsequent collisions however, will cause these modes to approach an equilibrium Maxwellian distribution with a limit on the kinetic energy set by the mean potential depth of the trap. Such an equilibrium distribution must clearly be truncated at the high velocity end and the total number of ions must slowly decay, as they escape from the trap.

For individual ions in ultra-high vacuum under laser irradiation, transitions involving center-of-mass motion as well as internal electronic states will dynamically alter the occupation number n-states of the harmonic potential, leading ultimately to a quantum statistical state which must depend on the characteristics of the laser field. A detailed study of the approach to equilibrium in the laser cooling of particles in a static trap has been made by Javanainen et al. [308], in which it is shown that the final distribution of *populations* over the (secular) vibrational states is one of thermal equilibrium. This is experimentally demonstrated in Fig. 10.1 where a single ion in a Pen-

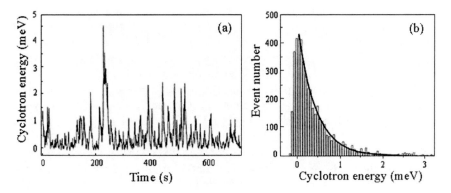

Fig. 10.1. (a) Thermal fluctuations of the cyclotron energy of a single stored ion coupled to a resonance circuit at $T = 4.2$ K. (b) Histogramm of cyclotron energies. The fit shows an exponential Boltzmann distribution corresponding to a temperature of 4.90 ± 0.08 K [208]

ning trap is kept in thermal equilibrium with the environment at liquid He temperature. A histogram of the thermal fluctuations of the cyclotron energy shows a one-dimensional Boltzman distribution to which a temperature can be assigned. This allows the application of the concept of temperature to single trapped ions, which, by analogy with Planck's oscillators in his theory of black body radiation, have a temperature T related to the average occupation number $\langle n \rangle$ according to

$$\langle n \rangle = \frac{1}{\exp\left(\frac{h\nu}{k_B T}\right) - 1} . \tag{10.1}$$

For practical purposes, however, in the following we will simply mean by "temperature" the average kinetic energy, according to what would be the classical relationship $T = \bar{E}_{kin}/k_B$ for each oscillator. For a single particle cooled nearly to the zero point of the harmonic potential, a common measure is taken to be the mean value $\langle n \rangle$ of the harmonic oscillator occupation number n.

10.1 Measurement of Ion Temperature

The method of measuring ion temperature most adapted to the ion trapping technique involves the same lasers already in place to fulfill other functions, such as preparing ions in particular quantum states or cooling them through transitions to lower n-states, the subject of the sections to follow.

The spectral profile of the resonance fluorescence from a convenient optical transition, as modified by the Doppler effect, can be used to determine the velocity distribution among the ions. In the simple derivation of the Doppler effect, the approximate assumption is usually tacitly made that the particle motion is such that it traverses many wavelengths of the laser field before changing direction. In the context of ion cooling this is often referred to as the "Doppler" region, in which the expected line shape, for particles with a Maxwellian velocity distribution, is a convolution of a classical Lorentz profile with a Gaussian. This applies during the cooling process when the ion temperature is yet far from the zero point and the condition $a \gg \lambda$ holds, where a is the amplitude of the ion motion, and λ is the wavelength of the exciting laser radiation. As the ions are cooled to the extreme, a point is reached where the amplitude is so far reduced that $a < \lambda$ and the optical field frequency as "seen" by the ion is modulated at its oscillation frequency, resulting in a spectrum consisting of a strong central line defined as the *carrier* frequency, with equally spaced *sidebands* stretching to infinity with decreasing amplitudes on both sides. This is referred to as the *Lamb–Dicke regime* and we will defer to the section on sideband cooling the discussion of the measurement of temperatures in this regime, which for optical wavelengths and typical practical

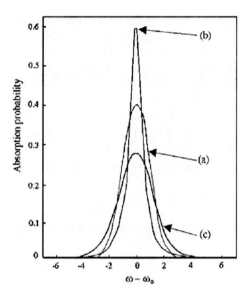

Fig. 10.2. The Voigt line shape (**a**) as the convolution of the Doppler (Gaussian) line shape (**b**) and the Lorentz line shape (**c**)

conditions of operation of ion traps, implies very low temperatures approaching the zero point where $\langle n \rangle$ is a small number, and a quantum description is required. At all but the lowest temperatures then, a laser is tuned through resonance with a suitable electronic transition, and the intensity of the fluorescence is observed as a function of frequency. The temperature is deduced from curve-fitting the observed profile to a theoretical line shape deduced from due consideration of all relevant sources of line broadening, principally the radiative lifetime of the excited state, Lorentz collision broadening, the laser spectral width and the expected Doppler effect.

It should be noted that in the process of resonance fluorescence, if the laser irradiates the ion with a spectrum that is constant over the spontaneous emission line of the ion, the fluorescence can be regarded as independent of the excitation. On the other hand if the excitation spectrum is much narrower than the natural line width of the transition, then the emitted photons will have nearly the same frequency distribution as the absorbed photons. If we neglect any velocity-dependence of ion collision cross sections, then the observed line shape in the Doppler regime is, as already stated, a convolution of a Lorentz with a Doppler (Gaussian) line shape, known as a *Voigt* line shape. The Voigt line shape, which occurs commonly in the context of radiation transfer theory and the study of stellar atmospheres [309, 310] may be expressed as follows:

$$\kappa(\omega - \omega_0) = \kappa_0 \frac{\gamma \Delta \omega_D}{4\pi \sqrt{\ln 2}} \int_{-\infty}^{+\infty} \frac{\exp(-\xi)^2}{(\omega - \omega_0 - k v_p \xi)^2 + (\gamma/2)^2} d\xi , \qquad (10.2)$$

where $v_p = (2k_BT/M)^{1/2}$ is the most probable velocity in the Maxwellian distribution, $\Delta\omega_D$ is the full Doppler width, and γ is the linewidth of the transition, which at the laser powers often applied may exhibit the effect of saturation of the ionic state populations, depleting the lower state and increasing the upper state populations. This saturation affects only ions which happen to be within the *natural* linewidth of exact resonance with the laser frequency, as the latter is swept across the Doppler broadened line. The effect on the linewidth γ is to broaden it according to $\gamma = \gamma_0(1+S)^{1/2}$ where S is the saturation factor, proportional to the intensity of the laser beam. By repeating the measurement of the line profile for a range of laser intensities, an extrapolation can be made to zero intensity. Detailed numerical data on the Voigt line profile are available in the literature to enable curve fitting to the experimental data, and extraction of the Doppler contribution. A comparison of the Voigt profile to the Doppler- and natural profile is given in Fig. 10.2.

11 Radiative Cooling

The emission of radiation by accelerating electric charges is an ever-present concern as a source of energy loss in the context of particle accelerators, where the emphasis is usually on behavior at relativistic energies. In the present context however, the interest rests at the opposite end of the energy scale, where, for all but the lowest energies, a classical analysis is adequate. The rate of energy emission by a particle carrying a charge Q and having a mass M moving with an acceleration α is given by Larmor's formula (in SI units)

$$\frac{dE}{dt} = -\frac{Q^2}{6\pi\varepsilon_0 c^3}\alpha^2 . \tag{11.1}$$

For a particle of *total* energy E, executing simple harmonic motion with the frequency ω, the mean rate of energy loss is given by

$$\left\langle\frac{dE}{dt}\right\rangle = -\gamma_r E , \tag{11.2}$$

from which follows

$$E = E_0 e^{-\gamma_r t} , \tag{11.3}$$

where the decay constant γ_r is given by

$$\gamma_r = \frac{1}{6}\frac{Q^2\omega^2}{\pi\varepsilon_0 M c^3} . \tag{11.4}$$

In a quantum description, the total rate of spontaneous photon emission in a transition from a quantum number state $|n\rangle$ to $|n'\rangle$ is given in the dipole approximation by [311]

$$\frac{dE}{dt} = -2\gamma_r M |\langle n'|r|n\rangle|^2 . \tag{11.5}$$

For an harmonic oscillator the nonzero matrix elements for $n > n'$ are:

$$\langle n-1|z|n\rangle = \sqrt{\frac{\hbar}{M\omega}\frac{n}{2}} , \tag{11.6}$$

leading to a value of γ_r which in the limit $n \gg 1$ agrees with the classical value for a linear oscillator, as derived above.

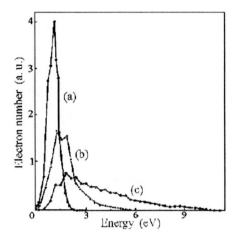

Fig. 11.1. Energy distribution of electrons in a Penning trap with $B = 6.5\,\text{T}$, for cooling times (**a**) 500 ms; (**b**) 300 ms; (**c**) 100 ms [312]. Copyright (2003), with permission from Elsevier

In the interesting case of particles executing cyclotron motion in a magnetic field B at a frequency $\omega_c = QB/(4\pi\varepsilon_0 M)$, the emitted *cyclotron radiation*, or in the context of relativistic energies *synchrotron radiation* has a decay constant $\gamma_c = 2\gamma_r$. Because of the dependence of γ_c on M^{-3} (for a given B), radiative energy loss by trapped particles is of interest as a cooling method only for light particles. For example, Kienow et al. [312] have demonstrated radiative cooling of electrons in a Penning trap operating with a magnetic field $B = 6.5\,\text{T}$, corresponding to a cyclotron frequency $f_c = \omega_c/2\pi = 182\,\text{GHz}$. This was done by measuring the energy distribution among the trapped electrons in a cloud for different cooling time intervals, and showed that as the cooling time is increased from 100 ms to 500 ms, the initially broad spatial distribution function becomes substantially narrower, as shown in Fig. 11.1, evidence of radiative cooling. The deduced cooling time constant of $\gamma \approx 13\,\text{s}^{-1}$ was found to be in reasonable agreement with the theoretical value.

In contrast to the radiation damping of a light particle such as an electron, an atomic ion such as Ca$^+$ in the same Penning trap would have $f_c \approx 2.5\,\text{MHz}$, and $\gamma = 4 \times 10^{-14}\,\text{s}^{-1}$, a rate clearly of no practical interest. Also Paul traps operate, as a practical matter, at frequencies that are far too small to give significant energy loss by radiation.

The final temperature of the trapped particles is determined by the equilibrium established through the exchange of radiant energy with the environment. When the temperature of the surrounding walls is kept at temperatures below 100 mK by contact to liquid He, the electron energy reaches a level at which the quantized nature of the cyclotron motion becomes visible. It has been demonstrated [313] that, at a B value of about 5 T, a trapped electron

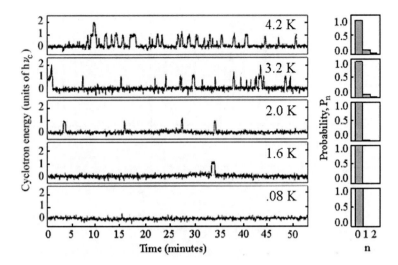

Fig. 11.2. Occupation of the lowest cyclotron oscillator states of a single electron radiatively cooled to different ambient temperatures [313]. Copyright (2003) by the American Physical Society

cools down to the ground state of the cyclotron oscillator states and that thermal fluctuations of the blackbody radiation have not sufficient energy to excite it from this state. At some elevated temperatures the distribution over the quantum states corresponds to the predicted form if thermal equilibrium between the electron and the environment is assumed (Fig. 11.2).

Consideration of the rate of the energy exchange with the environment raises the interesting question of the possible effect of the environment on the quantum transition probabilities of a radiating particle; for example, when confined inside a resonant cavity. It will be recalled that according to one of Fermi's "Golden Rules" the probability of a quantum transition, such as the spontaneous photon emission by an atom, depends on the density of available final photon states, which in free space is given by

$$\rho_\infty = \frac{2\omega^2}{\pi c^3} \; . \tag{11.7}$$

The radiation process of an atom will indeed be affected if the frequency falls within the range of the dominant modes of a cavity; this implies that the cavity dimensions must be on the order of the wavelength of the emitted radiation. An ion trap electrode structure obviously constitutes a conducting cavity, and the density of available photon states $\rho_c(\nu)$ differs from the free space value by the presence of cavity resonances. For example, in a perfectly conducting cylindrical cavity, singularities appear in the mode density, superimposed on the smooth mode density distribution in free-space. According to Kleppner [314] the density of states for a cylinder having walls of infinite

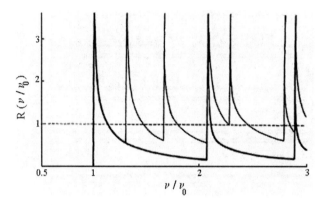

Fig. 11.3. The ratio of the waveguide mode density to the free space mode density [314]. Copyright (2003) by the American Physical Society

conductance is given by

$$\rho(\nu) = \frac{16\nu_0^2}{c^3} \sum_j \frac{\nu}{(\nu^2 - \nu_j^2)^{1/2}}, \tag{11.8}$$

where ν_0 is the fundamental cutoff frequency of the cavity, given by the relation

$$\nu_0 = \frac{c}{\lambda_0/2} = \frac{0.29c}{a}, \tag{11.9}$$

where a is the cavity radius, and ν_j is the cutoff frequency for higher modes. To apply realistically to walls having a finite conductance, an additional damping term is needed to limit the amplitude. These singularities are better expressed by the ratio of the waveguide mode density to that of free space as shown in Fig. 11.3. The enhancement in the spontaneous emission rate at the cavity resonances is accompanied by its suppression at other frequencies.

The observation of such *inhibited spontaneous emission* was achieved in 1985 [315] on a single electron in a Penning trap having a magnetic field of 6 T, at an equilibrium temperature of nearly 4 K. The quadrupole electrodes form a microwave cavity with dimensions comparable to the wavelength of the cyclotron radiation $\lambda_c = c/\omega_c$. The parameters of the experiment were such that the theoretical cyclotron energy decay time in free space was 0.086 s. The observed energy decay time constant was 0.35±0.06 s, when the magnetic field was slightly tuned to such a value that the cyclotron wavelength corresponds to one of the cavity modes (Fig. 11.4).

The ability to suppress spontaneous emission has obvious application to high resolution spectroscopy, where linewidths are often ultimately set by that process. More recently a 140-fold suppression of synchrotron radiation was reported in 1999 [313] on an electron *quantum cyclotron*, leading to extraordinary isolation of the electron from its environment. By manipulating

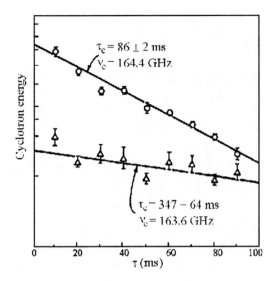

Fig. 11.4. Decay times of the cyclotron energy of an electron in a Penning trap when the cyclotron wavelength corresponds to a mode of the trap as microwave cavity (*triangles*), and as correspond to the decay time outside the mode wavelength (*circles*). The change in wavelength is achieved by a 0.5% change in the magnetic field strength [315]. Copyright (2003) by the American Physical Society

its quantum number (Fock) states using what they call a *quantum nondemolishing* technique, the quantum cyclotron was shown to remain in its ground state until a resonant photon is injected to excite it to a higher Fock state. This clearly represents an outstanding advance in the effort to further improve the precision of the measurement of the electron magnetic moment, and the fine structure constant.

12 Buffer Gas Cooling

The earliest cooling method for heavy trapped ions was by multiple weak collisions with a lighter inert gas, most commonly He introduced into the vacuum system as a buffer at typical pressures of 10^{-4}–10^{-1} Pa, at or below room temperature. The method relies on collisions between the ions and gas atoms to equilibrate the ion temperature to that of the gas, and ultimately to the trap walls, without affecting the internal quantum states of the ions. Of the types of collisions that can occur between the ions and buffer gas atoms, elastic scattering is expected to have the largest cross section; however, if the neutral background gas also contains the ions' parent atoms or molecules, then resonant charge exchange would also have a significant cross section, depending on the temperature. At high energies the elastic collision cross section is constant and about equal to the geometrical cross section; however, it increases at lower energies (Langevin theory). For collisions in all but the extremely low energy range, a classical treatment is adequate for elastic collisions between heavy particles. An approximate analysis of the statistical mean motion of a massive ion through a light buffer gas can be modeled as in the classical theory of *ion mobility* in a gas. In this model a constant ion drift velocity is derived under the action of an electric field, analogous with the action of a linear viscous drag proportional to the ion velocity, thus

$$M\frac{d\bm{v}}{dt} + \beta\bm{v} - Q\bm{E} = 0 , \quad \bm{v}_\infty = \frac{Q}{\beta}\bm{E} , \qquad (12.1)$$

where Q/β is defined as K, the mobility of the ion in the buffer gas at the given temperature and pressure.

There are limitations to this method of ion cooling which detract from its advantage of being simple. It is effective only for massive ions, and even for them the collisions with buffer gas atoms may, depending on the application, unacceptably perturb their internal energy levels, causing frequency shifts and broadening of their spectral lines. Also, while collisions with a lighter gas may lower the *average* energy of the ions, the discrete, random changes in the ions' motion will lead to broadening of motional resonances, and ultimately the loss of ions from the trap. The presence of a buffer gas in fact also affects the trapping process itself in several important respects: not only in the relaxation to an equilibrium ion temperature, but also in the initial ion entrapment process, and lifetime in the trap.

12.1 Paul Trap

On the basis of classical collision theory it has been shown [76] that, under the action of the rf-field, the average loss or gain in ion kinetic energy due to collisions with gas atoms is determined by the relative mass of the ion and buffer gas atom. This is fundamentally because the relative mass determines the degree to which the collisions interrupt the phase of the micromotion, introducing a degree of randomness which must be transferred to the secular motion, the micromotion remaining coherent with the rf-field. The amplitude of the micromotion remains at all times fixed by the amplitude of the field at the ion's position, and therefore the only way to reduce the micromotion energy is to reduce the amplitude of the secular motion, thereby causing the ions to remain near the origin where the rf-field approaches zero. If $M_{\text{ion}} \gg M_{\text{buffer}}$ the collisions only slightly deflect the ion, and the cumulative result of many collisions is to incrementally draw kinetic energy from the ion's secular motion, without randomizing the phase of the micromotion; the result *on the average* is cooling and a general concentration of the ion distribution about the center. The effect of the many weak collisions can be modeled, as shown above, as an average frictional force $\boldsymbol{F} = -\beta \boldsymbol{v}$, where $\beta = Q/K$. The small, but frequent, increments in the ion momentum, will of course also have a transverse component leading to diffusion away from the line of the trajectory. As already described in Part I, the presence of a linear damping term in the equation of motion of a single ion in a Paul trap, leads to significant changes in the boundaries of the stable regions in the a_j–q_j diagrams, $j = x, y, z$.

If on the other hand $M_{\text{ion}} < M_{\text{buffer}}$, each collision introduces randomness in the phase of the micromotion which is transferred to the secular motion resulting in an increase in the ion kinetic energy, and the rate of loss from the trap. In the absence of some other cooling mechanism being applied, the only way to ensure a long confinement time in this case is to exhaust the system to the highest vacuum possible.

With buffer gas cooling, the ultimate equilibrium ion temperature results from a balance between the collisional cooling and the rf-heating due to ion–ion collisions. This rf-heating has been shown by computer simulations of the classical dynamics of coupled nonlinear systems [316] to result from the existence of a chaotic regime for the interacting ions. Chaos leads to a continuous rather than a discrete power spectrum of the ion motion, that allows the ions to absorb energy from the field. Ultimately, however the ion distribution broadens and the density decreases, causing the motions of the ions to become independent of each other, and less chaotic; then the heating stops and the ion remains trapped for a long time.

Cutler et al. [317] have computed the number density distribution of a cloud of Hg^+ ions in a Paul trap subject to just a static harmonic pseudopotential and space charge, when cooled by a He buffer gas to different temperatures. The results, reproduced in Fig. 12.1, show that at the extreme

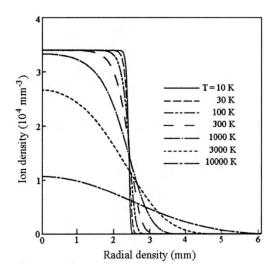

Fig. 12.1. Calculated radial density distribution of a cloud of 2×10^6 ^{199}Hg$^+$ ions in a Paul trap for different temperatures [317]. Copyright (2003) by the American Physical Society

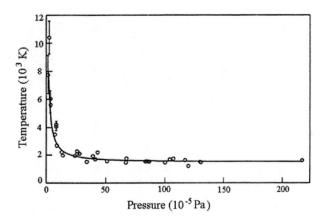

Fig. 12.2. Temperature of a cloud of 10^5 Ca$^+$ ions in a Paul trap at different N$_2$ buffer gas pressures [318]

limit of low temperatures, the ions tend to form a constant density core at the center of the trap.

Figure 12.2 shows an example of the approach to an equilibrium temperature of Ca$^+$ in a H$_2$ buffer gas, inferred from the Doppler width of an optical transition [318].

12.2 Penning Trap

The application of buffer gas cooling to ions in the Penning trap, while similar in regard to the axial oscillation to the secular motion in the Paul trap, is fundamentally different when it comes to the two modes of motion in the radial plane, the cyclotron- and the magnetron-motions. In the radial plane, the equations of motion of a single ion in the presence of a linear damping term are as follows:

$$\frac{d^2x}{dt^2} + \alpha \frac{dx}{dt} - \omega_0 \frac{dy}{dt} - \frac{1}{2}\omega_z^2 x = 0 ,$$

$$\frac{d^2y}{dt^2} + \alpha \frac{dy}{dt} + \omega_0 \frac{dx}{dt} - \frac{1}{2}\omega_z^2 y = 0 , \qquad (12.2)$$

where $\alpha = Q/(KM)$ and $\omega_0 = \frac{Q}{M}B$. Setting $u = x + iy$ we find

$$\frac{d^2u}{dt^2} + (\alpha + i\omega_0)\frac{du}{dt} - \frac{1}{2}\omega_z^2 u = 0 . \qquad (12.3)$$

The substitution $u = \exp(-i\omega t)$ yields the solutions for ω

$$\omega_\pm = \frac{1}{2}\left[\omega_0 \pm c_+ - i(\alpha \pm c_-)\right] , \qquad (12.4)$$

where

$$c_\pm = \frac{1}{\sqrt{2}}\left[\sqrt{(\omega_1^2 - \alpha^2)^2 + 4\omega_0^2\alpha^2} \pm (\omega_1^2 - \alpha^2)\right]^{1/2} , \qquad (12.5)$$

and $\omega_1 = \sqrt{\omega_0^2 - 2\omega_z^2}$. If we now assume $\omega_0, \omega_z \gg \alpha$, we find

$$\omega_\pm \simeq \frac{\omega_0 \pm \omega_1}{2}\left(1 \mp i\frac{\alpha}{\omega_1}\right) . \qquad (12.6)$$

The motion is in general a linear combination of the two modes, showing explicitly that whereas the amplitude $R_+(t)$ of the "reduced" cyclotron mode at frequency $\tilde{\omega}_+$ will decay in time, the amplitude of the magnetron mode $R_-(t)$ at frequency $\tilde{\omega}_-$ actually *increases* in time. Thus

$$x = R_+(t)\cos(\tilde{\omega}_+ t + \varphi_+) + R_-(t)\cos(\tilde{\omega}_- t + \varphi_-) ,$$
$$y = -R_+(t)\sin(\tilde{\omega}_+ t + \varphi_+) - R_-(t)\sin(\tilde{\omega}_- t + \varphi_-) , \qquad (12.7)$$

where

$$R_\pm(t) = R_\pm(0)\exp(\mp \alpha_\pm t) , \quad \alpha_\pm = \frac{1}{2}(c_- \pm \alpha) > 0 , \quad \tilde{\omega}_\pm = \frac{1}{2}(\omega_0 \pm c_+) . \qquad (12.8)$$

The buffer gas cooling of the axial and cyclotron ion motions therefore is unexceptional; however, the expansion of the magnetron orbit would lead ultimately to the loss of the ions from the trap. It is a difficulty that prompted the

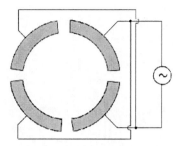

Fig. 12.3. Quadrupole electrodes for coupling the magnetron and cyclotron motions in a Penning trap

study of various means of coupling the magnetron motion to the other modes. One method of achieving this is by the introduction of an *rf-quadrupole* field in the radial plane [186]. The field is established by dividing the ring electrode of the trap into four equal parts, each pair of opposite sectors being short-circuited electrically, and between these two pairs, an additional rf-field at the combination frequency $(\tilde{\omega}_+ + \tilde{\omega}_-)$, is applied symmetrically with respect to ground (Fig. 12.3).

To describe the excitation of ion motion by the quadrupole field, consider the quantum levels n, $n \pm 1$, $n \pm 2$ of the modified cyclotron mode. The energies of these levels are given by

$$E_{n+} = \left(n + \frac{1}{2}\right)\hbar\omega_+ . \tag{12.9}$$

Since $\omega_0^2 > 2\omega_z^2$, and we have necessarily $\omega_+ > \omega_-$, each of these levels is split into sublevels characterized by the quantum number m, whose separations from the nth level are given by

$$E_{m-} = -\left(m + \frac{1}{2}\right)\hbar\omega_- . \tag{12.10}$$

Quadrupole transitions between different (m, n) levels are allowed in accordance with the selection rules for Δm and Δn; thus it can be shown that ions can absorb energy from an azimuthal quadrupole electric field at the frequencies $2\omega_+$, $2\omega_-$, and the cyclotron frequency $\omega_0 = \omega_+ + \omega_-$. At the first two frequencies transitions occur between states within each mode separately; however, at the combination frequency cross transitions occur between the two modes, making it possible to gradually transfer the excitation from the magnetron motion to the modified cyclotron. For a stored ion with a given magnetron radius R_-, this coupling between the modes will decrease R_- while increasing R_+ until the energy in the magnetron motion is transformed entirely into the modified cyclotron motion. If the excitation is continued, the

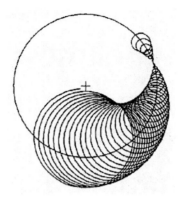

Fig. 12.4. Simulated motion of an ion in the radial plane of a Penning trap when an azimuthal quadrupole field at the frequency $\omega = \omega_+ + \omega_-$ is applied. Cyclotron and magnetron energies are continuously converted into each other [105]

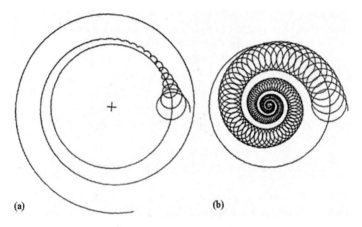

Fig. 12.5. Integration of the equation of motion in a plane perpendicular to the magnetic field for an ion in a Penning trap. A damping force proportional to velocity is taken into account which models the collisions with the buffer gas. The *cross* represents the center of the trap, the *circle* the initial magnetron radius. (a) A fast damping of the cyclotron motion and a slow increase of the magnetron motion are observed. (b) The effect of an additional quadrupole field resonant at ω_c is shown. Both cyclotron and magnetron radii are decreased [186]. Copyright (2003), with permission from Elsevier

direction of energy transfer is reversed between the modes and ultimately the energy is entirely in the magnetron mode. Figure 12.4 gives the results of a numerical solution of the equations of motion, including the quadrupole coupling field, showing the mutual exchange of energy between the modes.

In the presence of the coupling field, then, the cooling effect of collisions will now also cause the magnetron radius to shrink, resulting in the contraction of the ion distribution toward the center (Fig. 12.5). To succeed, this presumes that the energy exchange between the two modes is much faster than the collision process leading to the expansion of the magnetron motion. It can be shown that under these conditions the radial amplitudes decay exponentially, thus

$$R_\pm(t) = R_\pm(0)\exp\left(-\frac{\alpha\omega_\pm}{\omega_+ - \omega_-}t\right) . \tag{12.11}$$

The method of collisional cooling and of centering ions has been successfully applied in the fields of mass spectrometry [319] and optical spectroscopy [320]. Since the cyclotron frequency $\omega_0 = QB/M$ depends on the mass of the ion, it can be used to spatially separate ions of different mass; even where the difference is due only to the nuclei being in different nuclear states, such as the ground and isomeric states [321].

13 Resistive Cooling

In its earliest form, an external resistance was connected between the end caps of a trap so that charged particles oscillating along the axis induced charges on these electrodes, causing a current to flow through the resistance between them, a fact that was first made the basis of a *detection* technique for electrons by Dehmelt [205] called a "bolometric technique", and later also exploited by Wineland et al. [322, 323]. Unlike the original Paul detection method, in which the trapped cloud of ions was coherently driven by a dipole field generated by potentials applied between the end caps, this technique makes use of the noise current generated by the electron cloud in the external circuit. The cooling function derives from the dissipation of energy by the induced currents in the resistive load, which will continue until the particles reach thermal balance with the external resistance, whose Johnson noise is determined by its temperature. While the technique is particularly efficient for electrons, it is in principle applicable to any charged particle, and thus provides a method of cooling not only subatomic particles, but also ions whose properties make more efficient cooling by lasers impossible. As with collisional cooling, the magnetron degree of freedom cannot be cooled in this way, since as we have seen, a reduction in its (total) energy leads to an increase in its amplitude. The magnetron motion can be cooled, however, by coupling it to either the cyclotron, as already described, or the axial motion [324]. It is possible in principle to apply resistive cooling also to Paul traps; however, that is made difficult by the presence of the high rf-trapping voltages on the ring, which, if the rf-grounding of the end caps is not perfectly symmetric, will lead to unacceptable levels of rf on the high quality factor Q_{LC} tank circuit. Furthermore in a Paul trap, resistive cooling is of limited effectiveness, since only the secular macromotion can be cooled, not the micromotion, which is coherently driven by the trapping rf-field.

For a cloud of charged particles the *time average* image charge on the end caps and the average induced current in the external resistance are both zero; however, the *instantaneous* values of course fluctuate randomly, and the mean *square* current, that is, the mean power dissipated, is not zero. Conservation of energy then requires that the amplitude of the particle motion be damped. The electric field produced by the image charges acting on an individual particle consists of two terms: one depending on the particle position, and

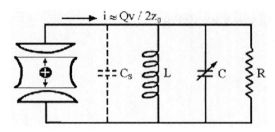

Fig. 13.1. The dissipation of ion energy through induced currents in an external resistive load formed by a tuned LC circuit

another, of the opposite sign, depending on its velocity. While the latter dissipates the particle energy, the former component primarily affects the oscillation frequency, and may have a significant effect in a microtrap on the resolution and accuracy of a resonance measurement.

Given the high frequency of the currents induced by the oscillating charges, the practical realization of a resistive load requires that the capacitance of the trap electrodes be "tuned out". Thus, as in the original Paul resonance absorption method of detecting trapped ions, the technique involves using a high quality factor Q_{LC} tuned LC circuit connected, for z-motion coupling, between the end caps. Its resonance frequency, $\omega_{LC} = (1/LC)^{1/2}$, is tuned to the particle z-oscillation frequency; for example, in a Penning trap we require

$$\omega_z = \sqrt{\frac{QU_0}{Md^2}} = \omega_{LC} \,. \tag{13.1}$$

At resonance, the impedance across the LC circuit is purely resistive, given by $Q_{LC}\omega_{LC}L$.

In the simplest model, we assume a cloud of N particles, each of charge Q, oscillating at frequency ω_z along the z-axis between two planar end caps that extend out to a radius much larger than their separation. Each charged particle oscillating with amplitude and phase (a_n, φ_n) is capacitatively coupled to the end caps, inducing image charges on them, causing a current I_n to flow through the external resistance given approximately by

$$I_n(t) = \frac{GQa_n\omega_z}{2z_0}\sin(\omega_z t + \varphi_n) \,, \tag{13.2}$$

where $G < 1$ is a geometric factor inserted to correct for the finite size and shape of the end caps, and assumed to be the same for all ions.

It will be assumed that the particles oscillate with random phases φ_n, have amplitudes a_n following the Boltzmann distribution, and have a Lorentz broadened oscillation frequency due to a limited average coherence time τ_0. To derive the power dissipated in an external resistance, we begin with the squared modulus of the Fourier transform of $I_n(t)$ assuming the oscillation continues for a time τ, and averaged over the phase, thus

$$\left\langle \left|\tilde{I}_n(\omega,\tau)\right|^2\right\rangle_\varphi = \frac{A_n^2}{2}\left\{\frac{[1-\cos(\omega_z-\omega)\tau]}{(\omega_z-\omega)^2}+\frac{[1-\cos(\omega_z+\omega)\tau]}{(\omega_z+\omega)^2}\right\}, \quad (13.3)$$

where $A_n = GQa_n\omega_z/(2z_0)$. If the density probability of the oscillations is assumed to be given by $p(\tau)d\tau$, where

$$p(\tau) = \frac{1}{\tau_0}\exp\left(-\frac{\tau}{\tau_0}\right), \quad (13.4)$$

the average over τ yields

$$\left\langle \left|\tilde{I}_n(\omega)\right|^2\right\rangle = \frac{A_n^2}{2}\left[\frac{1}{(\omega_z-\omega)^2+\gamma_0^2}+\frac{1}{(\omega_z+\omega)^2+\gamma_0^2}\right], \quad (13.5)$$

where $\gamma_0 = 1/\tau_0$. The average over a_n for particles in thermal equilibrium is simply obtained by setting $M\omega_z^2\langle a^2\rangle/2 = k_BT$. The power spectral density function $S(\omega)$ for individual particle oscillations averaged as a random time sequence of individual "events", each of average duration τ_0, is according to Carson's theorem given by

$$S(\omega) = \frac{1}{\pi\tau_0}\left\langle \left|\tilde{I}_n(\omega)\right|^2\right\rangle, \quad (13.6)$$

from which we obtain the total power dissipated in a resistance R by N particles as

$$\left\langle \frac{dE}{dt}\right\rangle = -NR\int_0^\infty S(\omega)d\omega, \quad (13.7)$$

and hence finally obtain

$$\left\langle \frac{dE}{dt}\right\rangle = -\frac{NR}{2}\left(\frac{GQ}{2z_0}\right)^2\omega_z^2\langle a_n^2\rangle. \quad (13.8)$$

Noting that $NM\omega_z^2\langle a_n^2\rangle/2 = \langle E\rangle$ we see that the total mean energy decays according to

$$\langle E\rangle = \langle E_0\rangle\exp\left(-\frac{t}{\tau_R}\right), \quad (13.9)$$

where, assuming, as is usual in practice, the particle oscillation spectrum is much sharper than the tank circuit, we can put $R = Q_{LC}\omega_0 L$, and the time constant τ_R is given by

$$\tau_R = \left(\frac{2z_0}{GQ}\right)^2\frac{M}{\omega_0 L}\frac{1}{Q_{LC}}, \quad (13.10)$$

which in terms of the trapping potential depth can be written as

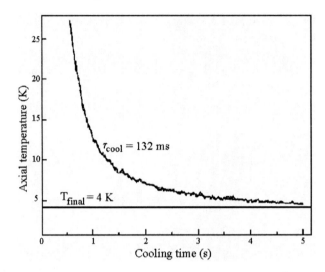

Fig. 13.2. The approach to thermal equilibrium of a single Ca$^+$ ion in a Penning trap

$$\tau_R = \frac{4\left(z_0\sqrt{M}\right)^3}{LQ_{LC}\sqrt{Q^5 U_0}}.\tag{13.11}$$

This result is accurate enough to indicate the way in which the time constant scales with respect to the various physical parameters; thus it shows the very strong dependence on the dimensions of the trap and the mass of the particles, being most effective for light particles such as electrons, or ions that are multiply charged. As a numerical example, for a single electron with $z_0 = 1.5$ mm, and $U_0 = 1.0$ V, corresponding to $\nu_z \approx 280/2\pi$ MHz, $L = 1\,\mu$H, $Q_{LC} = 100$, a time constant of $\tau_R \approx 11.4$ ms is obtained, whereas particles with the proton mass in the same field would have $\tau \approx 14.8$ min.

An example for highly charged ions is given in Fig. 13.2, where the exponential energy loss of a single C^{5+} ion in a Penning trap is demonstrated.

The observed time constant of 132 ms corresponds to the calculated one for the parameters $\nu_z = 270$ kHz$/2\pi$, $L = 10\,\mu$H, $Q = 2400$, and $z_0 = 1$ cm. The cooling time constant is, according to the simple theory, independent of ion population in the trap, although the effective capacitative coupling to the finite end caps may be expected to depend on the ion distribution in a complex way. The simple theory presumes the particle oscillation remains constantly in resonance with the LC circuit; however, the inevitable presence of imperfections and anharmonicities in the trapping field dictate that this condition can only be approximately met, particularly if there is a wide change in the particle oscillation amplitude. The situation is even graver for a large number of particles in a cloud, for which the space charge poten-

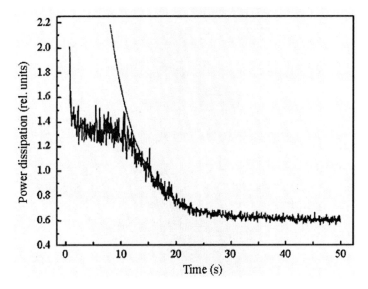

Fig. 13.3. Resistive cooling of the axial motion of an ion cloud. The time constant of the exponential fit to the tail is 5 s

tial introduces fields that depend on the spatial distribution, and change the frequency as the cloud contracts. It should be noted that for a cloud, the cooling of any coherent center-of-charge oscillation, if it exists, is far more rapid than the cooling due to noise current. This is demonstrated in Fig. 13.3, where a cloud of 30 C^{5+} ions is resistively cooled under the same conditions as those in Fig. 13.2. The initial fast time constant corresponds to that of a single particle and refers to the center-of-mass of the ion cloud. Then energy transfer from the incoherent oscillations of the individual particles to the center-of-mass mode having a much longer time constant of several seconds.

13.1 Negative Feedback

It has been reported by Dehmelt et al. [325] that the time constant and final temperature for passive resistive cooling can be changed by active *negative feedback* that is, by amplifying the signal appearing on the external resistance and reapplying it to the electrodes in opposite phase. In this case it can be shown that the cooling time constant with feedback τ_f is related to τ_R as

$$\tau_f = \frac{1}{1-G}\tau_R, \qquad (13.12)$$

where G is the loop gain and $T_f = (1-G)T_R$. It follows,

$$\tau_f T_f = \tau_R T_R. \qquad (13.13)$$

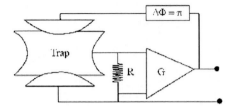

Fig. 13.4. Enhancement of the rate of resistive cooling by negative feedback in the external circuit

Gabrielse and coworkers [326] have demonstrated the reduction in temperature in the cyclotron motion of a single electron in a Penning trap. Using a gain of $G = 0.89$ a temperature of $0.85\,\text{K}$ has been reached while the effective electron temperature without feedback was $5.2\,\text{K}$.

13.2 Stochastic Cooling

Another method of cooling trapped particles involving feed back, that has been called *stochastic* cooling, is an adaptation of a method originally developed by van der Meer et al. [327] for the accumulation of relatively high energy antiprotons in a storage ring. Charged particles circulating in a storage ring exhibit longitudinal fluctuations of density as well as transverse fluctuations due to "betatron oscillations" of random phase. A central concern in the design of high energy particle accelerators and storage rings is the volume the particles occupy in *phase space*. The accumulation of subatomic particles diverging from a target necessitates as large an acceptance (as measured in phase volume) as possible, while the storage and interaction with other beams require a reduced phase volume, that is, smaller angular dispersion and energy spread. The compression of the distribution to occupy a smaller volume in phase space may seem to contradict Liouville's theorem for particles moving under conservative (electromagnetic) forces. However it has been argued that such is not the case, and indeed the experimental success of the method proves it. It assumes a finite number of interacting particles in a sample, so that there is a stochastic fluctuation in the ensemble average of the positions of a sample of the particles.

The way in which stochastic cooling is realized in a storage ring is most simply illustrated by the arrangement used to suppress the transverse betatron oscillations. In its essentials, the method naturally divides into three parts: the first is the measurement, using a sensor called the *pick-up*, of the mean transverse position of the beam at some point along its periphery; the second is to use the pick-up signal, appropriately adjusted in phase and amplitude, to generate a field pulse in a *kicker* placed farther along the beam, that cancels the mean displacement of the beam position from the central orbit.

Finally along the return path to the pick-up the particles must have sufficient time to *mix*, that is interact and reach a new equilibrium. The technique may evoke images of a "Maxwell demon", selectively measuring particular particle energies, and bringing about changes in apparent violation of the second law of thermodynamics. Again, it has been argued that no such violations are involved.

The principle on which this method is based can be simply stated: If there are N particles in a sample of the beam with transverse displacements x_i that are unresolved by the pick-up sensor due to a finite frequency bandwidth W, the signal produced is proportional to the mean displacement $\langle x \rangle$ of those N particles. This signal is adjusted in phase and amplitude and applied to the kicker to reduce this mean position to zero. This results in the positions of the individual particles becoming $(x_i - \langle x \rangle)$ and the resulting mean square displacement therefore reduced by $\langle x \rangle^2$, thus

$$\left\langle (x_i - \langle x \rangle)^2 \right\rangle = \langle x^2 \rangle - \langle x \rangle^2 . \tag{13.14}$$

Assuming that "good mixing" occurs on the return path to the pick-up, that is, randomness in the population is restored, free of correlations between successive transits through the pick-up, the fractional decrease in $\langle x^2 \rangle$ per turn is given by $\langle x \rangle^2 / \langle x^2 \rangle$. For a random distribution of x_i this ratio is $1/N$, or for a revolution time T, the time constant τ for "cooling", as measured by the dispersion in the transverse position, is proportional to NT. Thus, just from the point of view of the rate of cooling, the method is more effective for a small number N of particles. In the context of a storage ring, for a given total number of stored particles, this implies the widest possible bandwidth in the pick-up.

The adaptation of the principle of stochastic cooling to the Penning trap was motivated by the desire to capture and accumulate antiprotons in a Penning trap of elongated geometry. As already shown, the cooling time constant for simple resistive cooling increases rapidly with the dimensions of the trap, ruling out the use of the Penning trap unless some means is found of enhancing the cooling process. Unlike the storage ring however, where the pick-up and kicker can be spatially separated at will, in a Penning trap these must perform their functions on the particles while within the confines of the trap, separated only in time. The two end caps of a Penning trap may each be used to carry out one of these functions, provided some provision is made to eliminate feedback between them. Otherwise there is a close similarity in the theory for the two cases, particularly if the transverse particle displacements in the storage ring are analyzed as due to the betatron oscillations, which would then correspond to the oscillations in the trap.

Following Beverini et al. [328, 329] consider a simplified model consisting of a cloud of N identical noninteracting particles, each oscillating at a fixed (angular) frequency ω_k ($k = 1, 2, \ldots N$), distributed according to the spectral line shape of the real system. It is expected that the average values calculated using this "equivalent" model will yield identical average values as the real system. The axial displacement of the center-of-mass (CM) of the particles in the model is then given by

$$\langle z_c \rangle = \frac{1}{N} \sum_k A_k \sin(\omega_k t + \varphi_k), \qquad (13.15)$$

from which, by treating the terms in the sum as components of N rotating vectors, it is shown that the amplitude of oscillation of the CM (A_c) satisfies

$$A_c^2 = \frac{1}{N} \langle A^2 \rangle + \frac{1}{N^2} \sum_k \sum_{j \neq k} A_k A_j \cos(\theta_k - \theta_j), \qquad (13.16)$$

where $\langle A^2 \rangle$ is the ensemble average over k, and $\theta_{k,j} = \omega_{k,j} t + \varphi_{k,j}$. The time average of A_c^2 is simply given by

$$\langle A_c^2 \rangle = \frac{1}{N} \langle A^2 \rangle, \qquad (13.17)$$

since, although the sum fluctuates continuously in time, its average is zero. If the mean standard deviation in the ω_k's is assumed to be σ_ω it is shown that the time behavior of $\langle A_c^2 \rangle$ is governed by

$$\frac{d\langle A_c^2 \rangle}{dt} = \frac{1}{\tau_m} \langle A_c^2 \rangle, \qquad (13.18)$$

where the time constant $\tau_m = 1/\sigma_\omega$ is a measure of how fast the CM motion of the system of particles can evolve, that is, the phases to randomize.

As with the procedure in the storage ring, the amplitude A_c and phase φ_c of the axial CM motion are monitored, and this information used to synthesize a short pulse of amplitude proportional to A_c, that is applied at the proper phase of the CM oscillation in order to stop it. This obviously presumes that the amplitude and phase do not change appreciably during the time between the amplitude measurement and the pulse formation. The amplitude is measured as in the feed back version of the resistive cooling technique through the induced current in a resistive load placed between one end cap and ground. The pulse, chosen to be short compared with the period of oscillation, is applied to the other end cap; however because of the capacitative coupling between the two end caps, a special neutralizing circuit is required to prevent the pulse from appearing on the oscillation detector. It is shown that if a pulse of duration τ is timed so that it occurs when the CM velocity is at its maximum, that is at time t^* where

$$\omega_c t^* + \varphi_c = 2n\pi, \qquad (13.19)$$

and set to produce a field E given by

$$QE = -\frac{\omega_c}{\tau} M A_c , \qquad (13.20)$$

then the center-of-mass is momentarily stopped at the end of the pulse. The fractional loss of energy in this case is at its maximum and is simply equal to $1/N$ as was found earlier for the storage ring case. After a relaxation time on the order of τ_m the distribution of amplitudes will again randomize and a center-of-mass oscillation will again appear, but with slightly less energy. Thus if the process is repeated many times at intervals τ_0, energy of the system of particles will decay exponentially with a time constant $N\tau_0$, eventually reaching an equilibrium temperature determined by the external circuit. Evidence of the reduction of the cooling time constant by stochastic cooling has been obtained by Beverini et al. [329], who stored a cloud of about 10^6 ions in a Penning trap which was resistively cooled using room temperature electronics. Applying feedback waveforms with opposite phase to the detected ion oscillation resulted in a reduction of the cooling time constants that scales linearly with the mass of the ion and the square of the trap size according to (13.10). This method may be of advantage when very heavy ions are confined in large size traps.

14 Laser Cooling

The application of laser beam scattering by atomic particles to affect their state of motion is based on the long established concepts of radiation pressure, and photon recoil, that are manifested in various fields from comets to γ-ray scattering (the Mössbauer effect). Early attempts at the direct observation of the effect of radiation pressure on matter using thermal sources of light in the laboratory were difficult, and mostly inconclusive. However, the remarkable phenomenon of a form of *cooling* in atoms undergoing resonance fluorescence was noted as early as 1950 by the noted French spectroscopist Alfred Kastler, who described what he called the *lumino-frigorique* effect [330]. However, the extraordinary degree of cooling achievable with laser sources could not have been predicted even by Kastler, whose innovative life's work was steeped in classical spectroscopy, involving conventional light sources. It was not until a quarter century later with the availability of suitable lasers that radiation pressure cooling was proposed by T. Hänsch and A. Schawlow [30] for the cooling of free atoms and independently by D.J. Wineland and H. Dehmelt [322] for the cooling of trapped ions. The first successful laboratory attempt at laser cooling of stored ions was made in 1978 by Toschek et al. [32] on Ba^+ ions in an rf-trap, by Wineland et al. [331] on Mg^+ ions in a Penning trap and later on Mg^+ in a Paul trap by Nagourney et al. [332]. The extraordinary brightness of laser sources was shown to be sufficient to cool ions isolated in a trap toward a theoretical lower limit of temperature in the microkelvin range, virtually eliminating the second order Doppler shift in their spectrum. Since then, laser cooling and the direct optical observation of small numbers of trapped ions has been widely exploited, and extensive theoretical treatments of the laser cooling process have been given by several authors [272, 333].

14.1 Physical Principles

The constrained periodic motion of ions in a trap modifies their optical absorption spectrum in a way which distinguishes it, for example, from that of atoms that freely move in straight lines extending over many wavelengths between collisions. An oscillating ion irradiated by a monochromatic laser beam sees in its proper frame of reference a frequency-modulated wave, whose Fourier spectrum consists of a central *carrier* frequency at the laser frequency, and equally spaced *sideband* frequencies, extending above and below the carrier, at intervals equal to the oscillation frequency.

In practice, several defining parameters determine the model that is most appropriate to describe the cooling process under laser irradiation, and the limit to the achievable low temperature. These include ω_0 the particle oscillation frequency, γ the width of the particle oscillation resonance, γ_n the spectral width of the optical absorption line, γ_l the laser spectral width, the Lamb–Dicke parameter $\eta = k_l a$, and finally the Rabi frequency Ω_R.

To illustrate the principles involved, we consider the simple case where $\omega_0 \ll \gamma_n$ and $\eta \gg 1$, the so-called *weak binding* Doppler regime in which the ions execute large amplitude oscillations extending over many wavelengths, relatively slowly compared with the radiative process. Under these conditions we may approximate the velocity as being constant during the radiative process; furthermore, since $\eta \gg 1$, the Doppler effect may also be treated as if the motion were unconstrained and uniform.

Assume, then, the model of an atomic particle having an optically allowed transition between the ground state and an excited state, well separated from other states, of (angular) frequency ω_a, and moving with a constant velocity $\boldsymbol{v}$ in free space. It is irradiated by a laser beam, assumed to be a monochromatic plane wave propagating with a wave vector $\boldsymbol{k}_l$ and frequency ω_l, which in the particle's proper frame of reference is Doppler shifted to $(\omega_l - \boldsymbol{k}_l \cdot \boldsymbol{v})$, in the classical approximation. If for simplicity we assume the Doppler-shifted frequency $(\omega_l + k_l v)$ is on resonance, while $(\omega_l - k_l v)$ is off resonance with the atomic transition, then photons having a linear momentum of $\hbar \boldsymbol{k}_l$ are absorbed by particles moving only in the direction opposite to the laser beam, causing the particle to recoil in the direction of the beam, to conserve the total momentum of the system. The subsequent spontaneous re-emission of a photon will also cause a particle recoil; however, in this case the emitted radiation pattern in space has a symmetry such that the *average* recoil momentum over a large number of emissions, is zero. Therefore if the atomic particle undergoes N absorption-emission cycles, the average change in the particle's momentum is a net $N\hbar\boldsymbol{k}_l$. To judge the practical feasibility of this method of cooling, we note that, at most, the laser intensity can be raised to the point of saturating the transition, the ground and excited states becoming equally populated; beyond that the assumption of the re-emission being uncorrelated and randomly spontaneous would no longer be valid. In this limit, the mean force on the atomic particle, as given by the rate of change of momentum, is limited by the spontaneous re-emission lifetime $\tau_n = 1/\gamma_n$ of the excited state, thus

$$F_{\max} = -\frac{1}{2}\hbar k_l \gamma_n \ . \tag{14.1}$$

In the case of ions oscillating in a trap, cooling will occur even if the optical resonance profile is relatively broad, and the Doppler shifted laser frequencies fall within the absorption line, provided that the probability of absorption is greater when the ion is moving toward the laser in its oscillation, than when it is receding from it. Where this last condition obtains, it is called the *Doppler cooling* regime. By proper design it is in principle possible to cool all degrees of freedom of a trapped ion with only one properly directed laser beam, whereas for neutral particles in a gas, effective cooling would require *two* counter propagating laser beams along each coordinate axis.

The low temperature limit to this cooling process is set by the *heating effect* of the random discreteness of the absorption and emission processes. In the emission process, although the recoil momenta imparted to the atomic particle average to zero because of the spatial symmetry of the radiation pattern, the *square* of the momentum *change* which determines the *energy* does *not* average to zero. The particle executes a random walk in momentum space, each emission producing a single step. Thus there is a diffusion of the value of Δp, with $\langle \Delta p \rangle = 0$ and $\langle \Delta p^2 \rangle$ proportional to the number N of emitted photons: $\langle \Delta p^2 \rangle = N\hbar^2 k^2$, corresponding to an increase in kinetic energy of $N\hbar^2 k^2/2M$. The randomness of the discrete absorption process also contributes to the heating of the ion's motion. The minimum temperature is reached when the mean energy of oscillation is such that the cooling balances exactly the recoil heating. In the following sections a more detailed account will be given of the process, and the ultimate achievable temperatures under different conditions.

14.2 Doppler Cooling: Semi-classical Theory

As already noted, the term Doppler cooling refers to the cooling of either *free* particles, or in the case of trapped ions, to those in which the *weak binding* condition $\omega_0 \ll \gamma_n$ holds, and the Lamb–Dicke parameter $\eta \gg 1$. We assume that the oscillator quantum numbers of the trapped ions are so high as to permit a classical description of the particle motion. Suppose as in the previous section that an ion, moving initially with a uniform velocity $\boldsymbol{v}$, interacts with a plane, laser beam of frequency ω_l, and wave vector $\boldsymbol{k}_l$ directed opposite to $\boldsymbol{v}$, resulting in resonance excitation between two electronic levels, and subsequent reemission of photons. Because of the random timing of the discrete recoil "events" accompanying photon absorption, and the random direction of recoil in photon emission, (which cannot be presumed totally uncorrelated to the absorption recoil) a detailed treatment of the approach to equilibrium of the velocity distribution necessarily involves the solution of equations of motion that incorporate random functions, such as the Fokker–Planck type [272]. We limit the description here simply to the approach to final equilibrium of the mean energy, neglecting correlations.

Let the particle absorb a photon of wave vector $\boldsymbol{k}_l$ and subsequently re-emit a photon of wave vector $\boldsymbol{k}_s$, thereby changing its velocity from $\boldsymbol{v}$ to $\boldsymbol{v}'$. The conservation of momentum and energy in the nonrelativistic limit require

$$M\bm{v}' + \hbar\bm{k}_s = M\bm{v} + \hbar\bm{k}_l, \tag{14.2}$$

$$\frac{1}{2}Mv'^2 + \hbar\omega_s = \frac{1}{2}Mv^2 + \hbar\omega_l. \tag{14.3}$$

Combining these two equations we find for the change in kinetic energy of the particle the approximate expression

$$\Delta E_k = \hbar\bm{v}\cdot(\bm{k}_l - \bm{k}_s) + R\left(\hat{k}_l^2 - 2\hat{\bm{k}}_l\cdot\hat{\bm{k}}_s + \hat{k}_s^2\right), \tag{14.4}$$

in which it is assumed that $k_l \approx k_s = k$. Here $\hat{\bm{k}}_l$, $\hat{\bm{k}}_s$ are the unit vectors $\bm{k}_l/k$, $\bm{k}_s/k$ and $R = \hbar^2 k^2/2M$ equals the kinetic energy that a particle, initially at rest, would gain on absorbing a photon. To obtain the average over the radiation pattern of emitted photons, the type of quantum transition must be specified. Here it is assumed to be a dipole transition for which the appropriate angular distribution is as follows:

$$P_s(\hat{\bm{k}}_s) = (3/8\pi)\sum_i \int \left|\hat{\bm{d}}\cdot\hat{\bm{\varepsilon}}_i(\hat{\bm{k}}_s)\right|^2 d\Omega. \tag{14.5}$$

Because of the symmetry $P_s(\hat{\bm{k}}_s) = P_s(-\hat{\bm{k}}_s)$, one finds for the average net change in energy per scattering event

$$\langle \Delta E_k \rangle = R(1 + f_s) + \hbar\bm{k}_l\cdot\bm{v}, \quad f_s = \int P_s(\hat{\bm{k}}_s)\hat{k}_s^2 d\Omega. \tag{14.6}$$

The rate at which the absorption-re-emission cycles occurs is the product of the spontaneous decay rate γ_n and the population of the excited state, which, allowing for saturation gives

$$\frac{dE_k}{dt} = \gamma_n \frac{S}{2} \frac{\gamma_n^2 [R(1+f_s) + \hbar k_l v]}{(1+S)\gamma_n^2 + 4(\omega_l - \omega_a - k_l v)^2}, \tag{14.7}$$

where $S = 2|\Omega_R|^2/\gamma_n^2$ is the saturation factor and Ω_R is the Rabi frequency.

As the cooling process proceeds it is ultimately balanced by the heating that results from the random walk in the particle momentum space due to recoil during photon re-emission, and the cooling process approaches its limit. The initial particle velocity is reduced to the point where it is no longer dominant over the recoil velocities, and the initial Doppler shift has become much less than either the natural width or the detuning. Expanding with respect to the small velocity and using $P(\bm{v}) = P(-\bm{v})$ in averaging over many scattering events, one ultimately gets for the limiting mean energy [333]

$$\langle E_k\rangle_\infty = \frac{\hbar\gamma_n}{8}(1+f_s)\left\{(1+S)\frac{\gamma_n}{4\Delta} + \frac{\Delta}{\gamma_n}\right\}, \tag{14.8}$$

where $\Delta = \omega_a - \omega_l$ is the laser detuning, whose optimal value to obtain the lowest mean energy is given by

14.2 Doppler Cooling: Semi-classical Theory

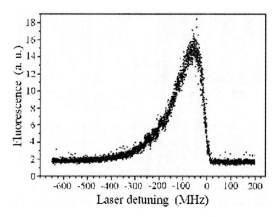

Fig. 14.1. Laser induced fluorescence from an electric dipole transition in a Ca$^+$ cloud in a Paul trap. As the laser is swept from the low frequency side towards the resonance center (zero detuning), it cools the ion cloud; above the resonance center the cloud is heated, expands and the fluorescence drops

$$\Delta_{\min} = \frac{\gamma_n}{2}\sqrt{1+S}, \qquad (14.9)$$

at which value the equilibrium temperature is given by

$$T_{\min} = \frac{\hbar\gamma_n}{8k_B}(1+f_s)\sqrt{1+S}. \qquad (14.10)$$

The process of cooling can then be defined as an approach to minimizing this quantity.

Reducing the amplitude of motion in all three degrees of freedom in the Penning trap is achieved by removing energy from the cyclotron and axial degrees of freedom, but *adding* energy to the magnetron motion. Cooling of the axial motion is similar to the cooling in the Paul trap and is achieved by detuning the laser below resonance. In order to cool the cyclotron motion, the laser must interact with the ions mostly when the cyclotron motion is directed towards the laser, again achieved by detuning the laser below resonance. However, for the magnetron motion, the laser must interact with the ions mostly when they are moving away from the laser. This is done by spatially offsetting an inhomogeneous laser beam to the side of the axis where the ions are receding from the laser. It has been shown [333] that if the laser beam intensity profile is assumed to vary linearly according to $I(y) = I_1(1+y/y_0)$, then the magnetron and cyclotron modes can be cooled with a single laser beam, oriented obliquely, provided the following condition is satisfied for $y_0 > 0$:

$$\omega_m < \frac{(\gamma_n/2)^2 + (\omega_a - \omega_l)^2}{2k_l y_0(\omega_a - \omega_l)} < \omega'_c. \qquad (14.11)$$

14.3 Resolved Sideband Cooling

This method applies to the cooling of an ion in a harmonic trap under conditions where its amplitude of oscillation does not exceed the laser wavelength ($\eta < 1$), so that a description of the fluorescence excited by a monochromatic laser beam involves resonances at a number of distinct Doppler sidebands, assuming the laser spectral width γ_l is narrower than the spectral width γ_0 of the particle oscillation frequency. The analysis will assume the Doppler sidebands in the fluorescence spectrum are well resolved, that is $\omega_0 > \gamma_n$, called *the strong binding condition*. Since γ_n for typical allowed electric dipole transitions is in the tens of megahertz range, meeting the strong binding condition entails extraordinarily weak dipole transitions or high oscillation frequencies. It is mitigated by using sharper quadrupole or stimulated Raman transitions [334]. The technique is most effective if $\eta \ll 1$, which defines the *Lamb–Dicke regime*, in which the dominant spectral component is the carrier, and the only significant sidebands are the ones of first order $\omega_a \pm \omega_0$, the higher order ones falling rapidly to zero. Hence the process must normally be preceded by other means of cooling, most appropriately Doppler cooling, to reduce the oscillation amplitude below the laser wavelength.

A simplified description of the cooling that accounts only for the energetics of the process is as follows: The laser is tuned to the first lower sideband at $(\omega_a - \omega_0)$ to induce transitions between the initial state $|g, n\rangle$ and the optically excited state $|e, n-1\rangle$, followed by spontaneous re-emission of a photon after a mean lifetime of $1/\gamma_n$ in the excited state. Under the assumed condition $\eta \ll 1$, the photon re-emission occurs almost totally at the carrier frequency ($\Delta n = 0$) (Fig. 14.2). This is a form of *optical pumping* of ions to lower n-states, in which the ion loses $\hbar \omega_0$ for every photon scattered. As the cooling proceeds, the absorption probability at the lower sideband decreases as the ground state ($n = 0$) is approached, until the cooling is ultimately balanced by the heating of the random photon recoil in the absorption and re-emission processes. The method can lower the temperature of an ion with practicable efficiency almost to the zero point where the particle is mostly in the ($n = 0$) oscillator state [335]. Of particular practical importance is the fact that the carrier signal, whose strength is dominant relative to the side bands, is totally free of the first order Doppler effect, and furthermore, as the cooling proceeds to the ultimate degree, even the second-order (relativistic) Doppler effect becomes negligible.

An approximate semi-classical description can be given of sideband cooling under conditions where $\eta \sim 1$, and the energy of oscillation is yet sufficiently high that n is large. Thus assume the motion of a single ion in the trap is simple harmonic, described by $z = z_0 \sin(\omega_0 t)$; then in the particle's frame of reference the laser field will be frequency modulated, as will be also the re-emitted photon as seen in the laboratory frame. Thus in the ion's proper frame, neglecting relativistic effects, we have

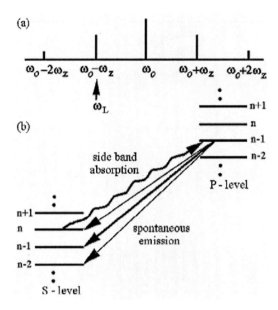

Fig. 14.2. Cooling by sideband absorption. (a) A laser is tuned to low-frequency sideband. Absorption of $E = \hbar(\omega_0 - \omega_z)$ is followed by spontaneous re-emission at $E = \hbar(\omega_0 \pm n\omega_z)$. A net energy loss of $\Delta E = \hbar\omega_z$ results. (b) Quantum description of sideband cooling

$$E(t) = E_0 \exp\left[i \int_0^t [\omega_l - k_l \omega_0 z_0 \cos(\omega_0 t')] \, dt'\right] . \tag{14.12}$$

This expression can now be expanded in a Fourier series, and the laser power spectrum in the ion's frame of reference, derived by taking the Fourier transform. This is combined with the photon spectral absorption probability, assuming the natural linewidth γ_n, to yield the desired spectral absorption cross section of the oscillating ion

$$\sigma(\omega) = \sigma_0 \sum_{n=-\infty}^{+\infty} |J_n(k_l z_0)|^2 \frac{(\gamma_n/2)^2}{(\gamma_n/2)^2 + [\omega_a - (\omega + n\omega_0)]^2} . \tag{14.13}$$

This absorption-emission spectrum is made up of a "carrier" or recoilless line ($n = 0$) at $\omega = \omega_a$, and equally spaced Doppler sidebands at frequencies $\omega = \omega_a \pm n\omega_0$, whose amplitudes depend on the argument of the Bessel function $k_l z_0$, the Lamb–Dicke parameter (η). If the condition $\eta < 1$ is satisfied, that is, the ion motion is constrained within a space no larger in extent than the (optical) wavelength of the laser light, the amplitude of the carrier ($n = 0$) is dominant relative to the sidebands, that rapidly approach zero for $|n| > 1$.

For a cloud of ions in thermal equilibrium at temperature T, in which the amplitudes of oscillation follow the Boltzmann distribution

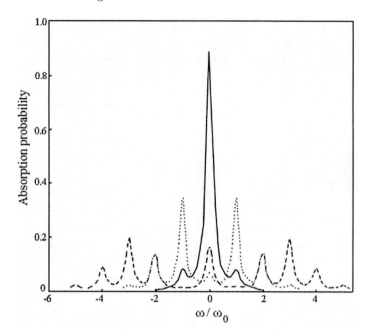

Fig. 14.3. The absorption spectrum of an ion, having a Lorentzian line shape at rest, executing simple harmonic motion with different amplitudes. The Lamb–Dicke parameter values are: *solid line* $\eta = 0.5$, *dotted line* $\eta = 2$, *dashed line* $\eta = 4$. The ion oscillation frequency is $\omega_0/2\pi$

$$dN = \frac{1}{k_B T} \exp\left(-\frac{E}{k_B T}\right) dE , \quad (14.14)$$

the ion absorption spectrum becomes considerably more complicated, thus

$$\sigma(\omega) = \sigma_0 \sum_{n=-\infty}^{+\infty} \frac{[\gamma_n/(2\pi)]^2}{[\omega_a - (\omega + n\omega_0)]^2 + (\gamma_n/2)^2}$$
$$\times \frac{1}{k_B T} \int_0^\infty \exp(-\beta E) J_n^2(2\alpha\sqrt{E}) dE , \quad (14.15)$$

where

$$\alpha = \frac{k}{\omega\sqrt{2M}} , \quad \beta = \frac{1}{k_B T} . \quad (14.16)$$

Fortunately the integral can be evaluated in closed form [336], thus we can write

$$\sigma(\omega) = \sigma_0 \sum_{n=-\infty}^{+\infty} \frac{(\gamma_n/2)^2}{[\omega_a - (\omega + n\omega_0)]^2 + (\gamma_n/2)^2} I_n(2\alpha^2/\beta) \exp\left(-\frac{2\alpha^2}{\beta}\right) , \quad (14.17)$$

where $I_n(x)$ is the modified Bessel function of the first kind. In the limit of high temperatures such that $ka \gg 1$, we expect the spectrum to approach the Voigt line shape. To show that it does, we substitute the following asymptotic form for the modified Bessel function for large values of its argument:

$$I_n(x) \to \frac{1}{2\pi x^{1/2}} \exp\left(x - \frac{n^2}{2x}\right), \quad x \gg n \gg 1, \quad (14.18)$$

and replace the sum by an integral, regarding now the order n as a continuous variable.

In a quantum perturbation analysis, drawing on the parallel with the theory of the Mössbauer effect, Wineland and Itano [337] have shown that taking quantum effects into account leads to the more accurate result, valid for lower temperatures:

$$\sigma(\omega) = \sigma_0 \sum_{m=-\infty}^{+\infty} \frac{(\gamma_n/2)^2}{[\omega_a - (\omega_l + m\omega_0)]^2 + (\gamma_n/2)^2} \sigma_m, \quad (14.19)$$

where σ_m is given by

$$\sigma_m = I_m\left[\exp\left(\frac{\hbar\omega_0}{k_B T}\right) \frac{2\alpha^2}{\beta}\right] \exp\left[\frac{1}{2}m\left(\frac{\hbar\omega_0}{k_B T}\right) - \frac{2\alpha^2}{\beta}\right]. \quad (14.20)$$

On the basis of these results the temperature of trapped ions can be obtained by fitting the data to the relative amplitudes of the Doppler sidebands, as illustrated for example by the work of Berquist et al. [338] on the fluorescence spectrum of a quadrupole transition in single laser-cooled Hg^+ ions, in which the Doppler sidebands were well resolved, and a temperature approaching the predicted limit was deduced as 1.7 mK.

In what follows the discussion will be limited to the resolved sideband cooling of *single* ions, and the achievement of cooling to the zero point energy of oscillation. At sufficiently low temperatures, where quantum effects become manifest, the condition for the Lamb–Dicke regime may be stated more precisely as $k_l^2 \sqrt{\langle z^2 \rangle} = k_l z_0 \sqrt{\langle 2n_z + 1 \rangle} < 1$, with $z_0 = \sqrt{\hbar/(2M\omega_z)}$ the spread of the zero-point oscillator wavefunction along the z-axis, for example. The quantum level structure of the ion at such low temperatures comprises the initial and final electronic levels of the transition each split into ladders of equally spaced harmonic oscillator levels. Under the assumed conditions $\gamma_n < \omega_0$ and $\gamma_l < \omega_0$, transitions between oscillator levels belonging to the two electronic levels are resolved as the sideband fluorescent peaks.

To describe the physical process in its simplest terms, assume a laser beam of intensity I and frequency $\omega_l = \omega_a - \omega_0$ irradiates the ion, thereby exciting it from an oscillator level $|n\rangle$ in the ground electronic state ladder to the $|n-1\rangle$ level in the exited state. In the subsequent spontaneous re-emission, the energy radiated is nearly $\hbar\omega_a$; hence there is a net reduction in the ion energy of $\hbar\omega_0$ per photon scattered. Neglecting the sources of heating due to photon recoil, the initial rate of sideband cooling is as follows:

$$\frac{dE}{dt} = -\frac{\omega_0}{\omega_a} I \sigma_0 J_1^2(k_l z_0) \ . \tag{14.21}$$

However an analysis of the ultimate temperature attainable requires a quantum description since the cooling can proceed toward the point where the ion energy amounts to but a few quanta $\hbar\omega_0$.

The quantum treatment of particles in electrodynamic traps irradiated by an optical field has received extensive study; for excellent reviews see Stenholm [272] and Leibfried et al. [306]. The Hamiltonian of interaction between a trapped ion and a laser field has been given in Sect. 9.1; see also Cirac et al. [214]. Near the low temperature limit, it is generally expected that the system is well into the Lamb–Dicke regime and one may safely neglect all sidebands beyond those of the first order. Assuming the cooling laser is tuned to $\omega_a - \omega_p = \omega_0$ the first red sideband, to first order in η the interaction Hamiltonian is reduced to the resonant red sideband (for a particle in a Paul trap $n = 0$, since the micromotion sidebands are far off resonance), the carrier which is detuned by ω_0, and the blue sideband which is detuned by $2\omega_0$ thus:

$$H_{\text{int}}^{\text{LD}} = \frac{\hbar\omega}{2}[\sigma_+ \exp(i\omega_0 t) + \sigma_- \exp(-i\omega_0 t)] + \frac{\hbar\Omega}{2} \eta \left(\sigma_+ a + \sigma_- a^+\right)$$
$$+ \frac{\hbar\Omega}{2} \eta \left[\sigma_- a \exp(-2i\omega_0 t) + \sigma_+ a^+ \exp(2i\omega_0 t)\right] \ . \tag{14.22}$$

A complete description of the evolution of the ionic motional and electronic states would begin with the master equation for the density matrix; however in what follows coherent effects, if present, are ignored and a rate equation for the populations of the states is sought. Following Leibfried et al. let $W(\Delta)$ represent the rate of photon scattering by an ion *at rest* for laser detuning $\omega_a - \omega_l = \Delta$. In terms of ϱ_e the steady state population of the excited state, and γ_n the spontaneous emission rate, $W(\Delta) = \gamma_n \varrho_e$. Recalling that in the Lamb–Dicke limit, the probabilities of the only transitions that are significant, namely the dominant carrier and the red and blue first order sidebands, are in the ratio $\Omega^2, \eta^2\Omega^2 n$ and $\eta^2\Omega^2(n+1)$, respectively, the rate equation for the oscillator $|n\rangle$-state populations P_n is shown to be

$$\frac{dP_n}{dt} = \eta^2 A_- \left[(n+1)P_{n+1} - nP_n\right] + \eta^2 A_+ \left[nP_{n-1} - (n+1)P_n\right] \ , \tag{14.23}$$

where

$$A_\pm = \gamma_n \left[\alpha W(\Delta) + W(\Delta \pm \omega_0)\right] \ . \tag{14.24}$$

From the population rate equation with $A_- > A_+$ may be derived the time evolution equation for the mean value of n, thus

$$\frac{d\langle n\rangle}{dt} = -(A_- - A_+)\langle n\rangle + A_+ \ . \tag{14.25}$$

14.3 Resolved Sideband Cooling

Then (14.25) leads to

$$\langle n \rangle = \langle n \rangle_0 \exp\left[-(A_- - A_+)t\right] + \langle n \rangle_\infty \left\{1 - \exp\left[-(A_- - A_+)t\right]\right\}, \quad (14.26)$$

with the asymptotic steady state n-value given by

$$\langle n \rangle_\infty = \frac{A_+}{A_- - A_+} = \frac{\alpha W(\Delta) + W(\Delta + \omega_0)}{W(\Delta - \omega_0) - W(\Delta + \omega_0)}, \quad (14.27)$$

where α is a numerical constant less than unity. Under conditions where the sidebands are well-resolved ($\omega_0 \gg \gamma_n$), the laser excitation weak ($\Omega_R \ll \gamma_n$) and the laser tuned to the first laser sideband ($\Delta = \omega_0$), the limit $\langle n \rangle_\infty$ achievable is given by [308]

$$\langle n \rangle_\infty = \left(\alpha + \frac{1}{4}\right) \frac{\gamma_n^2}{4\omega_0^2}, \quad (14.28)$$

corresponding to a total energy

$$E_\infty = \hbar\omega_0 \left[\left(\alpha + \frac{1}{4}\right)\left(\frac{\gamma_n}{2\omega_0}\right)^2 + \frac{1}{2}\right]. \quad (14.29)$$

This shows that the energy limit for resolved sideband cooling can reach closer to the zero point energy than Doppler cooling, however in neither case does the cooling pass the limit $\hbar\omega_0/2$. In the case of strong saturation ($\Omega_R \gg \gamma_n$) of the first sideband ($\Delta = \omega_0$), the final energy is proportional to the laser intensity, thus

$$\langle n \rangle_\infty \simeq \frac{\Omega^2}{8\omega_0^2}. \quad (14.30)$$

As stated earlier, if under the action of laser cooling the populations of the oscillator n-states reach thermal equilibrium, they can be characterized by a temperature T given in terms of $\langle n \rangle$ according to Planck's formula

$$k_B T = \frac{\hbar\omega}{\ln(1 + 1/\langle n \rangle)}. \quad (14.31)$$

The crucial question of just how far the cooling has gone, and what is a useful measure of the temperature under these conditions, fortunately finds an answer in the very fluorescence process used to produce these extremely low temperatures. The scattering probability is dominated by the $\Delta n = 0$ carrier frequency, with a rapidly decreasing intensity at the sideband frequencies $|\Delta n| > 1$. If the ion is irradiated with the laser frequency $(\omega_a + \omega_0)$ for a time τ causing a probability $P_+(\tau)$ of excitation, and $P_-(\tau)$ is the corresponding probability produced by the laser at frequency $(\omega_a - \omega_0)$, then it can be shown that

$$\langle n \rangle = \frac{r}{1-r}; \quad r = \frac{P_-(\tau)}{P_+(\tau)}, \quad (14.32)$$

where r, the ratio of probabilities is independent of the drive time τ, the Rabi frequency Ω_R, and the Lamb–Dicke parameter η. If the ion is at the zero point ($n = 0$), then setting the laser frequency at $(\omega_a - \omega_0)$, which can induce transitions only to an $(n-1)$ upper electronic state, will not cause any transition to occur, and $P_-(\tau) = 0$ in that case. For a given value of $\langle n \rangle$, the equilibrium distribution of the occupation probabilities of the oscillator states can be obtained from the following result:

$$P_n = \frac{\langle n \rangle^n}{(1 + \langle n \rangle)^{n+1}} \ . \tag{14.33}$$

To reach the region where $\langle n \rangle_\infty < 1$ usually requires two stages of laser cooling: the first stage is almost invariably Doppler cooling, where the ions are cooled by a dipole transition (usually $S_{1/2}$ to $P_{1/2}$) which causes strong fluorescence, and thus initially yields a high cooling rate, but which limits the ions' minimum energy to essentially the uncertainty in energy of the respective transition, $E_\infty = \hbar\gamma/2$. This predicts a minimum average quantum number for a given coordinate $\langle n_z \rangle$ on the order of 100, for typical trap parameters. To go beyond that with a weak quadrupole transition implies a much smaller cooling rate. The group at the NIST in the USA [41] was the first to come close to the zero point of energy using resolved sideband cooling on a *single* trapped ^{198}Hg$^+$ in a Paul trap. This was accomplished, after precooling, by sideband cooling using the $S_{1/2}$–$D_{5/2}$ quadrupole transition; but rather than begin at the slow 6 photons per second rate at which that transition naturally occurs, the radiative lifetime of the $D_{5/2}$ level was shortened by coupling it to an upper auxiliary $P_{3/2}$ level by another laser. For low saturation parameter on the $D_{5/2}$–$P_{3/2}$ transition, an analytical expression has been given for the effective linewidth of the cooling transition [339]

$$\gamma'_n = \frac{\gamma_n \Omega^2_{\text{aux}}}{(\gamma_n + \gamma_{\text{aux}})^2 + 4\Delta^2_{\text{aux}}} \ . \tag{14.34}$$

According to this formula it should be possible to obtain the desired linewidth by appropriately choosing the auxiliary laser power given by the Rabi-frequency Ω_{aux} and detuning Δ_{aux}. By adjusting the power and the detuning of the auxiliary laser, one can continuously tune from the Doppler regime to the resolved sideband regime. The NIST group has reported the achievement of 95% ground state in two dimensions for a single Hg$^+$ ion and a measured temperature of $47 \pm 3\,\mu$K. Also a single Be$^+$ has been cooled, in three dimensions, using resolved sideband Raman transitions, to 95% ground state in a miniature Paul trap [334] and similarly, on a single Ca$^+$ ion, with 99.9% ground state occupation, by [274] (Fig. 14.4), but the rates and limits of the cooling process are essentially the same as for the single-photon transitions.

The application of sideband cooling to small groups of ions has also been reported [46, 340, 341]. At these low temperatures, where the ion motion exhibits distinctly quantum behavior, there has been great interest in preparing

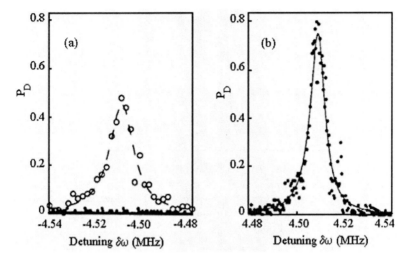

Fig. 14.4. Sideband absorption spectrum on the $S_{1/2}(m = +1/2) \to D_{5/2}(m = +5/2)$ transition after sideband cooling (full circles). The frequency is centered around (a) *red* and (b) *blue* sidebands at $\omega_z = 4.51$ MHz. *Open circles* in (a) show the red sideband after Doppler cooling. Each data point represents 400 individual measurements [274]. Copyright (2003) by the American Physical Society

eigenstates of the vibrational quantum number, that is, Fock states. Cirac et al. [214] have shown that under appropriate conditions it should be possible to prepare ions in Fock states, as indeed has recently been demonstrated [224].

14.4 EIT Cooling

In conventional sideband cooling, in addition to the desired cooling by excitation at the red sideband frequency with its attendant photon recoil heating, there is the heating due to recoil from excitation at the *carrier*, particularly if the sidebands are not widely resolved. A new method, exploiting *electromagnetic-induced transparency* (EIT) [342] was proposed by Morigi et al. [343] and has been experimentally demonstrated by Roos et al. [150]. By suppressing scattering at the carrier frequency using EIT, the method makes possible the efficient cooling of a trapped ion to the ground state, without requiring a weak transition; and by allowing a wider cooling bandwidth, enables the simultaneous cooling of more than one mode of oscillation. Thus longitudinal modes of vibration in a linear trap can be simultaneously cooled, a prerequisite for exploiting trapped ions as quantum logic gates. EIT broadly concerns the manipulation of the optical propagation properties of an atomic medium by quantum interference between transition amplitudes of internal quantum states in the atoms of the medium. It requires coherent excitation of

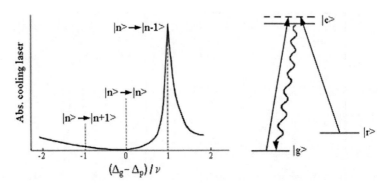

Fig. 14.5. The Lambda energy level scheme for EIT cooling, and the absorption of the cooling laser beam in the neighborhood of $\Delta_g - \Delta_r = 0$ [343]. Copyright (2003) by the American Physical Society

two transitions having a common level, typically in a 3-level system called a Λ-system (Fig. 14.5), where optical transitions induced by one laser between a pair of the levels are *inhibited*, under specific conditions, by quantum interference from a second laser, coupling the common upper level to a third lower level.

This phenomenon is referred to as a *dark resonance* and is indicative of a drop in the population of the excited state, the ion residing in a coherent superposition of the two lower states, a condition that has been called *coherent population trapping*. EIT was initially observed as a drop in fluorescence under coherent excitation of hyperfine levels in sodium vapor by a multimode laser, when a multiple of mode separation in the laser coincided with the separation of the hyperfine levels [344]. Its application to trapped ions in a somewhat modified form was undertaken [274] on the Zeeman components of the $S_{1/2} \rightarrow P_{1/2}$ transition in $^{40}Ca^+$ ion, whose sublevels form a four-level system denoted as $|S, \pm\rangle$ and $|P, \pm\rangle$, respectively. Three of these levels, $|S, \pm\rangle$ and $|P, +\rangle$, using σ^+- and π-polarized laser beams, form a Λ-system of the kind proposed by Morigi et al. [343]. First the ion must be precooled on the $S_{1/2} \rightarrow P_{1/2}$ transition at 397 nm using laser detuning of about 20 MHz below resonance for optimum Doppler cooling. Then EIT cooling is accomplished by irradiating with two blue-detuned laser beams around 397 nm (one with linear, the other circular polarization) having a frequency difference equal to the Zeeman splitting of the $|S, \pm\rangle$ level in the magnetic field. Finally, the motional state after EIT cooling is analyzed using the $S_{1/2} \rightarrow D_{5/2}$ quadrupole transition at 729 nm. With this scheme, it was confirmed that the theoretically predicted final mean vibrational quantum number $\langle n_y \rangle \approx 1$ was achieved after 7.9 ms of EIT cooling, starting from a thermal distribution with $\langle n_y \rangle = 16$ [150].

For a more quantitative understanding of EIT cooling, following Morigi et al. consider an atomic ion having among its electronic level structure three

14.4 EIT Cooling

simple levels $|1\rangle$, $|2\rangle$, $|3\rangle$ in a Λ-formation. Let the frequencies of the transitions between the levels $|1\rangle \to |2\rangle$ and $|2\rangle \to |3\rangle$ be ω_{a1} and ω_{a2} respectively, and the ion irradiated by two monochromatic laser fields, the *coupling field* represented by $E_1(\omega_1)$ and the *cooling field* $E_2(\omega_2)$, such that

$$\omega_1 - \omega_{a1} = \Delta_1, \quad \omega_2 - \omega_{a2} = \Delta_2, \tag{14.35}$$

where it is assumed $\Delta_1 \ll \omega_{a1}$ and $\Delta_2 \ll \omega_{a2}$, and $\Delta_1, \Delta_2 > 0$. The dynamics of the system is described by the Hamiltonian $H = H_A + H_I$ where H_A is the atomic Hamiltonian and H_I describes the ion-field interaction. The time evolution of the density matrix ρ_{nm} of the system is determined by

$$\frac{d\rho}{dt} = -\frac{i}{\hbar}[H, \rho] + \Gamma, \tag{14.36}$$

where Γ characterizes relaxation by spontaneous emission. In the interaction representation we have

$$\hat{H}_I = \hbar\Omega_1 |2\rangle\langle 1| + \hbar\Omega_2 |2\rangle\langle 3|, \tag{14.37}$$

with Ω_1 and Ω_2 the Rabi frequencies for the transitions $|1\rangle \to |2\rangle$ and $|3\rangle \to |2\rangle$ respectively, where it will be assumed for the purposes of EIT cooling that $\Omega_1 \gg \Omega_2$, that is the coupling transition is excited more strongly than the cooling transition. Evaluation of the matrix elements leads to a set of differential equations, the *"Optical Bloch equation"*:

$$\frac{d\rho_{22}}{dt} = -\Gamma\rho_{22} - i\frac{\Omega_1}{2}(\rho_{12} - \rho_{21}) - i\frac{\Omega_2}{2}(\rho_{32} - \rho_{23}),$$

$$\frac{d\rho_{11}}{dt} = \Gamma_1\rho_{22} + i\frac{\Omega_1}{2}(\rho_{12} - \rho_{21}),$$

$$\frac{d\rho_{33}}{dt} = \Gamma_3\rho_{22} + i\frac{\Omega_2}{2}(\rho_{32} - \rho_{23}),$$

$$\frac{d\rho_{12}}{dt} = -\left(\frac{\Gamma}{2} + i\Delta_1\right)\rho_{12} - i\frac{\Omega_1}{2}(\rho_{22} - \rho_{11}) + i\frac{\Omega_2}{2}\rho_{13},$$

$$\frac{d\rho_{32}}{dt} = -\left(\frac{\Gamma}{2} + i\Delta_2\right)\rho_{32} - i\frac{\Omega_2}{2}(\rho_{22} - \rho_{33}) + i\frac{\Omega_1}{2}\rho_{13},$$

$$\frac{d\rho_{13}}{dt} = i(\Delta_2 - \Delta_1)\rho_{13} + i\frac{\Omega_2}{2}\rho_{12} - i\frac{\Omega_1}{2}\rho_{32} - \Gamma_{13}\rho_{13}. \tag{14.38}$$

Here Γ_1, Γ_3 are decay constants from states $|2\rangle \to |1\rangle$ and $|2\rangle \to |3\rangle$, respectively, with $\Gamma_1 + \Gamma_3 = \Gamma$. Of particular interest is the element ρ_{22} because $\Gamma_3\rho_{22}$ determines the rate of emission at the cooling frequency from the excited level $|2\rangle$. The nondiagonal elements, which give the coherence between pairs of levels are not of direct concern here. Under conditions where the motional state of the ion varies slowly compared with the internal electronic dynamics of the ion, a steady state solution for ρ_{nm} may be used at each instant in following the motion of the ion [345]. Numerical solutions have been studied by a number of groups [346].

A simplified form of the steady state result for the scattering rate given by $\Gamma_3 \rho_{22}$ is obtained by imposing the relative magnitudes of parameters in practice, that is, $\Gamma_3 \approx \Gamma/2$, $\Delta_1 \approx \Delta_2$, and $\Omega_2 \ll \Omega_1, \Delta_1$, then [306]

$$W(\Delta) \approx \frac{\Omega_2^2 \Gamma \Delta^2}{\alpha \left[\Gamma^2 \Delta^2 + 4\left(\Delta_2 \Delta - \Omega_1^2/4\right)^2 \right]}, \quad (14.39)$$

where $\Delta = \Delta_2 - \Delta_1$. This shows explicitly for $\Delta_2, \Delta_1 > 0$ how the absorption profile of the transition $|3\rangle \rightarrow |2\rangle$ is changed from a Lorentz to a Fano type when the ion is strongly excited on the $|1\rangle \rightarrow |2\rangle$ transition: In particular, by choosing $\Delta = \Delta_1 - \Delta_2 = 0$, we get a dark resonance, indicative of complete suppression of $\Delta n = 0$ transitions involving the cooling laser. Next to the dark resonance, occurs a sharp fluorescence maximum, a *bright resonance*, whose position at $\Delta = \delta$ depends on the coupling laser intensity, according to

$$\delta = \frac{1}{2} \left(\sqrt{\Delta_1^2 + \Omega_1^2} - |\Delta_1| \right). \quad (14.40)$$

To display the absorption spectrum for the cooling laser, the (positive) detuning Δ_1 of the coupling laser is held fixed, while Δ_2 is tuned through the broad Lorentz profile centered on $\Delta_2 = 0$, then through the dark resonance at $\Delta_2 = \Delta_1$ and finally the sharp bright resonance, as shown in Fig. 14.5. By requiring the bright resonance to occur at $\Delta = \omega_0$, it is possible to ensure that the $|n\rangle \rightarrow |n-1\rangle$ excitation rate dominates over both the $|n\rangle \rightarrow |n\rangle$ and $|n\rangle \rightarrow |n+1\rangle$ transitions rates. This is achievable simply by adjusting Ω_1 and Δ_1 (keeping $\Delta_1 = \Delta_2$) to satisfy the condition $\delta = \omega_0$.

The great experimental advantage is that the lower limit of zero point energy is achievable without the need to be in the Lamb–Dicke regime, and moreover with dramatically reduced heating of other modes of motion due to carrier and blue sideband transitions. The levels and transitions suitable for the EIT cooling scheme have been given by Morigi et al. [343] for many ion species of interest.

14.5 Sisyphus Cooling

This is a cooling technique which, as we shall see, is appropriately named after Sisyphus, a figure in ancient Greek mythology whose punishment in Hades was for ever to keep pushing a stone uphill that immediately rolled down again. The origin of the technique lay in the search for an explanation of anomalous experimental results obtained in 1988 by Phillips et al. [347] on the temperature limit attained using Doppler cooling of neutral Na atoms. As shown in a previous section, that limit has a theoretical value for laser detuning $\Delta = -\gamma/2$, given by $k_B T_{\min} = \hbar \gamma / 2$, a value which for the Na D-line

14.5 Sisyphus Cooling

is computed to be $T_{\min} \simeq 240\,\mu\text{K}$. The experiments were carried out on Na atoms cooled by three orthogonal pairs of counter-propagating lasers forming an optical molasses of Na atoms. It was however optically confirmed that the minimum temperature was in fact $43\pm20\,\mu\text{K}$, far below the theoretically predicted limit of Doppler cooling. Clearly some other cooling mechanism must be involved to extend the cooling process beyond the two-level excitation Doppler limit.

Such a mechanism was proposed the same year independently by Dalibard et al. at the E.N.S. in Paris [348] and Chu et al. [349] at Stanford in the U.S. It is based on the process of optical pumping between ground state sublevels, a process known to induce *light shifts* (or AC Stark shifts) in atomic energy levels since the early work on Kastler optical pumping of Zeeman sublevels [350]. These shifts are determined by the intensity, polarization and frequency detuning from resonance of the light field to which the atoms are subjected. A thorough theoretical analysis of this new cooling mechanism was soon given by Dalibard and Cohen–Tannoudji [351].

In order that the light-induced energy shift in the quantum sublevels slow the atomic center-of-mass (CM) motion, clearly it is necessary that there be a positive potential gradient that the atoms must climb; such a potential gradient is provided by a laser optical field with a strong intensity/polarization gradient, such as might be produced in a standing wave. However, the atomic CM motion must not be so slow compared with the optical pumping rate that it sees a quasi-static periodic potential, since it will simply fall through the same potential as it climbs, with zero average change in kinetic energy. Fortunately the optical field detuning and intensity/polarization can be chosen to limit the optical pumping rate in practice, so the pumping lags behind the CM motion. This ensures that, as an atom climbs toward the top of a potential hill, thereby losing kinetic energy, it is pumped out of that light-shifted sublevel before the downward descent, and into another of the sublevels which has a much smaller (or zero) light shift. It is then returned to the initial state by some form of relaxation or optical pumping, and the process is repeated.

The initial form of Sisyphus cooling, as postulated to explain the anomalous Doppler cooling limit, was polarization-gradient cooling, since in the Na molasses experiments, the optical pumping involved magnetic hyperfine sublevels, whose excitation depends on the polarization of the optical field. Strong gradients in the polarization happen to exist in the Na molasses experiments since two counter-propagating laser beams with orthogonal linear polarizations (lin $\perp$ lin configuration) were used for cooling the free atoms. Thus if E_x and E_y are assumed to be given by

$$E_x = E_0 \cos(\omega t - kx)\,, \quad E_y = E_0 \cos(\omega t + kx)\,, \qquad (14.41)$$

we see that the difference in phase $\Delta\phi$ between the two components of the field is given by $\Delta\phi = 2kx$, and therefore varies linearly from 0 to 2π over the range $0 \leq x \leq \lambda/2$, passing through the values $0, \pi/2, \pi, 3\pi/2, 2\pi$. As

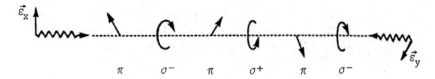

Fig. 14.6. Counterpropagating laser beams with orthogonal linear polarizations (lin ⊥ lin configuration) produce a total field whose polarization changes every eighth of a wavelength from linear to circular polarization

a result, atoms at different points along the optical field will be subjected to a resultant field alternating between linear and circular polarizations of opposite sense of rotation, as shown in Fig. 14.6.

The interaction of the atoms with this radiation field has two relevant effects: First, through absorption-spontaneous emission cycles, the atoms have their magnetic sublevels Kastler pumped into other sublevels, as determined by the polarization state of the field at their location. As a result, the atoms reach a steady state distribution among the sublevels with a characteristic time τ_p, which for low field intensities I_0 is proportional to I_0^{-1}. Second, the interaction induces energy shifts ΔE in the sublevel energies given approximately by $\hbar\Omega_R^2/2\delta$ where Ω_R is the Rabi frequency and δ is the optical detuning of the laser. The shift varies in general for different Zeeman sublevels and depends on the polarization of the light. It follows that in the counter-propagating laser beams, both the population distributions among the various sublevels and the energy shift of each level depend on the position of the atom in the optical field. Following Cohen–Tannoudji and Phillips we consider a simple illustrative example of an atom with a ground $S_{1/2}$ state with $m_J = \pm 1/2$ and an excited $P_{3/2}$-state with $m_J = \pm 1/2, \pm 3/2$ between which the relative probabilities for dipole transitions are as indicated in Fig. 14.7. At points where the optical field has circular polarizations $\sigma^\pm$, the selection rule $\Delta m_J = \pm 1$ leads to the atoms being pumped either into the $m_J = +1/2$ or $m_J = -1/2$ sublevel, with the light shift being three times greater for the receiving sublevel. At intermediate points where we have $\pm\pi$ (linear) polarization the sublevels tend to an equal population, with equal light shift.

As already indicated, if the CM motion of an atom is very much slower than the optical pumping rate, then the atom would move adiabatically up and down the potential hills with no average change in kinetic energy. But the pumping time τ_p, using for example negative (red) laser detuning, can be made long enough that an atom in the $m_J = -1/2$ sublevel, for example, at a potential minimum where the polarization is σ^-, will advance past the π position up the potential gradient toward the σ^+ point before it is pumped, without change of position, to the $m_J = +1/2$ sublevel, where the potential is lower. It is then returned to the $m_J = -1/2$ through some relaxation process, and the whole sequence is repeated, as shown in Fig. 14.8.

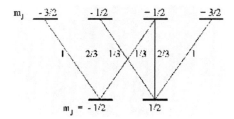

Fig. 14.7. The energy levels of an atomic particle with $S_{1/2}$ ground state and $P_{3/2}$ excited state. The *numbers* along the lines joining the various sublevels indicate the relative transition probabilities

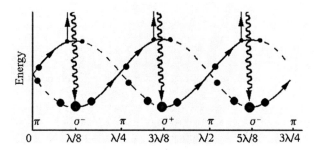

Fig. 14.8. Sisyphus effect in the lin ⊥ lin configuration. An atomic particle traveling in the laser field moves away from a potential valley and reaches a potential hill before being optically pumped to the bottom of another valley. The atomic particle sees on the average more uphill parts than downhill ones, losing energy. The size of the *full circles* indicate the steady state level population, which is maximum for the state with largest negative light shift

From this physical description it is evident that this method optimally slows the motion of the atoms when the velocity is such that during the characteristic pumping time τ_p the atom moves a distance on the order of $\lambda/4$. The energy loss by an atom in traversing that distance in the field is limited by the depth of the light shift, and while this decreases with laser intensity, the fractional cooling rate nevertheless tends to be maintained, since the pumping time is also lengthened.

The lowest temperature that is attainable is predicted to vary linearly with Ω^2/δ, a relationship that has been verified experimentally, lending support to the mechanism on which it is based. The theoretical limit to the temperature is expected to be on the order of the photon recoil energy $\hbar^2 k^2/(2M)$. The technique has been important in the cooling, trapping and manipulation of neutral atoms, such as the alkalis, particularly since the application of magnetic fields can manipulate the optical molasses in interesting ways, with applications in the formation of atomic beams and *fountains*.

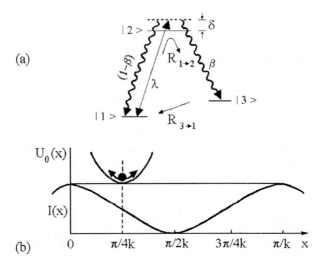

Fig. 14.9. A three level ion trapped in a harmonic potential. The ion has three internal energy levels $|1\rangle$, $|2\rangle$ and $|3\rangle$. (a) A standing-wave laser beam drives $|1\rangle \rightarrow |2\rangle$ above its resonance frequency. Level $|2\rangle$ decays to level $|3\rangle$ and $|1\rangle$ with branching fractions β and $1-\beta$, respectively. Transfer from level $|3\rangle$ to $|1\rangle$ occurs by relaxation at a rate $R_{3\rightarrow 1}$; (b) The maximum extent of the atom's motion in the trapping potential is assumed to be less than $\lambda/2\pi$ (Lamb–Dicke regime) [278]

The adaptation of the Sisyphus type of cooling to confined ions has been a subject of both theoretical [278] (hereinafter referred to as WDC-T) and experimental investigation [352]. In the case of weakly bound ions in a trap of frequency ω_0, that is in which $\gamma_n \gg \omega_0$, and outside the Lamb–Dicke regime ($ka \gg 1$), as we know from the analysis of Doppler cooling, the behavior of the confined ion with respect to its interaction with the laser field is approximately the same as for a free particle, as outlined above.

In order to see how to extend the application of the Sisyphus cooling mechanism to confined ions into the sub-Doppler, Lamb–Dicke regime, it must be reduced to its essential elements, namely, that through the AC Stark effect an ion may be subjected to a decelerating dipole force followed by a spontaneous Raman process leading to another state of lower potential and weaker dipole force. These elements can be achieved not only for an atom/ion linearly traversing an extended optical field periodic in space, but also to a trapped ion oscillating with an amplitude smaller than the spatial period of the field intensity. In either case the particle is continuously doing work going up a potential hill, and coming precipitously back down, although unlike Sisyphus's stone, it goes down with less kinetic energy gain, having made a transition to another sublevel in the downward phase.

Let us consider an ion confined by a static harmonic potential U_0 characterized by a frequency of oscillation ω_0 under the Lamb–Dicke regime, that is where the wave vector k and the oscillation amplitude $a_0 = \sqrt{\hbar/(2M\omega_0)}$ fulfill the relation $ka_0 \ll 1$, a condition equivalent to $\hbar^2 k^2/(2M) \ll \hbar\omega_0$. In order to facilitate the comparison of Sisyphus with Doppler cooling, it is further assumed that the spontaneous emission rate from an excited state $\gamma_n \gg \omega_0$. For simplicity the discussion is limited to one-dimensional motion of a single ion having three internal states labeled, following WDC-T, as $|1\rangle$ for ground state, $|2\rangle$ excited state, and $|3\rangle$ reservoir state, as shown in Fig. 14.9a.

It is assumed that the ion is in a standing-wave laser field whose frequency ω is tuned above the dipole transition between states $|1\rangle$ and $|2\rangle$, and of such intensity that the population of the excited state remains negligibly small. Ions in the excited state are assumed to decay radiatively to the lower states with a branching ratio of $\beta/(1-\beta)$. It is also necessary to assume that ions decay in a finite time τ_3 from the $|3\rangle$-state to $|1\rangle$-state and that during the $|1\rangle \to |2\rangle \to |3\rangle$ transition the ion's position is not significantly changed. The model is further restricted by the following assumed conditions:

$$R_{3\to 1}, \quad R_{1\to 3} \ll \omega_0 \ll \gamma_n \ll \delta, \qquad (14.42)$$

where $R_{3\to 1}$ and $R_{1\to 3}$ are respectively the rates of transfer from level $|3\rangle$ to level $|1\rangle$, and from level $|1\rangle$ to level $|3\rangle$. The first inequality on the left states that the ion will oscillate many times before a transition between the levels $|3\rangle$ and $|1\rangle$ can take place, allowing one to neglect the dependence of the populations of these levels on position, and to take the mean value over a complete oscillation. It further implies that the width of the harmonic oscillator levels is smaller than the spacing $\hbar\omega_0$ between these levels. The last inequality on the right ensures that the light-shift of level $|1\rangle$ is larger than the contribution to its width due to photon transitions involving it. Under these restrictions, and assuming the intensity is well below saturation, so that the $|2\rangle$ level population is negligibly small, it is possible to solve the master density matrix equation for the $|1\rangle$ and $|3\rangle$ level *populations*, neglecting terms involving nondiagonal elements (secular approximation).

If the motion of the ion is taken to be along the x-axis, then the light intensity on which the shift in the energy levels depends, will vary along the path of the ion according to

$$E^2 = \frac{1}{2} E_0^2 \cos^2(kx). \qquad (14.43)$$

However, in contrast to the neutral atom case, where extended spatial periodicity of the field is required, here the ion will go through the Sisyphus cycle periodically as it oscillates back and forth with an amplitude small compared with the laser wavelength. To optimize the cooling effect it is reasonable to expect that the ion should oscillate about a point of maximum gradient in the intensity: this implies that the center of the oscillation should be where $kx = \pi/4$ as shown in Fig. 14.9b. Since it is assumed that the amplitude of

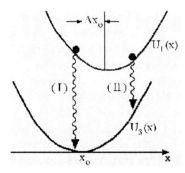

Fig. 14.10. The displacement Δx_0 of the harmonic confinement potential of an ion due to the dipole force of the laser field. It is assumed that the ion spents a negligible time in the excited state, and thus the spontaneous Raman transition may be represented by *vertical lines*, shown for example at phase (I) in the oscillation where the ion would lose energy, while at (II) it would gain energy [278]

oscillation is very much smaller than the wavelength (Lamb–Dicke regime), the variation of the intensity will be approximately linear with respect to x.

Following WDC-T, let $U_L(x)$ represent the light-shift in the $|1\rangle$ state energy level due to a positive detuning δ and U_0 the trapping potential, then the combined potential governing the motion of the ion in the $|1\rangle$ state is given by $U_0 + U_L(x) = U_1$. If the center of U_0 is assume to be at $kx = \pi/4$, then for positive δ the addition of U_L leads to U_1 being shifted in the $+x$-direction. Thus during the phase of its oscillation when the ion is moving in the $-x$-direction its motion is slowed by the rising potential as its spontaneous Raman transition $|1\rangle \to |2\rangle \to |3\rangle$ is becoming increasingly probable, and it returns to a lower point in the undisplaced U_0 potential, without appreciably changing its $x-$coordinate position, as shown in Fig. 14.10(I). After a mean time τ_3 it is returned to the $|1\rangle$-state through a relaxation process, and the cycle can be repeated. It should be clear from Fig. 14.10 that over a phase (II) of its oscillation an ion in the $|1\rangle$-state making the transition to the $|3\rangle$-state will in fact *gain* energy, but over that phase the probability of a transition is lower because the intensity is lower, and the loss of kinetic energy is favored.

In a semiclassical treatment of the dynamics of this cooling method, the CM motion of the ion is treated classically, an acceptable assumption provided the ion de Broglie wavelength is short compared to any characteristic length in the system. Let the coordinate of the center of the potential well in which the ion moves in the a zero light-shifted $|3\rangle$-state be x_0, so that the external trap potential has the form

$$U_0 = U_3 = \frac{1}{2}M\omega_0^2(x - x_0)^2 \ . \tag{14.44}$$

The light-shift induced by the laser field in the ion $|1\rangle$-state causes the ion to move in the effective potential

14.5 Sisyphus Cooling

$$U_1(x) = U_3(x) + U_L(x) , \qquad (14.45)$$

where U_L is approximately given by [351]

$$U_L \simeq \frac{\hbar\Omega^2}{4\delta} \cos^2(kx) , \quad \Omega, \gamma_n \ll \delta . \qquad (14.46)$$

Let x_0 be situated where the optical field has the maximum intensity gradient, that is let $kx_0 = \pi/4$. Since the amplitude of the ion oscillation is assumed to be small, we may expand U_L about $x = x_0$, retaining only the linear term to obtain

$$U_L \simeq \frac{\hbar\Omega^2}{4\delta} \left[\frac{1}{2} - k(x - x_0) \right] . \qquad (14.47)$$

It follows that in the light-shifted $|1\rangle$-state the potential well is shifted in the $+x$−direction (for $\delta > 0$) by an amount given by

$$\Delta x_0 = \frac{\hbar k \Omega^2}{4\delta M \omega_0^2} , \qquad (14.48)$$

which can be written more instructively as follows:

$$\Delta x_0 = 2\xi(ka_0)a_0 , \quad a_0 = \sqrt{\frac{\hbar}{2M\omega_0}} , \quad \xi = \frac{\Omega^2}{4\delta\omega_0} . \qquad (14.49)$$

If we write p_i, where $i = 1, 2, 3$ for the populations of the three states, then the rate of change of the total CM energy is given by

$$\frac{dE}{dt} = \frac{p_1}{\tau_{1\to 3}}[U_3(x) - U_1(x)] + \frac{p_3}{\tau_{3\to 1}}[U_1(x) - U_3(x)] + \frac{2p_1}{\tau_{1\to 2}}R . \qquad (14.50)$$

The time constants τ are the inverse of the corresponding transition rates. The first two terms are the contributions resulting from the transitions (occurring at fixed x-position) between the two potential wells U_1 and U_3, while the last term represents the rate of photon recoil energy gain R in the process. By averaging each term over a complete oscillation in the appropriate potential well, it is found that for times long compared with the time constant for the relaxation of the level populations

$$\frac{dE}{dt} = -\frac{1}{\tau_s}(E - E_0) . \qquad (14.51)$$

Thus the Sisyphus cycles result in the ion CM energy decaying exponentially toward a steady state value, in the limit $t \to \infty$, given by

$$E_\infty = \frac{\hbar\omega_0}{2}\left(\xi + \frac{1}{\beta\xi}\right) , \qquad (14.52)$$

with the following time constant

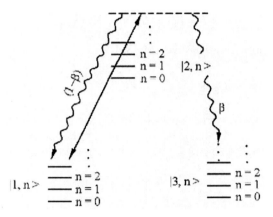

Fig. 14.11. Diagram for the cooling when the ion motion is quantized. Each electronic state has a sublevel structure of harmonic oscillator levels labeled by the quantum numbers $n(j)$, where $j = 1, 2, 3$. Cooling results from spontaneous Raman scattering process of the form $n(1) \to n(2) \to n(3)$, where $n(3) < n(1)$

$$\frac{1}{\tau_s} = \frac{4}{\tau_{3\to 1} + \tau_{1\to 3}} \left(\frac{R}{\hbar\omega_0}\right) \xi . \tag{14.53}$$

The final energy is minimized by choosing the laser intensity and detuning to make $\xi = 1/\sqrt{\beta}$, leading to the minimum energy attainable

$$E_{\min} = \hbar\omega_0/\sqrt{\beta} . \tag{14.54}$$

Thus according to this result, Sisyphys cooling, if continued is predicted to reach well into the region where the ion motion must be treated quantum mechanically. In such a treatment each electronic state of the ion has a sublevel structure of harmonic oscillator levels identified by the quantum numbers $n(j)$, where $j = 1, 2, 3$.

Cooling results from spontaneous Raman scattering processes involving transitions $|1, n(1)\rangle \to |2, n(2)\rangle \to |3, n(3)\rangle$ in which $n(3) < n(1)$. According to WDC-T using second order perturbation theory in the Lamb–Dicke regime, the populations of the different harmonic oscillator states ultimately obey

$$\frac{p_{n+1}}{p_n} = \frac{1 + \beta\xi(\xi - 1)}{1 + \beta\xi(\xi + 1)} , \tag{14.55}$$

which is a minimum for $\xi = 1/\sqrt{\beta}$ given by

$$\left(\frac{p_{n+1}}{p_n}\right)_{\min} = \frac{2 - \sqrt{\beta}}{2 + \sqrt{\beta}} . \tag{14.56}$$

Since this ratio p_{n+1}/p_n holds for all n, it is consistent with the system having reached thermodynamic equilibrium, and a temperature may be defined thus

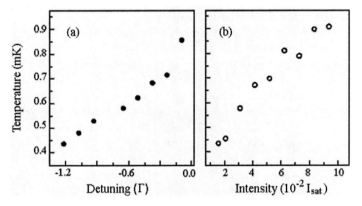

Fig. 14.12. Ion temperature with polarization gradient cooling as a function of laser detuning and intensity. (a) For a constant laser intensity below saturation the temperature decreases with increasing detuning which is given in units of the natural width Γ; (b) For a fixed detuning ($\delta = -0.8\Gamma$) the final temperature decreases with decreasing intensity [352]

$$\frac{p_{n+1}}{p_n} = \exp\left(-\frac{\hbar\omega_0}{k_B T}\right), \qquad (14.57)$$

from which we finally obtain

$$\langle n \rangle = \frac{1 + \beta\xi(\xi - 1)}{2\beta\xi}, \qquad (14.58)$$

and the minimum energy comes out to equal the semiclassical value of $\hbar\omega_0/\sqrt{\beta}$. This agreement with the semiclassical result, is a remarkable coincidence reflecting the special properties of particle dynamics in the harmonic potential.

An experimental realization of Sisyphus cooling in an ion trap has been reported by Birkl et al. on Mg$^+$ ions weakly bound in a circular Paul trap. Two counterpropagating linearly polarized laser beams overlap with a section of the storage volume, using the $3S_{1/2}$–$3P_{1/2}$ resonance transition at 280 nm (natural width 43 MHz). The Doppler cooling limit for this system is 1.0 mK. Depending on the amount of detuning δ and the size of the Rabi frequency Ω temperatures down to 300 μK have been obtained, well below the Dopler limit. As theoretically expected the temperature drops linearly with the detuning and with the laser intensity as shown in Fig. 14.12.

14.6 Stimulated Raman Cooling

Raman cooling is based on two-photon stimulated Raman transitions induced by two laser light waves between two states of an ion, generally involving the presence of allowed dipole transitions to a third common state. The two laser fields are detuned from resonance with the dipole transitions, but have a difference in frequency that matches the separation between the two states. Although the dipole transitions are strongly allowed, the laser fields are equally detuned far enough from resonance that the third state population remains small, and its presence can be eliminated theoretically in an adiabatic approximation. This leads in effect to a two-level description, analogous to having a single photon two-level system with the wave properties $\boldsymbol{k} \leftrightarrow \boldsymbol{k}_1 - \boldsymbol{k}_2$ and $\omega \leftrightarrow \omega_1 - \omega_2$, where $(\omega_1, \boldsymbol{k_1})$ and $(\omega_2, \boldsymbol{k_2})$ are the frequencies and wave vectors of the laser fields irradiating the ion. If the frequencies of both fields are detuned from resonance by an amount Δ, it can be shown that the equivalent coupling between the two states, Ω, is given by

$$|\Omega| = \frac{2}{\Delta} |\Omega_{12} \Omega_{32}|, \qquad (14.59)$$

where Ω_{12} and Ω_{32} are the dipole matrix elements coupling the two states, $|1\rangle$ and $|3\rangle$ with the upper state $|2\rangle$, as shown in Fig. 14.13.

The development of Raman cooling was motivated by the desire to overcome the limitation of the single photon, two-level technique imposed by the need to compromise ultimately between the cooling rate and temperature limit attainable. Thus a two-level system requires on the one hand, a single photon narrow transition, such as one arising from quadrupole coupling, in order to satisfy the condition $\omega_0 > \gamma_n$, while keeping the ion frequency ω_0 within practical limits, and on the other hand a fast transition for rapid cooling. Stimulated Raman transitions between ground state hyperfine sublevels

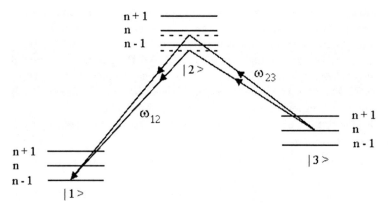

Fig. 14.13. Λ-type 3-level system with oscillator quatum states n and laser Raman transitions for cooling

for example, or the ground state and a neighboring metastable effectively provide such a narrow transition, making resolved sideband cooling more widely applicable to different ion species. Cooling of trapped ions by stimulated Raman cooling was first discussed for Ba$^+$ ions [353, 354].

Consider a three level Λ-system, consisting of the three levels, labeled $|1\rangle, |2\rangle, |3\rangle$, as shown in Fig. 14.7. The discussion will be limited to the one-dimensional problem of the cooling of the axial harmonic motion of an ion in a trap. It is assumed that the ion is subjected to two classical optical plane waves, given by

$$E = E_1 \exp\left[i(\boldsymbol{k}_1 \cdot \boldsymbol{z} - \omega_1 t)\right] + E_2 \exp\left[i(\boldsymbol{k}_2 \cdot \boldsymbol{z} - \omega_2 t)\right], \quad (14.60)$$

which couple the levels $|1\rangle, |3\rangle$ to $|2\rangle$ with Rabi frequencies Ω_{12}, Ω_{32}. The frequencies of the lasers are assumed to be such that their difference matches the first Doppler sideband below the carrier frequency of the "equivalent" two-level system, in order to stimulate Raman transitions in which the ion oscillation quantum number n decreases by one; thus we set

$$\omega_1 = \omega_{12} - \Delta - \omega_0 \; ; \quad \omega_2 = \omega_{32} - \Delta \; . \quad (14.61)$$

We assume that the system is first precooled well into the Lamb–Dicke regime, so that we have $\eta_1 = k_1 a_0 \ll 1$, and $\eta_2 = k_2 a_0 \ll 1$, that is $\hbar^2 k_{1,2}^2/(2M) \ll \hbar\omega_0$. Under these conditions, the spontaneous Raman transitions $|3\rangle \to |2\rangle \to |1\rangle$ occur mainly at the carrier frequency $\Delta n = 0$. The basic mechanism of this type of cooling then is simply the excitation of stimulated Raman transitions $|3, n\rangle \to |1, n-1\rangle$, followed by irreversible spontaneous Raman transitions of the type $|1, n-1\rangle \to |2, n-1\rangle \to |3, n-1\rangle$. To be effective as a cooling process, it is necessary that the population of $|1\rangle$ be larger than that of $|3\rangle$. This can be accomplished under continuous excitation by choosing the intensities of the laser beams so that $\Omega_{21} \gg \Omega_{23}$. A theoretical description has been given by Lindberg and Javanainen [355] of the dynamics of this system in terms of the density matrix, involving the application of degenerate perturbation theory to obtain a steady state solution of the Liouvillian equation, and thus the ultimate temperature. They show that the steady state solution for the populations of the oscillator states are expressible in terms of coefficients $A_\pm$ which determine the transition rates to and from a given harmonic oscillator state $|n\rangle$, as shown in Fig. 14.14. It is shown that if we write

$$q = \frac{A_+}{A_-} = 1 - \frac{A_- - A_+}{A_-}, \quad (14.62)$$

a normalized steady state solution is of the form

$$f(n) = (1-q)q^n, \quad q < 1. \quad (14.63)$$

This is consistent with a thermal equilibrium distribution characterized by a temperature T given by

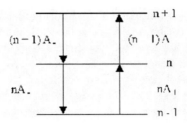

Fig. 14.14. Transitions between different oscillator states $|n\rangle$ with rate constants $A_\pm$

$$q = \exp\left(-\frac{\hbar\omega_0}{k_B T}\right). \qquad (14.64)$$

In the limit of large detuning Δ, Lindberg and Javanainen give the following result as a power series in $1/\Delta$:

$$A_\pm = \left(\Gamma_{21}\alpha_{21}\eta_1^2 + \Gamma_{23}\alpha_{23}\eta_2^2\right) P_2 + 2\Omega_{21}^2 \eta_1^2 L_{21}^\pm P_1 + 2\Omega_{23}^2 \eta_2^2 L_{23}^\pm P_3$$
$$+ 2\left(\frac{\Omega_{21}\Omega_{23}}{\Delta}\right)^2 \left(L_{31}^\mp P_3 + L_{31}^\pm P_1\right)(\eta_1 - \eta_2)^2 + \ldots, \qquad (14.65)$$

where Γ_{21}, Γ_{23} are spontaneous decay rates, α_{21} and α_{23} are geometrical factors of the spontaneous radiation patterns, P_j is the population density of level j, and $L_{ij}^\pm$ are Lorentz line shape factors given by

$$L_{ij}^\pm = \frac{\gamma_{ij}}{(\Delta_{ij} \pm \omega_0)^2 + \gamma_{ij}^2}. \qquad (14.66)$$

The first three terms on the right hand side of the expression for $A^\pm$ relate to single photon transitions: the first accounts for the redistribution of populations due to spontaneous emission, while the next two terms relate to the transitions $|1\rangle \to |2\rangle$ and $|3\rangle \to |2\rangle$. In the limit of large detuning Δ from resonance with the common state $|2\rangle$, these terms, as well as the spontaneous term, have $A_+ - A_- = O(\Delta^{-3})$ and the rates to higher and lower $|n\rangle$-states tend to become equal, and therefore would not produce net cooling. However the last term, which is proportional to the product of the intensities of the two laser fields, with the $L_{31}^\pm$ as the Lorentzian line shapes *near resonance* for a direct two-photon transition $|3, n\rangle \to |1, n-1\rangle$ has $A_+ - A_- = O(\Delta^{-2})$ and can lead to cooling. It is proportional to $(\eta_1 - \eta_2)^2$ and thus $(\boldsymbol{k}_1 - \boldsymbol{k}_2)^2$, and therefore is largest for counter-propagating laser beams.

Neglecting terms $O(\Delta^{-3})$ and higher, the leading term in the value of $(A_+ - A_-)/A_-$ will arise from the two-photon transition and is proportional to $(L_{31}^+ - L_{31}^-)(P_3 - P_1)$. A net cooling of the ion requires that the right hand side be positive, which imposes a requirement on the relative populations of states $|3\rangle$ and $|1\rangle$ for given relative tuning Δ_{31} with respect to the proper frequencies of these states, a relative tuning that determines the sign of $(L_{31}^+ - L_{31}^-)$. The cooling is favored by having a large difference in the populations, which in practice is accomplished by some form of optical pumping, which invariably involves irreversible spontaneous radiative processes, ultimately providing the means of reaching thermal equilibrium. The final temperature is determined by a balance between the rates of transitions which lead to an increase in the oscillator quantum number, and those that lower it. The most obvious transitions which will tend to raise the vibrational quantum number are $|3, n\rangle \to |1, n+1\rangle$ and $|1, n-1\rangle \to |3, n\rangle$ due to the finite combined spectral widths of the laser radiation, and the two-photon transition. The latter heating transition is reduced by optical pumping out of state $|1\rangle$ into state $|3\rangle$. It is shown that under the particular conditions where $\Gamma_{21} = \Gamma_{23} = \gamma$; $\alpha_{21} = \alpha_{23} = 2/5$; $\Delta_{31} = -\omega_0$; $\Omega_{21} \gg \Omega_{23}$; $\omega_0 \gg \gamma_{31}$, the q-parameter is given approximately by

$$q \simeq \frac{7}{10} \frac{\gamma \gamma_{31}}{\Omega_{12}^2} + \frac{19}{20} \left(\frac{\gamma_{31}}{\omega_0}\right)^2 + 2 \frac{\gamma_{31} \Omega_{23}^2}{\gamma \omega_0^2} + \left(\frac{\Omega_{23}}{\Omega_{21}}\right)^2. \qquad (14.67)$$

This shows that the predicted cooling limit is set either by the linewidth γ_{31} of the two-photon transition or, for sufficiently small linewidth, by the last term in the expression for q, the ratio of the optical pumping rates, and well capable of approaching the zero point energy of oscillation.

A pulsed mode of excitation was proposed by Heinzen and Wineland in 1990 [213]. Their analysis, given in terms of amplitudes, was limited to excitation times t short compared with any spontaneous decay times in the system, and was further simplified by imposing several other conditions on the physical parameters, including

$$k_j a_0 \sqrt{n} \ll 1, \quad \Omega_{21} = \Omega_{23} = \Omega, \quad \frac{\Omega^2}{\Delta} \ll \omega_0, \quad \Delta \gg \frac{\pi \gamma}{2\sqrt{n} a_0 (\delta k)}, \qquad (14.68)$$

where a_0 is the extent of the zero point motion of the ion as a harmonic oscillator, and $\delta k = k_1 - k_2$. The condition on Ω^2/Δ ensures that the population transferred by stimulated Raman transitions from $|1, n\rangle$ to states *other* than $|3, n-1\rangle$ remains small, while the last condition on the right ensures that spontaneous decay from $|2\rangle$ remains negligible. Within these limitations a solution was found valid for $t < 1/\Omega_R$, in which only the transition $|1, n\rangle \to |3, n-1\rangle$ is appreciably excited, in which the populations of these states evolve according to the following:

$$p_{1,n} \simeq \cos^2 \Omega_R t; \quad p_{3,n-1} \simeq \sin^2 \Omega_R t, \qquad (14.69)$$

where

$$\Omega_R = \sqrt{n} a_0 \Omega^2 \frac{|\delta k|}{\Delta}. \qquad (14.70)$$

The form of this result suggests a Rabi type of coherent oscillation of the populations between the two states at the frequency Ω_R, and that for $\Omega_R t \to \pi/2$ the populations tend to be exchanged between the states, with the population in the initial state $|1, n\rangle$ being transferred to the $|3, n-1\rangle$-state. Thus a pulse of duration $\tau_\pi = \pi/(2\Omega_R)$ may be regarded as a Raman π-pulse. Based on these considerations, a sequence of pulsed laser beams was proposed to achieve and monitor stimulated Raman cooling of an ion toward the zero point. The two Raman beams $(\omega_1, \mathbf{k}_1)$ and $(\omega_2, \mathbf{k}_1)$, adjusted to have $\Omega_{21} = \Omega_{23}$, are turned on for a period adjusted to equal $\tau_\pi(n=1)$, thereby transferring ions from the state $|1, n\rangle$ to $|3, n-1\rangle$, a state having oscillatory energy one quantum $\hbar\omega_0$ lower. Next the beam $(\omega_1, \mathbf{k}_1)$ exciting the transition $|1, n\rangle \to |2, n\rangle$ is turned off for a sufficient time to allow the ions in the state $|3, n-1\rangle$ to be pumped back to $|1, n-1\rangle$ by spontaneous Raman transitions. The beam $(\omega_1, \mathbf{k}_1)$ is turned back on again for a Raman π-pulse, and the sequence is repeated, with each cycle the ion losing one quantum of oscillatory energy.

This approach was implemented by the NIST group at Boulder [334] using a variant of a Paul trap having a coaxial resonator geometry [169] with compensating electrodes, capable of achieving the high secular oscillation frequencies required to resolve the Doppler sidebands in the optical spectra of such light ions as Mg^+ and Be^+. In the experiment on the $^9Be^+$ ion, the states $|1\rangle$ and $|3\rangle$ are two hyperfine sublevels of the electronic $^2S_{1/2}$ state and $|2\rangle$ is a sublevel of the $^2P_{1/2}$-state. The sequence for pulsed operation begins with Doppler precooling of the ion to the Lamb–Dicke regime. Tuning the laser to the oscillation sideband, the ion temperature is reduced, after repetitive cycles, well below the Doppler limit. From the change in laser fluorescence intensity, after a Raman π-pulse, the mean occupation number $\langle n \rangle$ of the quantum oscillator levels was determined.

By tuning of the Raman laser beams successively to match the frequencies along the three axes of the trap, three dimensional cooling was achieved, and from the constancy of the observed ratio $\langle n \rangle/(1+\langle n \rangle)$ concluded that thermal equilibrium was reached through Raman cooling. The limiting values of $\langle n_x \rangle$, $\langle n_y \rangle$, $\langle n_z \rangle$ deduced from the experimental data were 0.033, 0.022, 0.029, very close to the zero point of energy.

14.7 Sympathetic Cooling

The term *sympathetic* cooling refers to a method in which one species of particles, for which there is no known efficient method of cooling directly, is cooled by thermal exchange of energy with a second, for which an efficient method of cooling *does* exist. It assumes of course that the two species of particles can be brought into thermal contact, that is, with overlapping distributions, for the time required to achieve equilibrium.

14.7 Sympathetic Cooling   251

The transforming advance in the cooling of trapped ions made in 1978 of demonstrating that laser scattering provides a highly effective method of cooling ions to extremely low temperatures, was unfortunately limited to ions which have suitable quantum level schemes, and dependent on the availability of laser sources at the appropriate frequencies. Most atomic and molecular ions, however, have complicated level structures, but perhaps more significantly, elementary charged particles and antiparticles have of course no optical transitions at all. It was suggested from the beginning [356] that these limitations could be greatly relaxed if the particles to be cooled are simultaneously trapped with the directly laser-cooled species, and through collisions, allowed to "sympathetically" be cooled to an equilibrium common temperature. To achieve this, the trap parameters would have to be chosen in such a way that both the directly and the sympathetically cooled species fulfilled the stability criteria for the trap, whether it be a Paul or a Penning trap. We note that the use of electrons/positrons for sympathetic cooling precludes the use of traps involving rf-fields as part of the design, because such light particles would be subject to rf-heating. In general simultaneously trapped ion species with overlapping distributions will interact via Coulomb collisions, which will result in an exchange of thermal energy and approach to thermal equilibrium [356, 357].

In addition to broadening the range of applications of the laser cooling technique, sympathetic cooling can avoid the strong optical perturbation of the quantum levels of an atomic ion; otherwise, the cooling laser must be turned off during the interrogation time to avoid the light shifts and broadening [358] of these levels. The benefits of long-term confinement, high densities, and low temperatures of the laser cooled ion species is transferred to the sympathetically cooled species [359], resulting in ions well localized, and kept at temperatures of some tens of mK for long periods of time, without being perturbed by optical laser fields.

The rate of the temperature reduction of the sympathetically cooled species depends on the number of cooled ions and the overlap between the two species; however, the long range of Coulomb scattering allows effective cooling even in the case of only partial overlap of the ion clouds. An example is the cooling of protons by radiatively cooled electrons in a Penning trap (Fig. 14.15) [141]. Since electrons are efficiently cooled by synchrotron radiation in a magnetic field, they are natural candidates for the sympathetic cooling of protons simultaneously trapped in a Penning trap. As particles of opposite charge, electrons and protons cannot be stored simultaneously in an ordinary Penning trap, whose z-axis confinement depends on the polarity of the charge. This can be overcome only in a special extension of the Penning trap geometry having two potential minima, called a *nested Penning trap*. Although spatially separated there is evidently enough overlap nevertheless for the protons to approach a low temperature close to that of the electrons. While momentum transfer (Coulomb) collisions will efficiently thermalize the

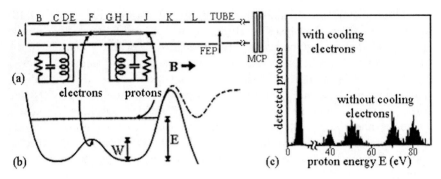

Fig. 14.15. (a) Scale outline of the inner surface of the nested Penning trap electrodes (b). The potential wells for the nested Penning trap (c). Energy spectrum of the hot protons (*right*) and the cooled protons (*left*), obtained by ramping the potential on electrode K downward and counting the protons that spill out to the channel plate. The hot and cooled spectra for 4 initial proton energy are summed. The electron well depth is $W = 7.4\,\text{V}$ [141]. Copyright (2003) by the American Physical Society

populations of electrons and ions, other types of collisions, such as radiative recombination and recombination in the presence of a third particle are rare at the particle densities involved. The rate of energy exchange in the cooling process depends on the ratio of the number of electrons to the number of protons: simultaneous with a decrease in the mean proton energy, there is naturally an increase in the electron energy dependent on the relative numbers of electrons and protons present. We note that the same process has been applied for cooling antiprotons by an electron cloud [139] which is of substantial interest for fundamental particle physics and forms the basis of the successful observation of antihydrogen by recombination of cold antiprotons with cold positrons [143, 188].

This type of cooling is clearly limited by the rate at which the energy of the electrons can be removed by synchrotron radiation and/or resistance cooling. The electron cooling process clearly does not depend on any resonance condition.

Consider a cloud of ions containing two species of different Q/M-ratios confined in a Penning trap [360], in thermal equilibrium. They both undergo an average uniform rotation at the same frequency ω_m, which tends to produce a separation of the two ion species due to the centrifugal barrier. Thus separation will occur if $(M_1 - M_2)\omega_m^2 r^2 > k_B T$. If $Q_1/M_1 > Q_2/M_2$ species 1 will have a larger density near the trap center, while species 2 will more likely be spread out (Fig. 14.16). In the low temperature limit, species 1 forms a column of plasma centered on the z-axis and species 2 forms a cylindrical shell around it with a gap between them. For finite temperatures there is a theoretically predicted overlap between the two ion columns (Fig. 14.16),

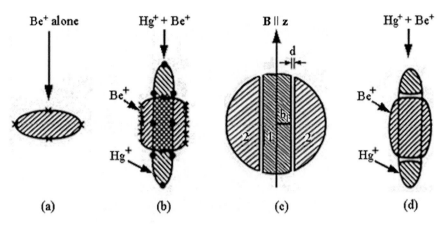

Fig. 14.16. (a) Measured Be$^+$ ion-cloud shape containing about 800 ions, without the Hg$^+$ present. (b) Measured Be$^+$- and Hg$^+$-ion-cloud shapes. The magnetic field is along the z-axis. (c) Approximate model for a sympathetically cooled ion sample in a Penning trap in the low temperature limit. It is assumed that $Q_1/M_1 > Q_2/M_2$ so that species 1 is approximated by a uniform density column of radius b_1. Species 2 is assumed to be continuously laser cooled and by the Coulomb interaction continuously cools species 1. All ions are assumed to be in thermal equilibrium. (d) Theoretically predicted shapes for cold Be$^+$ and Hg$^+$ clouds under conditions similar to those in (a) [358]. Copyright (2003) by the American Physical Society

as confirmed by experiments on laser cooled Be$^+$ ions and sympathetically cooled Hg$^+$ ions [358].

Charged particles exchanging energy through Coulomb collisions with other particles in a Penning trap, would come into equilibrium with a time constant given in classical plasma theory as [361]

$$\tau = (4\pi\varepsilon_0)^2 \frac{M_1 M_2 c^3}{10 n_2 Q_1^2 Q_2^2} \left(\frac{k_B T_1}{M_1 c^2} + \frac{k_B T_2}{M_2 c^2} \right)^{3/2}, \quad (14.71)$$

where Q_1, T_1 are the charge and temperature of the particle to be sympathetically cooled, n_2, T_2 the number density and temperature of the directly cooled particles, and complete overlap is assumed between the two species.

The final temperature of the sympathetically cooled species should be the same as for the directly cooled particles, when no heating mechanism is present, as in the case of a Penning trap. This is confirmed by the measured temperatures of different highly charged ions confined simultaneously in a Penning trap, when only one species is resistively cooled (Fig. 14.17). The first experiment on sympathetic cooling was performed in 1980 between different *isotopes* of Mg$^+$ [356] simultaneously trapped in a Penning trap, in which temperatures below 0.5 K were reached. Sympathetic cooling between ions of different *elements* was first demonstrated in 1986 [358] on ^{198}Hg$^+$ ions, that were sympathetically cooled to temperatures of 0.4 K–1.8 K by directly

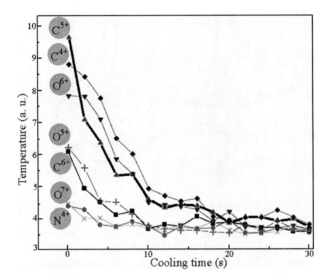

Fig. 14.17. Sympathetic cooling of simultaneously confined highly charged ions in a Penning trap. Only C^{5+} ions are cooled directly (resistively). The axial energy for all types of ions is reduced with about the same time constant

laser-cooled $^9Be^+$ ions at 50 mK–0.2 K. The difference in the temperatures may have resulted from the weak thermal coupling between the Be^+ and Hg^+ ions. Later, measurements on various Cd^+ isotopes, sympathetically cooled to temperatures of 0.7 K by $^9Be^+$ ions directly cooled to 0.4 K, have also been reported [362]. More recently the sympathetic cooling of highly charged ions Xe^{44+} by laser cooled Be^+ ions has been demonstrated [363], showing clearly the spatial separation of the two species (Fig. 14.18). Applications

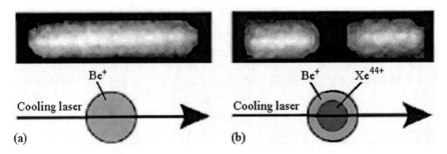

Fig. 14.18. (a) Side view and top-down diagram of a Be^+ ion cloud directly cooled to about 9 K by a laser beam passing through the center of the cloud. (b) The same Be^+ ion cloud displaced radially by a cloud of highly charged (nonfluorescing) Xe-ions, sympathetically cooled by Be^+ ions [368]. Copyright (2003) by the American Physical Society

of sympathetic cooling to molecular ions has been demonstrated for HCO^+, N_2H^+ [364], and $^{24}MgH^+$, $^{24}MgD^+$ [365, 366].

In a Paul trap the micromotion of the ions prevents spatial separation of sympathetically cooled ion clouds, unless the temperature reaches the value at which crystal formation starts; otherwise, the cooling mechanism works in the same way as in Penning traps. Crystallization at low temperatures achieved sympathetically has in fact been observed [366, 367].

Sympathetic cooling involving ions in different internal atomic states has been demonstrated in an experiment where some trapped $^{40}Ca^+$ ions are decoupled from the cooling laser by being optically pumped into a metastable dark state [148].

15 Adiabatic Cooling

The principle on which this method is based may be simply stated in terms of the classical behavior of a one-dimensional harmonic oscillator when its oscillation frequency in the trap is slowly lowered by reducing the trapping fields. Such a change will be *adiabatic*, that is the entropy does not change, if it is so slow that it requires much more time for the frequency to change an appreciable fraction of itself than the oscillation period or relaxation time. According to classical mechanics the action integral is one of the adiabatic invariants.

To illustrate, consider the equation of motion in the axial direction of a particle in a Penning trap with a time-dependent trapping voltage $U(t)$, thus

$$\frac{d^2 z}{dt^2} + \omega_z^2(t) z = 0, \quad \omega_z(t) = \sqrt{2QU(t)/Md^2}. \tag{15.1}$$

We assume that $\omega_z(t)$ is a weak function of t such that

$$\left| \frac{1}{\omega_z} \frac{d\omega_z}{dt} \right| \ll \omega_z. \tag{15.2}$$

Then there exists an adiabatic invariant I_0 given by [86]

$$I_0 = \frac{1}{\omega_z}\left[\left(\frac{dz}{dt}\right)^2 + \omega_z^2(t) z^2 \right] = \omega_z A^2, \tag{15.3}$$

where A is the amplitude of the oscillation. The classical and quantum invariants for the time-dependent harmonic oscillators and for the charged particles in time-dependent electromagnetic fields have been presented in [86] and [88].

It follows that the energy in the axial motion is $E_z = M\omega_z I_0/2$, and is thus reduced by a factor $k^{1/2}$ when the trapping voltage is adiabatically reduced by a factor k, while the amplitude *increased* by $k^{1/4}$.

Li et al. [369] have developed the theory of adiabatic cooling for an ion cloud in a Penning trap. The ions are treated as a classical gas in quasi-thermal equilibrium due to Coulomb interactions, as the trapping fields are adiabatically reduced in strength. Expressions are given showing the dependence of the volume occupied by the ion cloud and its temperature. From the above simple argument it is clear that as the frequency is lowered, the volume increases and the temperature falls. It follows that this method must begin with the ion cloud concentrated in a smaller volume than the capacity of the trap.

It has been suggested [370] that, in principle, an ion refrigeration cycle can be based on the adiabatic principle, in which the trap voltage is initially *increased* to compress the ions, allow thermalization to a lower temperature, for example, through collisions with a buffer gas, and then complete the cycle by lowering the voltage to its initial value.

Part V

Trapped Ions as Nonneutral Plasma

In describing a cloud of confined ions at temperatures where they are widely separated and diffuse, it is valid to assume that the particles follow essentially independent, single particle trajectories, with at most discrete collisions and an average space charge potential due to the long range nature of the Coulomb field. Under these conditions a description of the thermodynamic equilibrium state can still be given in terms of the single particle (Boltzmann) distribution in which the electrostatic potential must be obtained through a self-consistent solution of the Poisson equation.

However, when the density of the ion cloud is high and/or the temperature reduced, a point may be reached where the Coulomb interaction can no longer be adequately described by this independent particle model, but requires the correlation of neighboring discrete charges to be properly taken into account, as well as the long range effects. If the Debye length

$$\lambda_D = (\varepsilon_0 k_B T / n_0 Q^2)^{1/2} , \qquad (15.4)$$

is small compared to the dimensions of the cloud, the latter takes on the attributes of a plasma, since it exhibits many of the collective properties usually associated with a common (electron–ion) plasma, such as Langmuir oscillations and Debye shielding. This condition ensures that external electric fields are screened over a distance of the order of λ_D by the ions at the edge of the cloud. For ion densities of $n = 10^6 \, \text{cm}^{-3}$ and $T = 300 \, \text{K}$ we have $\lambda_D = 0.7 \, \text{mm}$. With cloud dimensions typically on the order of several millimeters, the criterion is clearly fulfilled. If, as will be the case in most systems of interest, laser cooling is applied, particularly in Penning traps, the criterion is met at much lower densities.

However, in a plasma at sufficiently low temperature and high density, the Debye length λ_D can become less than the mean particle separation and loses its meaning. In such plasma the Langmuir oscillations are ultimately replaced by the modes of vibration in a crystal lattice. The theoretical treatment of a plasma in which correlation between the particles becomes of the essence, requires an N-particle (Gibbs) description of the equilibrium distribution. It is expected that the long range mean Coulomb field will determine the gross shape of the plasma, while the correlation of neighboring particles will determine the detailed order within that overall shape. Because of the nonlinearity of the Coulomb interaction between particles, in a strongly coupled system the dynamical behavior even of only two particles can be quite complicated; in fact, chaotic conditions can result.

16 Plasma Properties

16.1 Coulomb Correlation Parameter

An appropriate way to characterize the plasma is by the *Coulomb correlation parameter*, Γ defined as follows (in SI units):

$$\Gamma = \frac{1}{4\pi\varepsilon_0} \frac{Q^2}{a_{\text{WS}} k_B T}, \qquad (16.1)$$

where a_{WS} is the Wigner–Seitz radius defined by $(4\pi/3) n_0 a_{\text{WS}}^3 = 1$, n_0 is the particle number density, and k_B is Boltzmann's constant. The correlation parameter is simply the ratio of the electrostatic energy of neighboring charges to the thermal energy $k_B T$. It alone determines the thermodynamic equilibrium properties of plasmas in a harmonic trap. The values of Γ encountered in practice range from weakly correlated systems $\Gamma \ll 1$ where the particles execute independent motions, to highly correlated $\Gamma \gg 1$ systems whose dynamics present a many-body problem more akin to condensed matter. More generally, values of $\Gamma > 1$ indicate "strong coupling" with significant correlation between the dynamical variables of the particles, which is manifested as collective behavior. The term nonneutral plasma has been used [371] to describe a confined cloud of unneutralized charged particles. The general properties of such plasmas have been extensively discussed in the literature [371–373].

In this chapter we will discuss features of nonneutral plasmas in Penning and Paul traps under the condition of weak coupling, $\Gamma \ll 1$. The strong coupling case, $\Gamma \gg 1$, leads to crystalline structures and will be the subject of the following chapter.

16.2 Weakly Coupled Plasmas

16.2.1 Penning Traps

The initial impetus to pursue the study of nonneutral plasmas in the context of ion traps can be traced to a 1977 proposal [374] relating to a cold electron plasma confined in a Penning type of trap, and the possibility of realizing experimentally the formation of a pure electron Wigner crystal. The equilibrium distribution of the electrons in the field configuration of a Penning trap is equivalent to that which they would assume if the confining fields are replaced by a uniformly distributed positive charge throughout the trap. This permits the adoption of results from a theoretical model, referred to as a *one component plasma* (OCP), that has been much studied in many contexts, including such diverse fields as plasmas, metals, dielectric solutions, etc. In this model, the system is assumed to consist of interacting classical point charges immersed in a uniform continuous background of neutralizing charge of the opposite sign. Since the equivalence extends to the thermodynamic properties and spatial correlation functions between particles, it is also of considerable practical importance as providing a laboratory means of studying the thermodynamics of an OCP.

For a cloud of ions in which the mutual Coulomb interaction among the particles is not negligible, a discussion of the system of particles as a whole under confinement, begins with those dynamical quantities that are conserved in the presence of those interactions; that is, the constants of the motion. Since the Hamiltonian for the particles in the Penning configuration is independent of time, if we assume all collisions with neutral particles, and other dissipative processes are negligible, it follows that the total particle energy is a constant of the motion. Also since ideally the system is axially symmetric about the magnetic field axis, the total canonical momentum P_ϕ is also conserved, thus

$$H = E = \text{const} \; ; \quad P_\phi = \sum_j [Mr_j^2 \frac{d\phi_j}{dt} + QA_\phi(r_j)r_j] = L = \text{const.} \;, \quad (16.2)$$

where the vector potential $A_\phi = Br/2$. To bring out the confinement property and establish the criterion for stable trapping of the plasma, the Hamiltonian is referred to a reference frame rotating about the axis of symmetry with the angular frequency $-\omega$, thus

$$H_R = H + \omega L \;, \quad (16.3)$$

from which we derive the effective confinement potential in the rotating frame $\Phi_R(r,z)$

$$\Phi_R(r,z) = \Phi_T(r,z) + \frac{M}{Q}\omega(\omega_c - \omega)\frac{r^2}{2} \;, \quad (16.4)$$

where $\Phi_T(r,z)$ is the applied electrostatic potential of the trap and ω_c the cyclotron frequency of the trapping particles. For harmonic confinement in the r-direction this requires $\omega(\omega_c - \omega) > \omega_z^2/2$. Moreover, since the total energy is conserved the sum of the total kinetic energy of the particles and the repulsive Coulomb energies of interaction between the particles cannot exceed the maximum of $Q\Phi_T$. In the limit of large B it can be shown [372] that for a system of identical charged particles, the mean square radius of the plasma is constrained as follows

$$\sum_{j=1}^{N} r_j^2 = \text{const.} \tag{16.5}$$

While this acts as a strong constraint on the outward diffusion of the plasmas, it does not prevent collisions with neutrals and imperfections in the trap from causing a slow expansion and eventual loss of particles from the trap. No such constraint applies to a neutral plasma since the constancy of $\sum_j e_j r_j^2$ is preserved if the paths of an electron and ion together expanded out toward the wall.

The thermal equilibrium distribution for a confined plasma in which correlations are small, that is $\Gamma \ll 1$, is given by a (one-particle) Boltzmann distribution function involving the one-particle Hamiltonian and canonical angular momentum [371] thus

$$f(\mathbf{r},\mathbf{v}) = n(r,z) \left(\frac{M}{2\pi kT}\right)^{3/2} \exp\left[-\frac{M}{2kT}(\mathbf{v}+\omega r\hat{\boldsymbol{\phi}})^2\right], \tag{16.6}$$

where $\hat{\boldsymbol{\phi}}$ is the unit vector, and

$$n(r,z) \sim \exp\left\{-\frac{Q}{kT}\left[\Phi_T(r,z) + \frac{M}{Q}\omega(\omega_c - \omega)\frac{r^2}{2} + \Phi_p(r,z)\right]\right\}. \tag{16.7}$$

Here Φ_p is the space charge potential at the position of a particle due to all the other particles in the plasma, that is

$$\Phi_p(r,z) = \int d^3r' d^3v' f(\mathbf{r}',\mathbf{v}') G(\mathbf{r}|\mathbf{r}'), \tag{16.8}$$

where $G(\mathbf{r}|\mathbf{r}')$ is the Green's function which vanishes at the conducting boundary electrodes and includes images in that boundary. The velocity distribution is Maxwellian in a reference frame rotating with angular velocity $-\omega$ and thus the plasma rotates bodily without shear at the frequency angular ω. In the limit $T = 0$, a finite $n(r,z) \neq 0$ is possible provided ω satisfies the condition

$$Q(\Phi_T + \Phi_p) + \frac{1}{2}M\omega(\omega_c - \omega)r^2 = 0. \tag{16.9}$$

For a quadrupole trap potential field, application of Poisson's equation leads to a constant charge density given by

$$n_0 = \frac{2\varepsilon_0 M \omega(\omega_c - \omega)}{Q^2}, \tag{16.10}$$

extending to a spheroidal boundary surface where the density falls off in a characteristic distance of one Debye length, provided λ_D is smaller than the size of the trap. Thus for a finite plasma confined in a quadrupole trap of much larger dimensions, allowing image charges on the electrodes to be neglected,

Fig. 16.1. Sideview image data obtained on a plasma in thermal equilibrium confined in a Penning trap. The magnetic field is in the *horizontal direction* [375]. Copyright (2003) by the American Physical Society

the thermal equilibrium distribution is one of uniform density bounded by an ellipsoid of revolution. The aspect ratio $\alpha = a_p/b_p$ of the ellipsoid, where a_p, b_p are the major and minor axes, is determined by the expression [376]

$$\omega_z^2 = \omega_p^2 \frac{Q_1^0(\beta)}{\alpha^2 - 1}, \qquad (16.11)$$

where $Q_1^0(\beta)$ is the associated Legendre function of the second kind, and $\beta = \alpha/(\alpha^2 - 1)^{1/2}$.

There is a fundamental limit in single component plasmas to the density of particles that can be magnetically confined, first pointed out by Brillouin [377] on the basis of Larmor's theorem. That limit is $n_{\max} = \varepsilon_0 M \omega_c^2/(2Q^2)$, called the *Brillouin density*. It represents a serious restriction and challenge to attempts to confine high particle densities for diverse applications, such as in particle physics, antimatter research, and the adaptation of Penning style devices for the purposes of alternative fusion energy generation. It corresponds to the plasma rotating without shear at the Larmor frequency $\omega_c/2$ around the trap axis. In a frame of reference rotating with the plasma, according to Larmor's theorem, the magnetic field vanishes and the particles pursue straight line orbits in the interior of the plasma.

The effect of a magnetic field on the properties of a plasma is often characterized by a parameter $S = 2\omega_p^2/\omega_c^2$ where ω_c is the free ion cyclotron frequency, and $\omega_p = [n_0 Q^2/(\varepsilon_0 M)]^{1/2}$ is the plasma frequency. The Brillouin condition can then be written as $S = 1$, defining what is termed *Brillouin flow* in a plasma. In a field $B = 0.1\,\text{T}$ the maximum density of singly ionized atomic particles of mass around 50 is approximately 6.6×10^{11} ions/m^3.

16.2.2 Paul Traps

In defining the concepts of thermal equilibrium and temperature to a plasma confined in the rapidly time-varying field of a Paul trap, the presence of the coherent micromotion and the degree of its coupling to the slow secular motion presents an essentially different physical circumstance from the Penning case. Fortunately there is evidence based on molecular dynamics simulations that the secular motion reaches a thermal equilibrium distribution among the particles that defines a temperature more or less independent of the micromotion. In a numerical simulation by Schiffer et al. [378] of particle motion in a linear Paul trap done on a 1000-ion system, the secular component of the motion was in effect separated out by computing velocities from displacements between times that differ by a complete period of the rf-field. It was found that the coupling between the periodic driven motion and the random secular motion is very small at low temperature, being virtually undetectable for $\Gamma > 100$. However at higher temperatures the results showed that the fraction of the kinetic energy in the periodic motion that mixes into the secular motion per rf-period increases quadratically with the temperature. A similar functional dependence was found at different values of the Mathieu parameter q, but the scale factor strongly depends on the operating point in the (a, q)-plane. Furthermore it was found that the secular motion in the transverse r-direction equilibrated with a numerically "cooled" z-motion, while the energy in the coherent high frequency r-motion remained orders of magnitude larger. Of some significance is the fact that, unlike the rf-periodic motion whose amplitude increases with distance from the axis, there appeared to be no corresponding change in the secular kinetic energy.

Other studies by Blümel et al. [316], have shown that the question of the coupling between the secular motion and the periodic motion, which is the basis of the phenomenon of rf-heating, is a complicated one. To shed light on it, their analysis begins with a simplified one-dimensional string of ions, for which the relevant Paul equation, including the Coulomb interactions between all the charges, is of the form

$$\frac{d^2 x_i}{d\tau^2} + (a + 2q \cos 2\tau) x_i = \frac{1}{2}\left(a + \frac{q^2}{2}\right) \sum_{i \neq j}^{N} \frac{x_i - x_j}{|x_i - x_j|^3} , \qquad (16.12)$$

in which it is assumed $a \ll 1$, $q \ll 1$, and x is in units of d_0, the equilibrium separation of one ion from another ion placed at the center of the trap. The equation was numerically solved for up to five ions and the mean kinetic energy recorded as a function of time. Little heating was found, a fact confirmed by a fast Fourier transform of the positions of the ions as functions of time, the results of which showed a small number of discrete frequencies dominating the spectrum. However in a two-dimensional system constrained to move in the x–z- or y–z-plane (assuming no spherical symmetry in the trap) rapid heating occurs. This time, although the power spectrum of the positions showed a small quasiperiodic region about the point in phase space corresponding to a "crystalline" form, *continuous* bands of frequency were found when an initial state typical of cloud conditions was chosen. This, it is asserted shows evidence of *deterministic chaos* in the cloud phase. Calculations of the work done by the rf-field on the ions confirmed that at large interparticle distances little rf-heating occurs and the particle motions are uncorrelated; this, Blümel et al. call the "Mathieu regime". However as the interparticle distance is reduced, a regime is reached where chaotic heating does occur, followed by a narrow region where quasiperiodic motion is seen, prior to the ultimate formation of a crystal.

It is thus evident that within certain physical conditions, a state of thermal equilibrium among particles confined in time-dependent fields can evolve leading to a distribution similar to the one given for the static field case. In this case ϕ_R will now contain the pseudo-potential governing the secular motion, and there is generally no rotation, so that $\omega = 0$.

17 Plasma Oscillations

The result of small displacements of the ions from their positions of equilibrium is a subsequent motion dominated by the coupling between them. This requires a description in terms of normal modes of oscillations of the system as a whole, rather than in terms of individual ions. The normal modes are distinguished as having all particles move with the same frequency (but of course not the same phase). The mode in which all the ions move together in the same phase is the center-of-mass mode, whose axial frequency ω_z is the same as a single particle would have in the trap. A mode in which the displacement of each ion is proportional to the distance of its equilibrium position from the trap center is called a breathing mode with the next higher frequency at $\omega = \sqrt{3}\omega_z$. A plasma containing N ions has of course $3N$ normal modes of oscillation. A singular fact about the mode frequencies is that their ratios are nearly independent of the number of ions, provided it is small ($N < 10$).

In the special case of $N = 2$ ions having equilibrium positions along the z-axis, but free to make small oscillations in three dimensions, there are six degrees of freedom: the mode frequencies are $\omega_x, \omega_y, \omega_z$ for the center of mass along the three coordinate axes, $\sqrt{3}\omega_z$ "breathing" mode along the z-axis, $(\omega_y^2 - \omega_z^2)^{1/2}$ and $(\omega_x^2 - \omega_z^2)^{1/2}$ "rocking" in the y–z- and x–z-planes. A plasma ellipsoid confined in an ion trap can sustain oscillations in many different types of modes; a thorough treatment of linear normal modes are given by Bollinger et al. [379] and Prasad et al. [380].

Other modes include azimuthally propagating *diocotron* waves near harmonics of the plasma rotation frequency arising from shears in the rotational drift velocity [381], as well as higher frequency modes near the cyclotron frequency [382].

If we approximate the boundary of the confined plasma by an infinite circular cylinder of radius a, the oscillating potential modes, as derived by Trivelpiece and Gould have the form [383]

$$\varphi = AJ_m(p_{mn}r/a)\exp\left[i(m\theta + k_z z - \omega t)\right], \qquad (17.1)$$

with eigenfrequencies given by

$$k_z a = \pm p_{mn}\left[\frac{\omega^2(\omega^2 - \omega_p^2 - \omega_c^2)}{(\omega^2 - \omega_p^2)(\omega^2 - \omega_c^2)}\right]^{1/2}, \qquad (17.2)$$

where p_{mn} is the n-th root of the Bessel function J_m and k_z is the axial wave number.

Dubin [384] has published an analytical solution to the problem of the eigenmodes of a cold spheroidal plasma that assumes no perturbation approximation nor an infinite plasma column, but actually finds an unusual coordinate system in which the problem is separable. However the assumptions are still made that both λ_D and $n_0^{-1/3}$ are small compared with the size of the plasma, and that the effects of a finite temperature and correlation are negligible, so that the plasma may be regarded a cold-fluid ellipsoid of revolution of uniform density. Starting with the equation of continuity, conservation of momentum, and Poisson's equation, the normal mode problem is referred to the rotating frame of reference, and by transforming to an unusual set of coordinates it was reduced to a separable problem in the theory of electrostatics, involving an anisotropic plasma dielectric constant. The plasma modes are classified according to two integers (l, m), where $l \geq 1$ and $0 \leq m \leq l$, and have eigenfrequencies ω_{lm}. They are expressed as a perturbed potential functions Ψ_{lm} inside the plasma of the form

$$\Psi_{lm} \propto P_l^m\left(\frac{\bar{\xi}_1}{\bar{d}_1}\right) P_l^m(\bar{\xi}_2) \exp\left[i(m\phi - \omega_{lm}t)\right], \qquad (17.3)$$

where $\bar{\xi}_1$ and $\bar{\xi}_2$ are scaled spheroidal coordinates. The equation giving the mode frequencies may be regarded as a generalization of the Trivelpiece–Gould dispersion relation quoted above. In a subsequent paper [385] a comparison is drawn between the results of a molecular dynamics simulation and the cold fluid model in order to quantify the effects of the neglected correlations in the fluid theory. They found that the fluid theory predictions for low order mode frequencies are well described even for clouds as small as 100 particles. However, the MD simulation mode frequencies were found to be slightly shifted and dampened compared to the fluid theory, differences that were confirmed to be systematic, and explainable as physical effects caused by strong interparticle correlation, accompanied by viscous drag due to collisions.

In Fig. 17.1 are sketched some of the low order oscillation modes: The $(1, 0)$ mode represents a dipole oscillation of the plasma ellipsoid, the $(2, 0)$ mode is a quadrupole oscillation where the plasma density remains uniform but the aspect ratio oscillates in time, in the $(2, 1)$ mode the spheroid has a tilt angle with respect to the trap axis (produced, e.g., by trap misalignments) and precesses around the axis. In the $(2, 2)$ mode the plasma is distorted in the radial plane forming a triaxial ellipsoid which rotates around the z-axis.

Studies of modes in confined ions in a quadrupole Penning trap at the Brillouin limit [382] have confirmed that the azimuthally propagating mode frequencies agree with the predictions of a simple fluid model. Plasma modes also have been experimentally studied by laser Doppler imaging using phase coherent detection [375]. The Doppler images provide a direct measu-

17 Plasma Oscillations   271

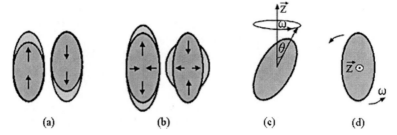

Fig. 17.1. Modes of a spheroidal plasma. (**a**) Dipole mode $(1,0)$; (**b**) Quadrupole mode $(2,0)$; (**c**) Tilt mode $(2,1)$; (**d**) Asymmetry mode $(2,2)$

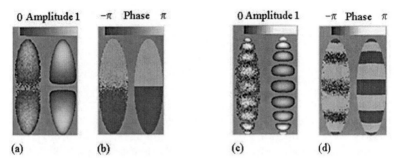

Fig. 17.2. Experimental (*left*) and simulated (*right*) amplitude distribution analyzed into amplitudes of $(2,0)$ mode (**a**) and $(9,0)$ mode (**c**); Experimental (*left*) and simulated (*right*) phase distribution in terms of the same modes (**b**) and (**d**) [375]. Copyright (2003) by the American Physical Society

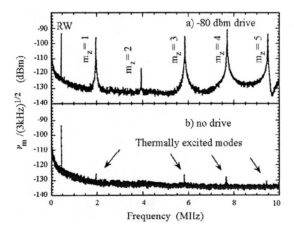

Fig. 17.3. Spectrum of $m_\theta = 0$, $m_z = 1, 2, \ldots 5$ Trivelpiece–Gould modes for 3 drive amplitudes. The axial wave numbers k_z are given by $k_z = \pi m_z / L_z$, where L_z is the effective length of the plasma. (**a**) -80 dbm drive; (**b**) no drive (thermally excited) [386]. Copyright (2003) by the American Physical Society

rement of the mode's axial velocity (Fig. 17.2), while also yielding an accurate measurement of the mode eigenfrequency. Using this technique measurements were made of mode frequencies and eigenfunctions of a number of axially symmetric "magnetized" plasma modes, defined as having frequencies $|\omega_{lm}| < |\omega_c - 2\omega|$. Modes as high as $(12, 1)$ and $(11, 0)$ have been measured by this technique. Agreement between the observed mode frequencies and the predicted values was typically 1%.

Modes in a pure electron plasma lend themselves to study by Fourier analysis of the voltage induced by plasma oscillations in the wall of the trap [386]. Figure 17.3 shows the spectra of axially symmetric Trivelpiece–Gould modes for a pure electron plasma.

17.1 Rotating Wall Technique

In order to control the rotation of an ion plasma in a Penning trap, an external torque may be applied to it. This can take the form of the radiation pressure exerted by the resonance scattering of a laser beam, directed off-axis in the radial plane of the trap [387–389] or an electric field that rotates around the symmetry axis, as introduced in 1997 by Huang et al. [390, 391]. The ring electrode of a Penning trap is split into different segments, and a sinusoidal voltage $V_{wj} = A_w \cos[m(\theta_j - \omega_w t)]$ is applied to the segments at $\theta_j = 2\pi j/n$, where n is the number of segments, and $m = 1$, $m = 2$ correspond to dipole and quadrupole fields, respectively.

Figure 17.4 illustrates an 8-segment configuration with $m = 2$. The plasma rotation frequency caused by the torque is in general somewhat less than ω_w. The slip decreases with increasing amplitude A_w, and scales approximately as the square root of the plasma temperature.

For steady state confinement, the additional centrifugal force from the plasma rotation leads to a change in plasma density and spatial profile. Thus by varying the frequency of the rotating wall, one is able to compress or expand the plasma profile, depending on whether the field is rotating with or

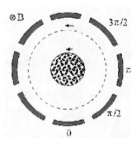

Fig. 17.4. Schematic of $m = 2$ rotating field wall sector signals with rf-phases applied to 8 segments of a cylindrical trap electrode

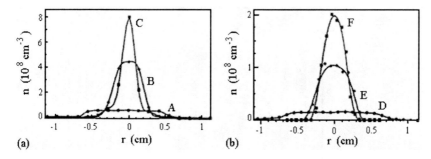

Fig. 17.5. Measured density profiles, demonstrating rotating wall compression and expansion. (a) For electron; (b) for Mg$^+$ ions. Profiles B and E describe the density distribution without a rotating wall, A and D expansion by a backward rotating field, C and F compression by forward rotating field [392]

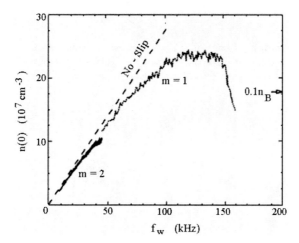

Fig. 17.6. Evolution of central ion density during gradual ramp of rotating field frequency. Density compression by an order of magnitude up to 13% Brillouin density limit n_B is shown [390]. Copyright (2003) by the American Physical Society

against the rotation of the plasma. Figure 17.5 shows the results on a Mg$^+$ plasma and a pure electron one.

Figure 17.6 shows the evolution of the density of a pure electron plasma upon variation of the wall frequency, for field configurations having $n = 8$ and $m = 1$ and $m = 2$. It is compared with the expected change if no slip occurs, and the density reaches 12% of the maximum possible value, the Brillouin density.

If the rotation frequency matches one of the plasma mode frequencies, energy from the field is coupled into the plasma, and heating occurs. By monitoring the plasma temperature, the spectrum of the different modes can be displayed (Fig. 17.7).

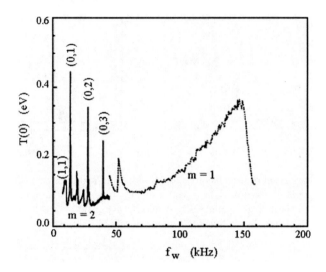

Fig. 17.7. Evolution of central ion temperature during gradual ramp of rotating field frequency. Heating resonances due to excitation of $k \neq 0$ plasma modes are observed [390]. Copyright (2003) by the American Physical Society

18 Plasma Crystallization

18.1 Phase Transitions

The thermodynamic equilibrium state of an infinite homogeneous OCP can be shown, after appropriate scaling of variables, to be determined by one parameter, the Coulomb coupling parameter Γ, as already defined in (16.1). Numerical simulations have shown that ordering characteristics of liquids appear at $\Gamma \approx 2$ [393], while crystallization into a bcc-crystal occurs at $\Gamma \approx 174$ [394, 395]. In reality the number of particles constituting the non-neutral plasma in a trap is almost never large enough to be considered infinite or homogeneous. It is useful to differentiate the plasma sizes into three categories:

1. Coulomb clusters with $N \leq 10$,
2. microplasmas, which have larger N but whose distribution still exhibits finite-size effects, and
3. plasmas sufficiently large that one might expect the behavior of an infinite plasma may be manifested, such as the formation of the bcc–crystal order.

Experimentally, phase transitions from a gaseous or liquid state of confined ions to a crystalline structure are observed as a kink in the fluorescence spectrum, when a cooling laser is swept from below the optical resonance frequency ω_0 towards ω_0 (Fig. 18.1). It indicates a spatial rearrangement of the ions leading to a sudden reduction in the Doppler width of the transition. The value of laser detuning from resonance at which crystallization occurs depends on the rf-amplitude V_0 of the trap voltage and the cooling laser power: In order to achieve the condition for crystallization, laser cooling has to overcome ion heating by the rf-field, which increases with V_0. This is illustrated in Fig. 18.2.

The first observations of such phase transitions were made by Walther's group in Munich in 1987 on Mg^+ ions [37] and in the same year on Hg^+ ions at NIST [396], both groups using Paul traps for confinement. The interpretation of the fluorescence feature as due to the formation of an ordered state was confirmed by direct observation with a photon counting imaging system. Small clusters of ions were seen to form regular figures in the plane perpendicular to the symmetry axis of the trap [397]. They were similar in structure

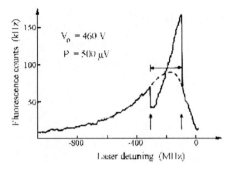

Fig. 18.1. Experimental excitation spectrum of five ions as a function of the laser detuning Δ. The *vertical arrows* indicate the detunings where phase transitions from cloud to crystal ($\Delta = -300\,\text{MHz}$) and from crystal to cloud ($\Delta = -100\,\text{MHz}$) occur. The *horizontal arrow* shows the range of detunings in which a stable five-ion crystal is observed. The spectrum was scanned from left to right [397]. Copyright (2003), with permission from Elsevier

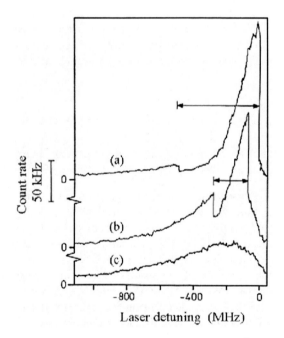

Fig. 18.2. Dependence of the excitation spectra of 5 Mg^+-ions simultaneously stored in a Paul trap upon the radio frequency amplitude V_0 for fixed laser power ($500\,\mu\text{W}$). (a) $V_0 = 360\,\text{V}$; (b) $V_0 = 460\,\text{V}$; (c) $V_0 = 570\,\text{V}$. The laser frequency region where an ordered structure of the ions is observed is marked by an *arrow* [37]. Copyright (2003) by the American Physical Society

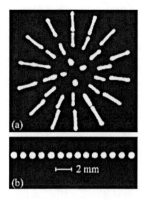

Fig. 18.3. Photos of microparticle crystals in Paul traps. (a) Charged Al microparticle in a three-dimensional trap at a background pressure of about 10^{-3} Pa [39]; (b) chain of charged borosilicate glass microspheres of 25 μm diameter in a linear trap at standard atmospheric temperature and pressure [60,61]

as configurations observed many years ago by Wuerker, Shelton and Langmuir [39] on charged Al particles trapped in air at pressures in the 10^{-3} Pa range in a three-dimensional device similar to a Paul trap (Fig. 18.3a), and more recently [60,61] in a linear Paul trap operated at atmospheric pressure (Fig. 18.3b). Simulations of crystal structures appearing in traps are in obvious agreement with the experimentally observed ones, whether operating in ultrahigh vacuum or at standard pressure and temperature (see Fig. 18.18).

The theory of phase transitions in trapped ion clouds draws on the extensive existing work [394] on the model of a classical one component plasma (OCP). Interest in the structural phases of an OCP, particularly the fluid-crystalline transition, precedes the latter's demonstration in ion traps, and was generally motivated by its relevance to such fields as the theory of stellar interiors. In the present context of stored ions in a trap, the transition to an ordered crystalline state has become of great practical interest since the development of laser cooling brought very large values of the Coulomb parameter Γ within experimental reach. Ions crystallized in a harmonic trapping potential, at the limit attainable with laser cooling, provide a laboratory means of studying a high density OCP, particularly in the quantum regime, which is of special interest in the realization of a quantum computer.

Through the work of Hansen et al. extensive Monte Carlo computations on the equilibrium properties of a classical OCP have been published. The initial configuration assumed in these computations was a bcc lattice, the one known to be formed by an infinite OCP. The approach was to find the critical value of Γ below which the crystal would spontaneously melt. To do this, the indicator used was the dependence of the mean square displacement of the charges from their equilibrium positions δ^2, on the number of configurations generated in the computation, where

$$\delta^2 = \left\langle \frac{\Delta r^2}{d^2} \right\rangle = \frac{1}{N} \left\langle \sum_i \frac{(r_i - R_i)^2}{d^2} \right\rangle, \tag{18.1}$$

and d is the nearest neighbor distance in the bcc lattice. The general trend in the dependence of δ on the value of Γ is that for $\Gamma > 160$, in successive configurations δ rapidly stabilizes to fluctuate about an average value which depends on the value of N. To determine exactly the transition point, requires the derivation of expressions for the Helmholtz free energy (F) for the two phases as a function of the temperature T or equivalently of (Γ^{-1}). For the crystal lattice, classical harmonic lattice vibration theory is used with an anharmonic correction to the order of Γ^{-2}, to obtain the desired free energy. An upper bound of $\Gamma = 170$ is set for the transition on the basis of what the value could reach in the absence of anharmonicity in the lattice vibrations, which lowers Γ.

Having computed the value of Γ for fluid-crystal transition, the dependence of the temperature of that transition on the number density of ions follows immediately from the definition of Γ, thus (in SI units)

$$T = \left(\frac{4\pi}{3}\right)^{1/3} \frac{Q^2}{4\pi\varepsilon_0 k_B \Gamma} n^{1/3}. \tag{18.2}$$

Some authors [398–400] interpreted the phase transitions as an order → chaos transition induced by ion–ion collisions. A transient chaotic regime precedes the transition point when the amplitude of the radio frequency voltage is increased and approaches the stability limit of the trap. Blümel et al. [316,401] yet showed that there is no sharp order-chaos transition point and that the crystalline regime may even persist until the Mathieu instability of the trap is reached. If transitions to chaos occur before the stability limit, they have been induced by external perturbations.

18.2 Chaos and Order

The complexities of the motion of confined particles when the nonlinear Coulomb interaction between them is fully taken into account, is already evident in considering just two identical ions in a quadrupole Paul trap. Detailed molecular dynamics simulations of such a system have predicted transitions from an ordered state to one of chaotic motion.

In one such computer study [400] the Mathieu equations of motion, including the Coulomb interaction, for two laser-cooled ions are numerically solved for different values of the Mathieu parameter q, with $a = 0$. The center-of-mass (CM) motion, of course, is not affected by the interparticle Coulomb force, however the interaction with the optical field of the cooling laser remains, which adds a random photon-recoil term. The study was carried out in terms of the two *relative* coordinates r, z, which satisfy equations of motions of the form

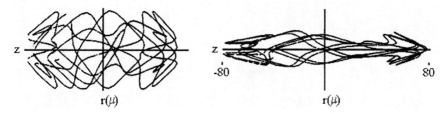

Fig. 18.4. Examples of computer generated plots of r, z spatial patterns for two ions for values of the Mathieu parameters $q = 0.73$, $a = 0$, showing chaotic behaviour in a Paul trap [400]. Copyright (2003) by the American Physical Society

$$\frac{d^2 r}{d\tau^2} + \beta \frac{dr}{d\tau} - \left[\frac{\alpha}{(r^2 + z^2)^{3/2}} + 2q\cos(2\tau) \right] r + n_r(\tau) = 0 , \qquad (18.3)$$

with a similar equation for the z-coordinate. The coefficient α is defined as $Q^2/4\pi\varepsilon_0 M \omega_{rf}^2$, and $n_r(\tau)$ represents the spontaneous emission noise due to the random photon recoil. The damping term, in the form given, adequately models the possible effect of a light buffer gas or, to a first approximation, the effect of the cooling laser. An accurate form for the latter has the familiar nonlinear dependence on the particle velocity given by

$$\mathbf{f}_{\text{laser}} \sim \frac{\mathbf{k}}{[(\Delta - \mathbf{k} \cdot \mathbf{v})^2 + \gamma^2 + \Delta^2]} . \qquad (18.4)$$

The salient results of the computer solutions of these equations, in the adiabatic approximation, are that there exists a certain critical value of the parameter $q = q_c$ below which the motion is quasi-periodic, and the Poincaré section of the particles' phase space has as an attractor a simple limit cycle. The motional spectrum of the ion pair contains the driving frequency ω_{rf} as well as the secular frequencies ω_r and ω_z, which the Coulomb force in this case fortuitously makes degenerate. At the critical value q_c there is a sudden onset of chaotic behavior with erratically changing motion, an extraordinary increase in r- and z-displacements, and evidence of radial bifurcation with motion also centered on a second equilibrium position. An expansion of the Coulomb term about the equilibrium position in fact shows a double dip in potential. This bifurcation breaks the degeneracy of the r and z secular oscillation frequencies The Liapunov exponent [402] changes sign at q_c, signaling an exponentially increasing divergence in the motion for a small change in initial conditions, further evidence of the onset of chaos. In Fig. 18.4 are shown examples of computer generated plots of chaotic z, r patterns for $q = 7.3 > q_c$.

The system also exhibits other properties characteristic of nonlinearity, such as hysteresis [403] made evident by monitoring the photon count rate as a function of the direction in which the laser power, detuning or the parameter q is varied (Fig. 18.5). Frequency locking can occur on the return branch of a hysteresis loop, as exemplified in Fig. 18.6 which shows a dumb-bell-shaped

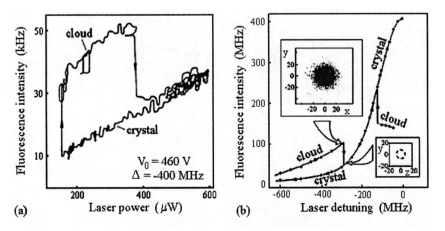

Fig. 18.5. Phase transition curves showing hysteresis and bistability. (a) Experimental hysteresis loop in the fluorescence intensity of five ions as a function of the laser power; (b) theoretical excitation spectrum of five ions as a function of the laser detuning Δ. The *solid arrows* on the fluorescence curves indicate the scanning direction. Two phase transitions (and bistability) are apparent. The *insets* demonstrate directly the existence of a transition from a cloud phase to a crystalline state. They are the results of three-dimensional molecular dynamics calculations for detuning below and above $\Delta = -300$ MHz. The *axes of the insets* are in units of µm [397]. Copyright (2003), with permission from Elsevier

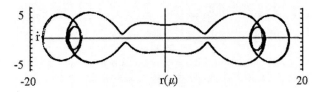

Fig. 18.6. Calculated phase diagram for two ions showing the dumb-bell shaped attractor [400]. Copyright (2003) by the American Physical Society

attractor in the $r - \dot{r}$ phase plane, where the requencies $\omega_{\text{rf}} : \omega_r : \omega_z$ are locked in the ratio $10 : 2 : 1$.

Also observed was the hysteresis both with respect to variations in q, and cooling laser power and detuning as expected from the simulations (Fig. 18.5).

18.3 Crystalline Structures

It has been known for some time that, at temperatures sufficiently low that the correlation energy exceeds the thermal energy, an *infinite* one component plasma forms a body centered cubic (bcc) crystal. And indeed it has been confirmed by numerical simulations on very large confined ion clouds that the bcc ordering has the least energy [404]. However it had been pointed out [405] that the bulk energies per ion of fcc and hcp lattices differ very little from bcc, and therefore one should expect that an imposed boundary geometry may favor these other lattices in that region. As already stated the transition from a fluid to a crystalline state will occur when the Coulomb correlation parameter $\Gamma \geq 170$. However, for a *finite* plasmas, such as are confined in real Penning or Paul traps, and particularly microplasmas or large clusters, boundary conditions impose a modified structure to conform to the geometry of the confining potential field. Numerical simulations by numerous authors have shown that in large confined plasmas, concentric shells are formed [406–408], with approximately constant spacing between shells, whose surface structure has plane hexagonal symmetry. For sufficiently large plasmas, the interior evolves into the structure characteristic of a bcc crystal.

18.3.1 Crystals in Paul Traps

As described in the previous section the formation of ionic crystals in traps is indicated by a characteristic kink in the fluorescence distribution (see Fig. 18.1). The crystals, however, can be also directly observed replacing the photomultiplier tube for fluorescence detection by a CCD camera. A simple optics having a magnification factor of the order of 10 is sufficient to resolve the individual ions separately. This gives direct information on the different crystalline structures which depend on the trapping parameters, the cooling power, and the number of stored ions. The first observation of crystalline structures – so called Wigner crystals [409, 410] in a three-dimensional Paul trap was reported by a group at Garching [37], using $^{24}Mg^+$ ions and by a group at NIST [396] using Hg^+ ions. Up to 50 ions form geometrical structures determined by the equilibrium between the trapping forces and the Coulomb repulsion. The average inter-particle distance is of the order of $20\,\mu m$. It is, however, difficult to obtain larger crystals since energy gain from the time-varying electric trapping field ("rf-heating") requires increased cooling forces to obtain the required strong confinement conditions.

The linear Paul trap (see Sect. 4.5) offers a larger region of low electric field strength along the trap axis and therefore is better suited to produce large ion crystals. If the number of ions is small, they arrange themselves in a chain along the trap axis (Fig. 18.7). An analysis of the ion distribution in a linear Paul trap given by Dubin [411] assumed that the ions are far enough apart that there is negligible overlap of their wavefunctions, and therefore a classical description of the particle dynamics is appropriate. Unlike ion distributions of higher dimensions, the correlation energy arising from the discreteness of the charges extends beyond nearest neighbors to the limits of the distribution. Numerical results are obtained using a "local density approximation" for the energy (including correlation), minimized by a variational method.

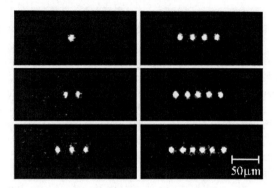

Fig. 18.7. Strings of ions confined along the axis of a linear trap [412]

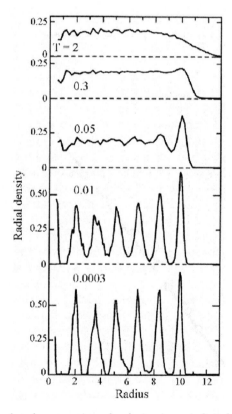

Fig. 18.8. MD-simulated progression of ordering in an infinitely long, radially harmonically confined ion plasmas of cylindrical symmetry, at decreasing temperature T, corresponding to increasing plasma coupling parameter Γ [414]

18.3 Crystalline Structures

Molecular dynamics (MD) calculations of spherical ion clouds containing a few hundred ions confined in a harmonic potential, predict that with increasing value of the coupling parameter Γ the ions first separate in concentric spheroidal shells and then order within shells [406, 407]. Similar quantitative features have been predicted from MD simulations of infinitely long, radially harmonically confined ions of cylindrical symmetry [413]. These simulations were performed in the pseudopotential approximation, not including rf-heating from the micromotion of the ions in the time dependent trapping field of a Paul trap. Similar MD simulations including time-dependent forces show a similar behaviour (Fig. 18.8) [414].

The formation of shell structures in linear Paul traps appearing at low temperatures have been experimentally observed using up to 10^6 ions (Figs. 18.9, 18.10) [148, 149, 352]. Depending on the depth of the confining radial potential a well defined number of shells and strings appear (Fig. 18.11). The observations agree well with the predictions from MD simulations.

For ions confined in a linear quadrupole trap, for a given N, the crystal structure depends only on the anisotropy parameter $\alpha = \omega_z^2/\omega_r^2$. By variation of the axial dc confining potential in a linear Paul trap α can be tuned to any desired value, and an initially prolate crystal can be changed to a spherical or to an oblate one (Fig. 18.9). For sufficiently small values of α, the elongated prolate spheroid degenerates into a line of charge, and with increasing density of ions, transitions to zig-zag and helix forms may be observed (Figs. 18.12, 18.13).

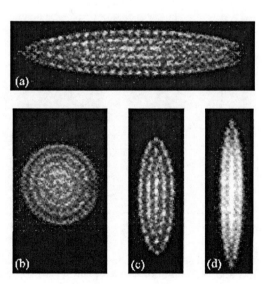

Fig. 18.9. Change of crystal shape when the static axial confinement voltage is increased. The asymmetry parameter $\alpha = \omega_z^2/\omega_r^2$ changes from values $\alpha \ll 1$ (**a**) to $\alpha = 1$ (**b**) and $\alpha < 1$ (**c, d**) [417]

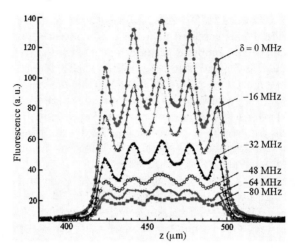

Fig. 18.10. Fluorescence peaks indicating shell structure of a large oblate crystal for various detunings of the cooling laser [417]

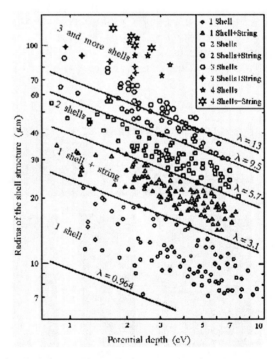

Fig. 18.11. Observed shell structures with up to four shells plus string, characterized by their radius and the potential depth. As in Fig. 18.16, the various observed structures are separated by lines of theoretically determined critical λ [397]. Copyright (2003), with permission from Elsevier

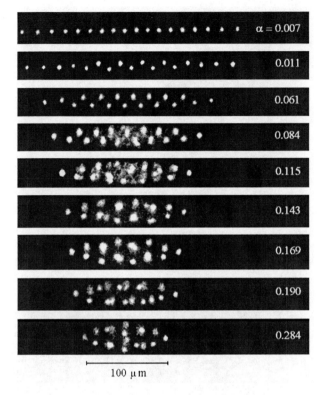

Fig. 18.12. Transition from string to a helix in a linear Paul trap. With increasing potential strength in axial direction indicated by the anisotropy parameter $\alpha = \omega_z^2/\omega_r^2$ the shape of the ion crystal changes from a linear string to a planar zig-zag structure and finally into a helix [412]

Since the linear density of ions is a function of z with a maximum at the center, the transitions are expected to begin at that point. Furthermore since the maximum density depends on α, the transition points may be stated in terms of it, thus [415]

$$\alpha_i(N) = \left(\frac{8}{3Nx_i}\right)^2 \left[\ln\left(\frac{3Nx_i}{2^{3/2}}\right) - 1\right], \quad i = 1, 2, \quad (18.5)$$

where $x_1 = 2.05\ldots$ for the zig-zag transition, and $x_2 = 1.29\ldots$ for the helix. The theoretical result is at odds with a power law dependence of α_1 on N [416]. More recently, experimental confirmation of the power law dependence was reported for strings of Ca^+ ions in a linear Paul trap. Measurements are reported on up to 10 ions to be in good agreement with analytical as well as numerical predictions.

Linear strings of ions can exhibit different modes of vibrations. They can be excited by an additional rf-field applied at the trap electrodes. Most

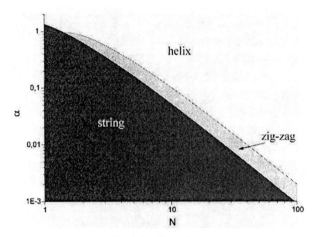

Fig. 18.13. Theoretical types of crystalline order for trapped ions as a function of ion number N and the ratio between the axial and radial trapping potential $\alpha = \omega_z^2/\omega_r^2$ [412]

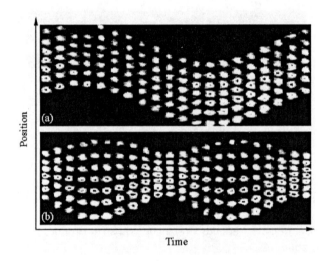

Fig. 18.14. Common mode excitation of a string of seven ions. (**a**) Center-of-mass mode of frequency ν; (**b**) breathing mode of frequency $\sqrt{3}\nu$ [421]

important are the center-of-mass mode along the trap axis which has the same frequency as a single ion, and the breathing mode, which appears at $\sqrt{3}$ times that frequency. Figure 18.14 shows an example of 7 ions.

Transitions from strings into zig-zags, helices and shells for larger numbers of ions have been investigated experimentally (Fig. 18.15) [397] and well defined regions for the different structures, depending on the depth of the confining potential have been found.

18.3 Crystalline Structures   287

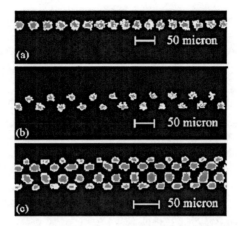

Fig. 18.15. Ion crystals observed in ultra high vacuum having (a) linear structure; (b) zig-zag structure; (c) helicale structure [398]

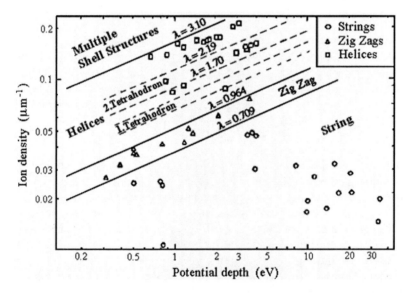

Fig. 18.16. Observed structures of ion crystals are characterized by the ion linear density and the depth of the potential well. The *straight lines* show critical λ values separating the regions corresponding to the various theoretically expected structures. The observed configurations are labelled with different symbols for each structure [397]. Copyright (2003), with permission from Elsevier

Figure 18.16 shows the observed geometrical structures for different linear particle densities λ as a function of the potential depth. The observed distribution of structures agrees well with Molecular Dynamics (MD) calculations [411].

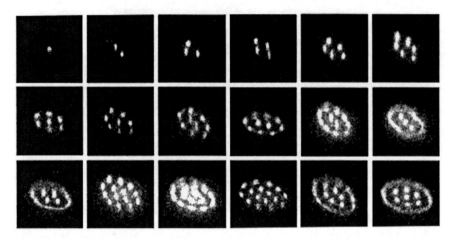

Fig. 18.17. Structure of two-dimensional Ca ion crystals in a linear Paul trap, with progressively larger numbers of ions. Stable ground state configurations show regular division into shells, reminiscent of the periodic system of the elements [412]

An oblate spheroidal plasma collapses into a planar formation when the anisotropy parameter $\alpha = \omega_z^2/\omega_r^2$ is made very large [412, 418]. The critical value of the anisotropy factor at which this will occur is given by [415]

$$\alpha_3(N) = \left(\frac{96N}{\pi^3 w_1^3}\right)^{1/2}, \qquad (18.6)$$

where $w_1 = 1.11\ldots$ for the formation of the first planar lattice; beyond this value, the minimum energy state consists of three closely spaced two-dimensional hexagonal lattice planes. This configuration evolves from an instability in the single plane configuration as the charge density increases.

The ground state of finite classical two-dimensional configurations of ions confined in a harmonic potential have been extensively studied [419, 420]. These studies have a broader application than just ions in a Penning or Paul trap; for example, electrons above a liquid He-surface and electrons in *quantum dots* in semiconductor devices. Bedanov and Peeters [419] used a Monte Carlo method, in which the initial configurations assumed were fragments of a perfect Wigner triangular lattice, with a judicious choice of interparticle spacing appropriate to the number of particles. The ground state configuration having a global minimum energy is then derived through the Monte Carlo procedure at zero temperature. Geometrical analysis of the crystalline order was carried out to determine the coordination number. Some of the results are shown in Figs. 18.17 and 18.18, and displayed in Table 18.1, which shows that at low temperatures the ions arrange themselves into rings, the numbers of particles in different rings being tabulated in the manner of the Mendeleev periodic table of the elements.

18.3 Crystalline Structures

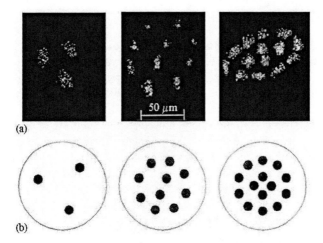

Fig. 18.18. (a) Observed Ca$^+$ ion crystals of 3, 10, and 14 ions in a Paul trap; (b) simulated crystals formed with the same numbers of particles in a Paul trap [422]

Table 18.1. Observed and simulated shell structure of ions in a two-dimensional harmonic potential [412]

Total ion number	Observed ion number			Simulated ion number		
	1st shell	2nd shell	3rd shell	1st shell	2nd shell	3rd shell
1	1			1		
2	2			2		
3	3			3		
4	4			4		
5	5			5		
6	6			1	5	
7	1	6		1	6	
8	1	7		1	7	
9	2	7		2	7	
10	2	8		2	8	
11	2	9		2	9	
12	3	9		3	9	
13	4	9		4	9	
14	4	10		4	10	
15	5	10		5	10	
16	6	10		1	5	10
17	6	11		1	6	10
18	1	6	11	1	6	11

18.3.2 Crystals in Penning Traps

The absence of time-varying fields and the associated rf-heating found in Paul traps allows one to obtain in a Penning trap crystals containing large numbers of ions. They arrange themselves in shell structures, as first observed by the NIST group [423], and have since been studied in detail as nearly planar (aspect ratio 0.05) crystalline formations [375].

Remarkable success has been achieved in the diagnostics of large ion crystals confined in a Penning trap using Bragg diffraction patterns created by the regular array of ion scatterers [391, 424]. Since the plasma rotates, the pattern observed on a plane perpendicular to the axis will consist of concentric rings, much like the Debye–Scherrer "powder technique" used in X-ray diffraction analysis. Such patterns are difficult to analyze, particularly if there is a mixture of different crystalline parameters in the ion formation. Fortunately there exist gated imaging detectors in the form of CCD cameras and photomultipliers which allow the diffraction pattern to be observed stroboscopically and "stop the motion". This will work provided the frequency of rotation of the plasma is sufficiently stable. To make sure that is the case, it has been found possible to phase lock the plasma rotation to a stable frequency signal generator suitably applied to the plasma ("rotating wall technique", see Sect. 18.1). Figure 18.19 shows a schematic diagram of the aparatus for studying the diffraction pattern produced by a confined ion crystal and Fig. 18.20 a representative result.

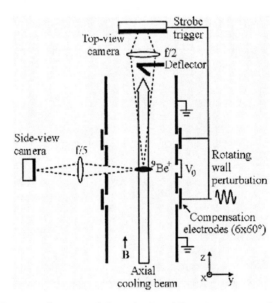

Fig. 18.19. Schematic drawing of the cylindrical Penning trap used to obtain Bragg diffraction pattern of bulk ion crystals, using phase lock of plasma to a rotating field to stroboscopically "stop" the rotation [425]

Fig. 18.20. Bragg diffraction pattern from a crystal in a Penning trap phase locked to a rotating quadrupole field. (**a**) Time averaged pattern. The *center rod* is to block the direct laser light, the *dashed lines* indicate shadows of a wire mesh in the observation line; (**b**) time-resolved pattern of (**a**). The *rectangular grid lines* are for a bcc-crystal with a (110) axis aligned with the laser beam [425]

On the basis of observed Bragg diffraction patterns, it has been established that three-dimensional long range order in the form of a bcc lattice begins to appear when the number of ions exceeds about $N \approx 5 \times 10^4$, except at the boundary of the plasma. At $N > 3 \times 10^5$ the plasma shows exclusively bulk behavior as a bcc-crystal. In some cases, two or more bcc-crystals having fixed orientations with respect to each other were observed. In some oblate plasmas, a mixture of bcc- and fcc-orders was seen [426]. Also structural phase transitions have been observed when the axial density is changed into a two-dimensional extended plane of confined ions [427]. Five different phases occured which correspond to the energetically favoured structure, in agreement to theoretical predictions (Fig. 18.21) [416, 428].

18.3.3 Crystals in Storage Rings

For experiments performed with charged particles in storage rings, an important requirement is a high phase space density, that is, low momentum spread and tight spatial confinement. A phase transition from a cloud-like behaviour of the particles to an ordered state, as observed in Paul and Penning traps, would provide the highest available brilliance. The ions would reside in well defined lattice positions and no dissipative close Coulomb collisions between particles (intra-beam scattering) would occur.

Test experiments on a small table-top storage ring ("PALLAS") at the University of München, based on previous experiments at the Max–Planck-Institut for Quantum Optics, Garching [160], have demonstrated that such crystalline beams can in fact be realized [429]. The ring, very similar in construction to that shown in Fig. 4.19 has a radius of 57.5 mm. Radial confinement is provided by a quadrupole potential created by a radio-frequency voltage applied to the electrodes as in a linear Paul trap. Two counter-propagating laser beams properly detuned from resonance cool the circulating ion beam. When the strong confinement condition is met ($\Gamma > 170$)

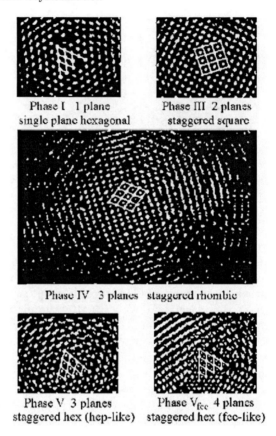

Fig. 18.21. Images of five different structural phases observed on Be$^+$ ions in a Penning trap. The trap axis is perpendicular to the observation plane. The *lines* show fits to the indicated structure [427]. Copyright (2003) AAAS

the ions undergo a phase transition into a crystalline state. Strings, zig-zag structures and helices are observed as in linear Paul traps, depending on the linear ion density. The ring can be additionally equipped with drift tubes to generate longitudinal electric fields which can be used to transport and position stationary ions along the orbit or to influence the velocity of the ion beam.

In the absence of heating forces the crystalline beams are found to be stable even if the cooling lasers have been turned off. For ion crystals at rest, times exceeding 90 s (corresponding to 6×10^8 periods of the trapping field) have been measured during which helical structures remained stable. For fast moving beams, the stability was assured for about 3000 round trips in the storage ring, corresponding to $t > 0.4$ s [430].

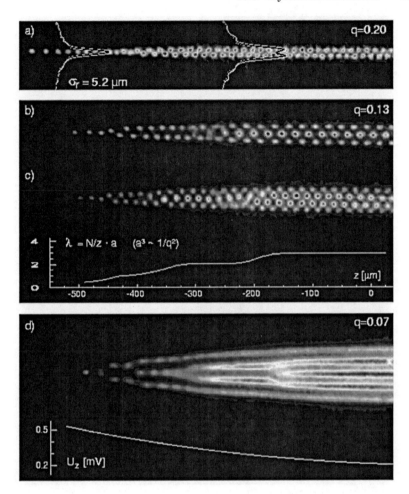

Fig. 18.22. Images of ion crystals at rest in PALLAS. The ions are longitudinally confined in the ring by weak electrostatic potentials. The crystal form is a string at low values of the linear ion density λ and becomes more complex when λ increases. The part of the *ring* which is seen by the observation laser acts approximately like a linear Paul trap [431]

19 Sympathetic Crystallization

The crystallization of stored ions through sympathetic cooling may prove to be of critical importance in broadening their application, since only for a small number of species can the required wavelength for cooling be provided by available laser sources. This is particularly true when dealing with highly charged ions. If one species can be cooled by lasers, a second one simultaneously confined in the trap, will be cooled also by Coulomb interaction, as previously discussed in Sect. 14.7. Sympathetic crystallization was first observed in crystals of one ion species exhibiting dark lattice points arising from simultaneously created impurity ions (Fig. 19.1).

A systematic study of sympathetic cooling has been performed by a group at the University of Aarhus [367] using a linear Paul trap. It was found that in a linear string of ions the cooling of a single ^{24}Mg$^+$ ion was sufficient to keep as many as 14 additional sympathetically cooled impurity ions in a crystalline state. This is of particular interest not only for extending the range of species that can be simultaneously cooled and spectroscopically studied, but also for permitting most of the ions in a chain to perform quantum information processing, while reserving the function of cooling to other ions.

With larger crystals, showing a shell structures, a fraction of 20–30% of the ions in the crystal have to be cooled directly by laser radiation to keep the total crystal in a stable configuration. This is, of course, because the rf-heating of ions off the trap axis must be offset by larger cooling power. Although the impurity ions cannot be observed directly their mass can be measured by resonant excitation at their oscillation frequency in the trap. The additional energy of resonant excitation is transferred to the directly cooled species and as a consequence the amount of fluorescence from the cooled ions

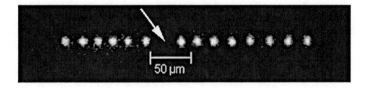

Fig. 19.1. Linear Ca$^+$ crystal with sympathetically crystallized impurity ion [148]

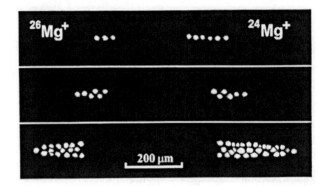

Fig. 19.2. A sequence of images recorded during loading of a multi-component crystal by electron impact ionization. A laser beam resonant with ^{26}Mg$^+$ ions is incident from the *right*, while the laser beam resonant with the ^{24}Mg$^+$ ions is incident from the *left*. The figure shows the ^{24}Mg$^+$ ions being pushed to the *right end* and ^{26}Mg$^+$ ions to the *left* by their respective near resonant cooling lasers. In the *center region* the nonresonant impurity ions are collected [367]. Copyright (2003) by the American Physical Society

will change, as has been shown in the case of impurity ions O_2^+ in a Mg$^+$ crystal.

Apart from reducing the ion temperature, the radiation from the cooling laser exerts a light pressure on the ions. This can be used to spatially separate ions cooled directly from those sympathetically cooled, as shown in Fig. 19.2.

A different example is the spatial separation of ions in an excited energy level: In a laser cooled Ca$^+$ crystal, some of the ions were excited by an additional laser into a long lived metastable state, and hence decoupled from the cooling laser radiation. Since the directly coupled ions are pushed by the light forces in the direction of the laser beam, the metastable ions are left segregated on that part of the crystal opposite to the beam direction (Fig. 19.3).

When two different ion species can be simultaneously cooled and observed by their fluorescence radiation, the spatial properties of such two-component crystals can be investigated. Due to the mass dependence of the radial confining potential in a Paul trap, spatial separation appears. The lighter ion is situated near the trap axis while the heavier one forms a crystal in the outer regions of the trap. While the shape of the heavier ion formation resembles that of a single component crystal, the lighter ones form a nearly cylindrical structure, which continues to the edge of the surrounding crystal (Fig. 19.4).

19 Sympathetic Crystallization   297

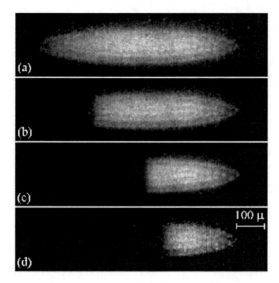

Fig. 19.3. Spatial separation of atomic states by sympathetic crystallization. In a Ca$^+$ crystal (**a**) containing a few 100 ions a laser excites some ions into a long lived metastable state. These ions are decoupled from the laser radiation but remain crystallized through Coulomb interaction. Radiation pressure from the cooling laser forces the directly cooled ions to the right side. From (**b**) to (**d**) the number of ions excited to the metastable states increases [148]

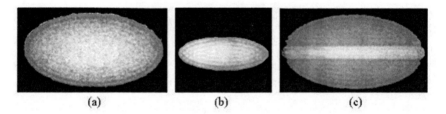

Fig. 19.4. (**a**) Crystal containing approximately 3,200 ^{40}Ca$^+$ ions; (**b**) crystal containing approximately 650 ^{24}Mg$^+$ ions; (**c**) a bicrystal formed by loading 320 ^{24}Mg$^+$ ions into the ^{40}Ca$^+$ ion crystal shown in (**a**) [417]

A Mathieu Equations

A.1 Parametric Oscillators

An ion confined within a quadrupole Paul trap can be considered as a three-dimensional parametric oscillator described by three Mathieu equations. The Mathieu equation is an ordinary differential equation with real coefficients:

$$u'' + (a - 2q \cos 2\tau)u = 0 \ . \tag{A.1}$$

It was introduced by Mathieu in the investigation of the oscillations of an elliptic membrane [432]. From Floquet's theorem it follows that the Mathieu equation (A.1) has a solution of the form $e^{\mu\tau}\Phi(\tau)$, where Φ is a periodic function with period π, and μ depends on a and q. The parameter μ is called the *characteristic exponent*. Clearly $e^{-\mu\tau}\Phi(-\tau)$ is also a solution of (A.1). If $i\mu$ is not an integer, then $e^{\mu\tau}\Phi(\tau)$ and $e^{-\mu\tau}\Phi(-\tau)$ are linearly independent. In this case, the general solution of (A.1) can be written as

$$u(\tau) = A_1 e^{\mu\tau} \sum_{n=-\infty}^{\infty} c_{2n} e^{2ni\tau} + A_2 e^{-\mu\tau} \sum_{n=-\infty}^{\infty} c_{2n} e^{-2ni\tau} \ . \tag{A.2}$$

Substitution of (A.2) in (A.1) gives the following recurrence relationship for c_{2n}:

$$\gamma_n(\mu)c_{2n-2} + c_{2n} + \gamma_n(\mu)c_{2n+2} = 0 \ , \tag{A.3}$$

where

$$\gamma_n(\mu) = \frac{q}{(2n - i\mu)^2 - a} \ . \tag{A.4}$$

The characteristic exponent μ is determined by the equation

$$\Delta(\mu) = 0 \ , \tag{A.5}$$

where $\Delta(\mu)$ is the determinant of system (A.3). $\Delta(\mu)$ is convergent and (A.5) can be reduced to the remarkable equation [433]

$$\cosh(\pi\mu) = 1 - 2\Delta(0) \sin^2\left(\frac{1}{2}\pi\sqrt{a}\right) \ . \tag{A.6}$$

From the recurrence relationship (A.3), we obtain

$$\frac{c_{2n}}{c_{2n\pm 2}} = -\frac{-q(2n-i\mu)^{-2}}{1 - a(2n-i\mu)^{-2} + q(2n-i\mu)^{-2} \frac{c_{2n\mp 2}}{c_{2n}}}, \quad (A.7)$$

repeated applications of which lead to the convergent continued fractions $R_n^+(\mu)$ and $R_n^-(\mu)$

$$\frac{c_{2n}}{c_{2n\mp 2}} = R_n^\pm(\mu) . \quad (A.8)$$

Then μ is given by

$$R_0^-(\mu) R_1^+(\mu) = 1 , \quad (A.9)$$

and the coefficients $c_{\pm 2n}$ are obtained as

$$c_{\pm 2n} = c_0 R_1^\pm(\mu) R_2^\pm(\mu) \cdots R_n^\pm(\mu) . \quad (A.10)$$

From (A.10)

$$\lim_{n \to \infty} \frac{n^2 c_{2n}}{c_{2n\pm 2}} = -\frac{q}{4} , \quad (A.11)$$

such that each series in (A.2) converges.

In a stable domain, $\beta = -i\mu$ is real. Then all the c_{2n} are real provided c_0 is real. In this case, the general solution u of the Mathieu equation can be written as a linear combination, with real coefficients A and B, of the fundamental solutions u_1 and u_2

$$u(\tau) = A u_1(\tau) + B u_2(\tau) , \quad (A.12)$$

$$u_1(\tau) = \sum_{n=-\infty}^{\infty} c_{2n} \cos(2n+\beta)\tau , \quad u_2(\tau) = \sum_{n=-\infty}^{\infty} c_{2n} \sin(2n+\beta)\tau . \quad (A.13)$$

From (A.12) and its derivative with respect to τ results

$$A = \frac{1}{W}[u_2'(\tau)u(\tau) - u_2(\tau)u'(\tau)] ,$$

$$B = \frac{1}{W}[u_1(\tau)u'(\tau) - u_1'(\tau)u(\tau)] , \quad (A.14)$$

$$W = u_1(\tau)u_2'(\tau) - u_1'(\tau)u_2(\tau) . \quad (A.15)$$

For $\tau = 0$ the Wronskian is

$$W = \sum_{m,n=-\infty}^{\infty} (2n+\beta) c_{2m} c_{2n} . \quad (A.16)$$

On the other hand, from (A.12) results $|u(\tau)| \leq u_m$ for any τ, where

$$u_m = \sqrt{A^2 + B^2} \sum_{n=-\infty}^{\infty} |c_{2n}| . \quad (A.17)$$

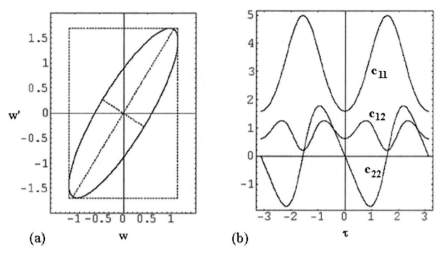

Fig. A.1. Phase space analysis of Mathieu equation. (a) Ion trajectory ("emittance" ellipse) in phase space; (b) The evolution of the functions c_{11}, c_{12} and c_{22} for $a = 0$, $q = 0.5$

Introducing (A.14) and (A.15) in (A.17), we obtain

$$c_{11}u^2 + 2c_{12}uu' + c_{22}u'^2 = c_0 , \qquad (A.18)$$

where

$$c_{11} = \frac{1}{W}\left[(u_1'(\tau))^2 + (u_2'(\tau))^2\right] , \quad c_{22} = \frac{1}{W}\left[u_1^2(\tau) + u_2^2(\tau)\right] , \qquad (A.19)$$

$$c_{12} = -\frac{1}{W}\left[u_1(\tau)u_2'(\tau) + u_2(\tau)u_1'(\tau)\right] , \qquad (A.20)$$

$$c_0 = u_m^2 W \left[\sum_{n=-\infty}^{\infty} |c_{2n}|\right]^{-2} . \qquad (A.21)$$

Using (A.19), (A.20), (A.14) and (A.15), we obtain the constraint

$$c_{11}c_{22} - c_{12}^2 = 1 , \qquad (A.22)$$

showing that (A.18) is the equation of an ellipse with the area πc_0 in phase space.

In Fig. A.1a is represented the ellipse

$$c_{11}w^2 + 2c_{12}ww' + c_{22}w'^2 = 1 , \qquad (A.23)$$

where $a = 0$, $q = 0.5$, $\tau = 0.75$, and

$$w = \frac{1}{\sqrt{c_0}}u(\tau) , \quad w' = \frac{1}{\sqrt{c_0}}u'(\tau) . \qquad (A.24)$$

In Fig. A.1b the evolution of the functions c_{11}, c_{12} and c_{22} on the interval $-\pi \leq \tau_0 \leq \pi$ for $a = 0$, $q = 0.5$ can be seen.

B Orbits of Trapped Ions

Here some examples are given of periodic and quasiperiodic ion trajectories in Paul traps (Fig. B.1-Fig. B.4), and in Penning traps (Fig. B.6-Fig. B.8) in different planes, in three dimensions, and in the phase space, for specific trapping parameters, time intervals, and initial conditions.

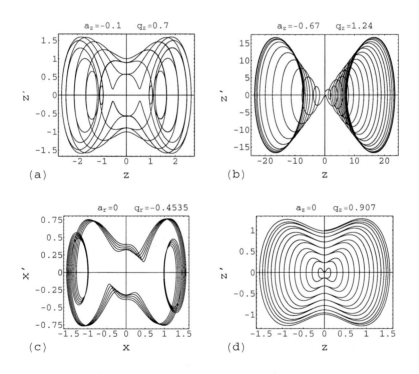

Fig. B.1. Phase space trajectories for an ideal Paul trap with $0 \leq \tau \leq 30\pi$. (a) $\beta_z = 0.071$, $\beta_r = 0.035$ in the (z, z')-plane. (b) $\beta_z = 0.969$, $\beta_r = 0.335$ in the (z, z')-plane. (c) $\beta_z = 0.054$, $\beta_r = 0.984$ in the (x, x')-plane. (d) $\beta_z = 0.054$, $\beta_r = 0.984$ in the (z, z')-plane

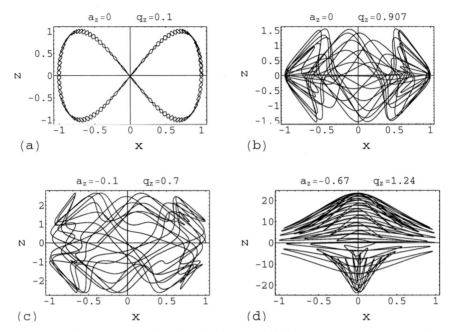

Fig. B.2. Ion trajectories in the (x, z)-plane. (a) $\beta_z = 0.071$, $\beta_r = 0.035$ for $0 \leq \tau \leq 120\pi$. (b) $\beta_z = 0.969$, $\beta_r = 0.335$ for $0 \leq \tau \leq 60\pi$. (c) $\beta_z = 0.416$, $\beta_r = 0.344$ for $0 \leq \tau \leq 60\pi$. (d) $\beta_z = 0.054$, $\beta_r = 0.984$ for $0 \leq \tau \leq 60\pi$

The radial ion trajectories in a Penning trap can be written as

$$x = R_+ \cos(-\omega_+ t + \theta_+) + R_- \cos(-\omega_- t + \theta_-) , \tag{B.1}$$

$$y = R_+ \sin(-\omega_+ t + \theta_+) + R_- \sin(-\omega_- t + \theta_-) . \tag{B.2}$$

If $\theta_+ = \theta_- = 0$, then (B.1) and (B.2) can be written as the parametric equations of an epitrochoid:

$$x = (a + b) \cos \tau + h \cos\left(\frac{a+b}{b}\tau\right) , \tag{B.3}$$

$$y = (a + b) \sin \tau + h \sin\left(\frac{a+b}{b}\tau\right) , \tag{B.4}$$

where

$$a = \frac{\omega_1}{\omega_+} R_- , \quad b = \frac{\omega_-}{\omega_+} R_- , \quad h = R_+ , \quad \tau = -\omega_- t . \tag{B.5}$$

The epitrochoid given by (B.1) and (B.2) is the trace of a point P that is rigidly attached to a circle of radius b rolling without slippage on the outside of a fixed circle of radius a. The distance from the tracing point P to the center O of the rolling circle is equal to the cyclotron radius R_+. The trace of O is the magnetron circle.

B Orbits of Trapped Ions   305

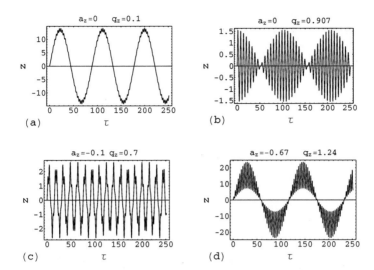

Fig. B.3. The axial coordinate z for $0 \leq \tau \leq 80\pi$. (a) $\beta_z = 0.071$, $\beta_r = 0.035$; (b) $\beta_z = 0.969$, $\beta_r = 0.335$; (c) $\beta_z = 0.416$, $\beta_r = 0.344$; (d) $\beta_z = 0.054$, $\beta_r = 0.984$

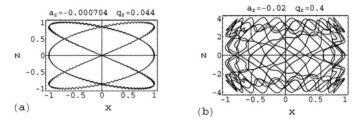

Fig. B.4. Periodic and quasiperiodic orbits in the (x,z)-plane. (a) Orbit of period 360π for $0 \leq \tau \leq 360\pi$. (b) Quasiperiodic orbit for $0 \leq \tau \leq 360\pi$

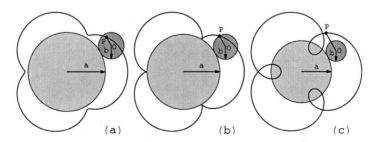

Fig. B.5. Epitrochoids in the radial plane of the ideal Penning trap for $\omega_+ = 4\omega_-$, $a = 3b$, $b = R_-/4$. (a) Shortened epitrochoid with $R_+ = 3b/4$; (b) epicycloid with $R_+ = b$; (c) elongated epitrochoid with $R_+ = 2b$

Table B.1. Periodic orbits for $T \leq 12T_z$

m	n	ω_z/ω_-	ω_+/ω_z	ω_c/ω_z	T/T_z
2	1	2	1	3/2	2
4	1	4	2	9/4	4
3	1	3	3/2	11/6	6
6	1	6	3	19/6	6
8	1	8	4	33/8	8
5	1	5	5/2	27/10	10
10	1	10	5	51/10	10
12	1	12	6	73/12	12
3	2	3	3/4	17/12	12

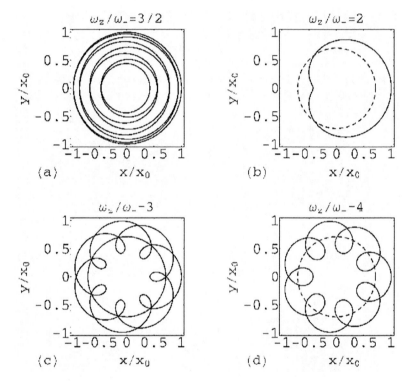

Fig. B.6. Radial projections of periodic orbits with the initial conditions $x_0 = R_+ + R_-$ and $R_- = 5R_+/2$. (a) Orbit of period $17T_c/12$ for $\omega_+/\omega_- = 9/8$; (b) orbit of period $T = 3T_c$ for $\omega_+/\omega_- = 2$; (c) orbit of period $T = 11T_c$ for $\omega_+/\omega_- = 9/2$; (d) orbit of period $T = 9T_c$ for $\omega_+/\omega_- = 8$

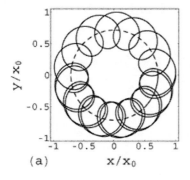
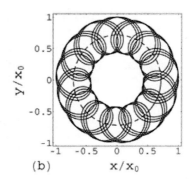

Fig. B.7. Radial quasiperiodic orbits for $\omega_+/\omega_- = 10\sqrt{2}$ with the initial conditions $x_0 = R_+ + R_-$ and $R_- = 5R_+/2$. (a) $0 \le t \le 24T_c$. (b) $0 \le t \le 72T_c$

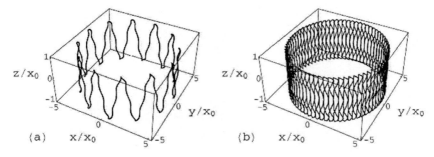

Fig. B.8. Trajectories in three dimensions with the initial conditions $R_+ = 100R_- = 5R_z$. (a) Orbit for $\omega_+/\omega_- = 72$. (b) Quasiperiodic orbit with periodic radial projection for $\omega_+/\omega_- = 70$

The generation of epitrochoids is illustrated in Fig. B.5. If $R_+ < b$, then the epitrochoid is called shortened, and when $R_+ > b$, elongated. The epicycloid is a special case of the epitrochoid for $b = R_+$.

For $a/b = p/q$, where p and q are coprime positive integers, the epitrochoid has a time period of $2\pi q/\omega_-$. If a/b is irrational, the motion is quasi-periodic.

If ω_z/ω_- is a rational number, then there exist positive integers m and n such that

$$\frac{\omega_z}{\omega_-} = \frac{m}{n}, \qquad (B.6)$$

where the numerator m and denominator n have no common factors. Using (3.14), we obtain the following rational numbers

$$\frac{\omega_+}{\omega_z} = \frac{m}{2n}, \quad \frac{\omega_c}{\omega_z} = \frac{m^2+2n^2}{2mn}, \quad \frac{\omega_1}{\omega_z} = \frac{m^2-2n^2}{2mn}. \qquad (B.7)$$

Then an orbit in an ideal Penning trap is periodic if and only if $\omega_z/\omega_- = m/n$ is an irreducible fraction and the positive integers m and n satisfy the condition $m > \sqrt{2}n$. Moreover, the period T of a periodic orbit is given by $T = mnT_z$ for m even, and $T = 2mnT_z$ for m odd, where $T_z = 2\pi/\omega_z$. Using (B.6) and (B.7), we obtain the Table B.1. Some trajectories with the initial conditions $x(0) = x_0 \neq 0$, $y(0) = 0$, and $v_x(0) = 0$ are illustrated in (B.6)–(B.8).

C Nonlinear Oscillator

C.1 Multipole Expansions

The trap electric potential is given by

$$\Phi = \sum_{n=0}^{\infty} \sum_{m=-n}^{n} c_{nm} \left(\frac{\rho}{d}\right)^n P_n^m(\cos\theta) \exp(im\varphi) , \qquad (C.1)$$

where $\rho = (x^2 + y^2 + z^2)^{1/2}$ and

$$x = \sin\theta \cos\varphi , \quad y = \sin\theta \sin\varphi , \quad z = \cos\theta . \qquad (C.2)$$

The associated Legendre functions are defined by

$$P_n^m(x) = \frac{(-1)^m}{2^n n!} (1-x^2)^{m/2} \frac{d^{n+m}}{dx^{n+m}} (x^2-1)^n . \qquad (C.3)$$

We now introduce the harmonic polynomials

$$H_{nm}(x,y,z) = \rho^n P_n^m(\cos\theta) \exp(im\varphi) . \qquad (C.4)$$

Using (C.2)–(C.4), we obtain

$$H_{nm}(x,y,z) = (n+m)! \sum_{p,q,s} \frac{2^{-p-q}}{p!q!s!} (-x-iy)^p (x-iy)^q z^s , \qquad (C.5)$$

where the sum is over nonnegative integers p, q, s obeying $p - q = m$ and $p + q + s = n$. Then Φ can be expanded in the form

$$\Phi = \sum_{n=0}^{\infty} \sum_{m=-n}^{n} \frac{c_{nm}}{d^n} H_{nm} . \qquad (C.6)$$

In the case of rotational symmetry arround the z-axis, a basis of harmonic polynomials consists of

$$H_n(r,z) = H_{no}(x,y,z) = \rho^n P_n(\cos\theta) , \qquad (C.7)$$

where $P_n(x) = P_n^0(x)$ is the Legendre polynomial. Using (C.5), we have

$$H_n(r,z) = \sum_{k=0}^{n/2} (-4)^{-k} b(n,k) r^{2k} z^{n-2k} , \qquad (C.8)$$

where $r = (x^2 + y^2)^{1/2}$ and

$$b(n,k) = \frac{n!}{(n-2k)!(k!)^2} . \qquad (C.9)$$

Then the multipole expansion of the electric field is

$$\Phi(r,z) = \sum_{n=0}^{\infty} \frac{c_n}{d^n} H_n(r,z) , \qquad (C.10)$$

where $c_n = c_{n0}$, and the solid harmonic polynomials are

$$H_2(r,z) = \frac{1}{2}(-r^2 + 2z^2) , \qquad (C.11)$$

$$H_3(r,z) = \frac{1}{2}(-3r^2 z + 2z^3) , \qquad (C.12)$$

$$H_4(r,z) = \frac{1}{8}(3r^4 - 24r^2 z^2 + 8z^4) , \qquad (C.13)$$

$$H_5(r,z) = \frac{1}{8}(15r^4 z - 40r^2 z^3 + 8z^5) , \qquad (C.14)$$

$$H_6(r,z) = \frac{1}{16}(-5r^6 + 90r^4 z^2 - 120r^2 z^4 + 16z^6) , \qquad (C.15)$$

$$H_7(r,z) = \frac{1}{16}(-35r^6 z + 210r^4 z^3 - 168r^2 z^5 + 16z^7) , \qquad (C.16)$$

$$H_8(r,z) = \frac{1}{128}(35r^8 - 1120r^6 z^2 + 3360r^4 z^4 - 1792r^2 z^6 + 128z^8) . \qquad (C.17)$$

C.2 Normal Forms

We can apply the method of normal forms [107] to the small anharmonic perturbations of a *quadrupole* electromagnetic trap. Then the resonance condition for a nonlinear combined trap can be written as

$$n_x \omega_x + n_y \omega_y + n_z \omega_z = k\Omega , \qquad (C.18)$$

where n_x, n_y, n_z, k are integers, and

$$\omega_j = \frac{1}{2}\Omega \beta_j , \quad j = x, y, z , \qquad (C.19)$$

with β_j the characteristic exponent in the solution of the Mathieu equation corresponding to the parameters $a_j + \omega_c^2/4$ and q_j.

For a dynamical trap with rotational symmetry around the z-axis, we may write $\omega_x = \omega_y = \omega_r$ and $n_x + n_y = n_r$, leading to

$$n_r \omega_r + n_z \omega_z = k\Omega \ . \tag{C.20}$$

In the case of a Penning trap, we have the resonance condition

$$n_+ \omega_+ + n_- \omega_- + n_z \omega_z = 0 \ , \tag{C.21}$$

which for an ideal Penning trap, reduces to

$$n_1 \omega_1 + n_z \omega_z = 0 \ , \tag{C.22}$$

where n_+, n_-, n_1 and n_z are integers. Then (C.22) can be written in the form

$$\omega_c = \omega_z \sqrt{s^2 + 2} \ , \tag{C.23}$$

in which s is rational.

The Hamiltonian, under appropriate canonical transformations with the new coordinates ξ_{j+3} and momenta ξ_j, can be expressed in the form [434, 435, 437]

$$K(\boldsymbol{\xi}, \tau) = \frac{1}{2} \sum_{j=1}^{3} \beta_j (\xi_j^2 + \xi_{j+3}^2) + W(\boldsymbol{\xi}, \tau) \ , \tag{C.24}$$

where $\boldsymbol{\xi} = (\xi_1, \xi_2, \xi_3, \xi_4, \xi_5, \xi_6)$ and W is the anharmonic multipole part.

Consider a system of new variables $\boldsymbol{\eta} = (\eta_1, \eta_2, \eta_3, \eta_4, \eta_5, \eta_6)$, and the generating function

$$S(\xi_1, \eta_1, \xi_2, \eta_2, \xi_3, \eta_3, \tau) = \sum_{j=1}^{3} \xi_j \eta_j + \sum_{n \geq 3} S_n(\xi_1, \eta_1, \xi_2, \eta_2, \xi_3, \eta_3, \tau) \ , \tag{C.25}$$

where every S_n is a homogeneous polynomial of degree n. Then

$$\eta_{j+3} = \frac{\partial S}{\partial \eta_3} \ , \quad \xi_{j+3} = \frac{\partial S}{\partial \xi_3} \ , \quad 1 \leq j \leq 3 \ . \tag{C.26}$$

The new Hamiltonian G after the nonlinear transformation is

$$G(\boldsymbol{\eta}, \tau) = K(\boldsymbol{\xi}, \tau) + \frac{\partial}{\partial \tau} S(\xi_1, \eta_1, \xi_2, \eta_2, \xi_3, \eta_3, \tau) \ , \tag{C.27}$$

$$\frac{\partial S_n}{\partial \tau} + \sum_{j=1}^{3} \beta_j \left(\eta_j \frac{\partial S_n}{\partial \xi_j} - \xi_j \frac{\partial S_n}{\partial \eta_j} \right) = R_n(\xi_1, \eta_1, \xi_2, \eta_2, \xi_3, \eta_3) \ . \tag{C.28}$$

It is convenient to introduce the complex variables

$$\zeta_{j\pm} = \xi_j \pm i\eta_j \ , \quad 1 \leq j \leq 3 \ , \tag{C.29}$$

and to expand the π-periodic functions S_n and R_n into Fourier τ-series and into power series in $\zeta_{j\pm}$:

$$S_n = \sum_{\boldsymbol{m},|\boldsymbol{m}|=n} \sum_{\nu=-\infty}^{\infty} \left[S_{n\nu\boldsymbol{m}} e^{2i\nu\tau} \left(\prod_{j=1}^{3} \zeta_{j+}^{m_{j+}} \zeta_{j-}^{m_{j-}} \right) \right], \quad (C.30)$$

$$R_n = \sum_{\boldsymbol{m},|\boldsymbol{m}|=n} \sum_{\nu=-\infty}^{\infty} \left[R_{n\nu\boldsymbol{m}} e^{2i\nu\tau} \left(\prod_{j=1}^{3} \zeta_{j+}^{m_{j+}} \zeta_{j-}^{m_{j-}} \right) \right], \quad (C.31)$$

where $m_{j\pm}$ are integers, $\boldsymbol{m} = (m_{1+}, m_{1-}, m_{2+}, m_{2-}, m_{3+}, m_{3-})$ and

$$|\boldsymbol{m}| = \sum_{j=1}^{3} (m_{j+} + m_{j-}). \quad (C.32)$$

Substituting (C.30) and (C.31) into (C.28), we obtain

$$S_{n\mu\nu} = \frac{iR_{n\mu\nu}}{\sum_{j=1}^{3} \beta_j (m_{j+} - m_{j-}) - 2\nu}. \quad (C.33)$$

It is convenient to introduce the integers

$$n_j = m_{j+} - m_{j-}, \quad 1 \leq j \leq 3. \quad (C.34)$$

Then the condition for small divisors in (C.33) can be written as

$$\beta_1 n_1 + \beta_2 n_2 + \beta_3 n_3 = 2\nu,$$
$$n_1 + n_2 + n_3 = n. \quad (C.35)$$

C.3 Nonlinear Resonances

The condition for nonlinear resonances in a Penning trap observed in [113, 436] and theoretically discussed in [437], can be written as

$$n_+ \omega_+ + n_- \omega_- + n_z \omega_z = 0, \quad (C.36)$$

where n_+, n_-, n_z are integers with $n_+ \geq 0$.

The orbits of the charged particles in the nonlinear traps can be classified as periodic and nonperiodic.

In the Table C.1 some examples of periodic orbits for $T \leq 12 T_z$ ($T_z = 2\pi/\omega_z$) are given. For it, $n_+ = 0$ and we have

$$\frac{\omega_z}{\omega_-} = \left| \frac{n_-}{n_z} \right|. \quad (C.37)$$

C.3 Nonlinear Resonances

Table C.1. Examples of periodic orbits in a nonlinear Penning trap

n_-	n_z	ω_+/ω_z	ω_+/ω_-	ω_z/ω_-	T/T_z
2	−1	1	2	2	2
4	−1	2	8	4	4
3	−1	3/2	9/2	3	6
6	−1	3	18	6	6
8	−1	4	32	8	8
5	−1	5/2	25/2	5	10
10	−1	5	50	10	10
12	−1	6	72	12	12
3	−2	3/4	9/8	3/2	12

If the nonperiodic orbits have periodic radial projections, ω_z/ω_- is irrational, $n_z = 0$ and $\omega_+/\omega_- = |n_-/n_+|$.

In the Table C.2 some examples of nonperiodic orbits in a nonlinear Penning trap are given.

If ω_z/ω_- and ω_+/ω_- are irrational, using

$$s = \frac{\omega_z}{\omega_-} = 2\frac{\omega_+}{\omega_z} , \qquad (C.38)$$

(C.36) can be written as

$$n_+ s^2 + 2n_z s + 2n_- = 0 . \qquad (C.39)$$

Then

$$s = \frac{1}{n_+}(-n_z \pm \sqrt{n_z^2 - 2n_+ n_-}) , \qquad (C.40)$$

Table C.2. Examples of nonperiodic orbits in a nonlinear Penning trap

n_+	n_-	ω_+/ω_z	ω_+/ω_-	ω_z/ω_-
3	−1	$\sqrt{6}/2$	3	$\sqrt{6}$
4	−1	$\sqrt{2}$	4	$2\sqrt{2}$
5	−1	$\sqrt{10}/2$	5	$\sqrt{10}$
6	−1	$\sqrt{3}$	6	$2\sqrt{3}$
7	−1	$\sqrt{14}/2$	7	$\sqrt{14}$
9	−1	$3\sqrt{2}/2$	9	$3\sqrt{2}$
10	−1	$\sqrt{5}$	10	$2\sqrt{5}$
3	−2	$\sqrt{3}/2$	3/2	$\sqrt{3}$
5	−2	$\sqrt{5}/2$	5/2	$\sqrt{5}$

C Nonlinear Oscillator

Table C.3. Nonperiodic orbits for nonlinear resonances of the order $0 < N \leq 4$

N	n_+	n_-	n_z	ω_+/ω_z	ω_+/ω_-	ω_z/ω_-
2	1	0	−1	1	2	2
3	2	0	−1	2	8	4
3	1	1	−1	$(1+\sqrt{3})/2$	$2+\sqrt{3} \approx 3.73$	$1+\sqrt{3}$
4	3	0	−1	3	18	6
4	2	−1	−1	$(1+\sqrt{5})/4$	$(3+\sqrt{5})/4 \approx 1.31$	$(1+\sqrt{5})/2$
4	1	−2	−1	$(1+\sqrt{5})/2$	$3+\sqrt{5} \approx 5.24$	$1+\sqrt{5}$
4	1	1	−2	$(2+\sqrt{2})/2$	$3+2\sqrt{2} \approx 5.83$	$2+\sqrt{2}$
4	1	−1	−2	$(2+\sqrt{6})/2$	$5+2\sqrt{6} \approx 9.9$	$2+\sqrt{6}$

for $n_+ > 0$, $n_z^2 - 2n_+ n_- > 0$, $0 < s < \sqrt{2}$. We define $N = |n_+| + |n_-| + |n_z|$ as the order of resonance.

In Table C.3 are given some examples of nonperiodic orbits in a nonlinear Penning trap for resonances of the order $0 < N \leq 4$.

D Generating Functions for Quantum States

D.1 Uncertainty Relations

Consider a quantum mechanical system with two observables represented by the Hermitian operators A and B. The variance σ_{AA} of A, the variance σ_{BB} of B, and the covariance σ_{AB} of A and B given by

$$\sigma_{AA} = \langle \tilde{A}^2 \rangle - \langle \tilde{A} \rangle^2, \quad \sigma_{BB} = \langle \tilde{B}^2 \rangle - \langle \tilde{B} \rangle^2,$$

$$\sigma_{AB} = \frac{1}{2} \langle \tilde{A}\tilde{B} \rangle - \langle \tilde{A} \rangle \langle \tilde{B} \rangle, \tag{D.1}$$

where $\langle \rangle$ means the expectation value in the state vector Ψ and

$$\tilde{A} = A - \langle A \rangle, \quad \tilde{B} = B - \langle B \rangle. \tag{D.2}$$

The Schrödinger inequality can be written as

$$\sigma_{AA}\sigma_{BB} \geq \sigma_{AB}^2 + \frac{1}{4} |\langle [A,B] \rangle|^2. \tag{D.3}$$

We have

$$\langle \tilde{A}\tilde{B} \rangle = \left\langle \frac{1}{2}[\tilde{A}, \tilde{B}] + \frac{1}{2}(\tilde{A}\tilde{B} + \tilde{B}\tilde{A}) \right\rangle = \sigma_{AB} + \frac{1}{2} \langle [A,B] \rangle. \tag{D.4}$$

Since σ_{AB} is real, (D.4) implies

$$\left| \langle \tilde{A}\tilde{B} \rangle \right|^2 = \sigma_{AB}^2 + \frac{1}{4} |\langle [A,B] \rangle|^2. \tag{D.5}$$

According to the Schwarz inequality, we have

$$\langle \tilde{A}^2 \rangle \langle \tilde{B}^2 \rangle \geq \left| \langle \tilde{A}\tilde{B} \rangle \right|^2. \tag{D.6}$$

Inserting (D.1) and (D.5) into (D.6) we obtain (D.3). The uncertainty ΔA of A, the uncertainty ΔB of B, and the correlation coefficient r_{AB} of A and B are given by

$$\Delta A = \sqrt{\sigma_{AA}}, \quad \Delta B = \sqrt{\sigma_{BB}}, \quad r_{AB} = \frac{\sigma_{AB}}{\Delta A \Delta B}. \tag{D.7}$$

The Schrödinger inequality can be rewritten as

$$\Delta A\, \Delta B \geq \frac{1}{2\sqrt{1 - r_{AB}^2}} |\langle [A, B] \rangle|. \tag{D.8}$$

If $\sigma_{AB} = 0$, the Schrödinger inequality is reduced to the Heisenberg inequality

$$\Delta A\, \Delta B \geq \frac{1}{2} |\langle [A, B] \rangle|. \tag{D.9}$$

If $\tilde{A}\Psi \neq 0$ and $\tilde{B}\Psi \neq 0$, then the equality in (D.6) is satisfied by the state vector Ψ if and only if there exists a nonzero complex number λ such that

$$\tilde{A}\Psi = i\lambda \tilde{B}\Psi. \tag{D.10}$$

Multiplying (D.10) on the left first by $\tilde{A}$ and then by $\tilde{B}$ and taking the expectation values, we obtain

$$\left\langle \tilde{A}^2 \right\rangle = i\lambda \left\langle \tilde{A}\tilde{B} \right\rangle, \quad \left\langle \tilde{B}\tilde{A} \right\rangle = i\lambda \left\langle \tilde{B}^2 \right\rangle, \tag{D.11}$$

$$\sigma_{AA} = |\lambda|^2 \sigma_{BB}, \quad \sigma_{AB} = \frac{1}{2}(\lambda + \lambda^*)\sigma_{BB}.$$

If $\sigma_{AB} = 0$, then λ is real and the equality in the Heisenberg relation (D.9) is satisfied.

D.2 Generating Functions

D.2.1 Hermite functions

The Hermite polynomials H_n are defined by the generating function

$$\exp(2z\xi - \xi^2) = \sum_{n=0}^{\infty} \frac{z^n}{n!} H_n(\xi). \tag{D.12}$$

The Hermite polynomials can be expanded as

$$H_n(\xi) = \sum_{k=0}^{[n/2]} \frac{(-1)^k (2\xi)^{n-2k}}{k!(n-2k)!}. \tag{D.13}$$

The first five polynomials are

$$H_0(\xi) = 1, \quad H_1(\xi) = 2\xi, \quad H_2(\xi) = 4\xi^2 - 2,$$
$$H_4(\xi) = 8\xi^3 - 12\xi, \quad H_5(\xi) = 16\xi^4 - 48\xi^2 + 12. \tag{D.14}$$

The polynomial H_n is a solution of Hermite's differential equation

$$\frac{d^2 H_n}{d\xi^2} - 2\xi \frac{dH_n}{d\xi} + 2nH_n = 0 \ . \tag{D.15}$$

The Hermite functions φ_n are defined by

$$\varphi_n(\xi) = \left(\sqrt{\pi}2^n n!\right)^{-1/2} H_n(\xi) \exp\left(-\frac{1}{2}\xi^2\right) , \tag{D.16}$$

and satisfy the differential equation associated with the quantum one-dimensional harmonic oscillator

$$-\frac{1}{2}\frac{d^2 \varphi_n}{d\xi^2} + \frac{1}{2}\xi^2 \varphi_n = \left(n + \frac{1}{2}\right)\xi_n \ . \tag{D.17}$$

The set of functions φ_n, $n = 0, 1, \ldots$, forms a complete, orthonormal system on the interval $(-\infty, \infty)$:

$$\int_{-\infty}^{\infty} \phi_{n\alpha}(\xi)\phi_{n'\alpha}(\xi)d\xi = \delta_{nn'} \ . \tag{D.18}$$

The generating function of the Hermite functions φ_n is

$$\pi^{-1/4} \exp\left(\sqrt{2}z\xi - \frac{1}{2}z^2 - \frac{1}{2}\xi^2\right) = \sum_{n=0}^{\infty} \frac{z^n}{\sqrt{n!}} \varphi_n(\xi) \ . \tag{D.19}$$

Using the generating function (D.19), Schrödinger constructed a Gaussian wave function from a suitable superposition of the stationary wave functions of the harmonic oscillator [123]. With time, the center of the Gaussian follows the classical motion and does not change its shape.

D.2.2 Laguerre Polynomials

The associated Laguerre polynomials L_n^α are defined by the generating function

$$(\xi z)^{-\alpha/2} \exp(z) J_\alpha(2\sqrt{\xi z}) = \sum_{n=0}^{\infty} \frac{z^n}{\Gamma(n+\alpha+1)} L_n^\alpha(\xi) , \tag{D.20}$$

where the Bessel function J_α is given by

$$J_\alpha(x) = \sum_{m=0}^{\infty} \frac{(-1)^m}{m!\Gamma(m+\alpha+1)} \left(\frac{x}{2}\right)^{\alpha+2m} . \tag{D.21}$$

318   D Generating Functions for Quantum States

The Laguerre polynomials can be expanded as

$$L_n^\alpha(\xi) = \sum_{k=0}^{n} \frac{\Gamma(n+\alpha+1)}{(n-k)!\Gamma(k+\alpha+1)} \frac{(-\xi)^k}{k!} . \tag{D.22}$$

The first three polynomials are

$$L_0^\alpha(\xi) = 1 , \quad L_1^\alpha(\xi) = \alpha + 1 - \xi ,$$
$$L_2^\alpha(\xi) = \frac{1}{2} \left[\xi^2 - 2\xi(\alpha+2) + \alpha^2 + 3\alpha + 2 \right] . \tag{D.23}$$

The polynomials L_n^α are solutions of Laguerre's differential equation

$$\xi \frac{d^2 L_n^\alpha}{d\xi^2} + (\alpha + 1 - x) \frac{dL_n^\alpha}{d\xi} + 2n L_n^\alpha = 0 . \tag{D.24}$$

Define the functions $\phi_{n\alpha}$ as

$$\phi_{n\alpha}(\xi) = \sqrt{2 \frac{n!}{\Gamma(n+\alpha+1)}} \xi^{\alpha+1/2} L_n^\alpha(\xi^2) \exp\left(-\frac{1}{2}\xi^2\right) . \tag{D.25}$$

The set of functions $\phi_{n\alpha}$, $n = 0, 1, \ldots$, forms a complete, orthonormal system on the interval $(0, \infty)$:

$$\int_0^\infty \phi_{n\alpha}(\xi) \phi_{n'\alpha}(\xi) d\xi = \delta_{nn'} . \tag{D.26}$$

The functions $\phi_{n\alpha}$ are solutions of the differential equation associated to the quantum singular oscillator:

$$\frac{1}{2} \left[-\frac{d^2}{d\xi^2} + \xi^2 + \left(\alpha^2 - \frac{1}{4}\right) \frac{1}{\xi^2} \right] \phi_{n\alpha} = (2n + \alpha + 1) \phi_{n\alpha} . \tag{D.27}$$

Moreover, the functions $\varphi_{n\alpha}$ defined by

$$\varphi_{n\alpha}(\xi) = \xi^{-1/2} \phi_{n\alpha}(\xi) \tag{D.28}$$

are solutions of the radial differential equation associated to the quantum two-dimensional isotropic harmonic oscillator:

$$\frac{1}{2} \left[-\frac{d^2}{d\xi^2} - \frac{1}{\xi} \frac{d}{d\xi} + \xi^2 + \frac{\alpha^2}{\xi^2} \right] \varphi_{n\alpha} = (2n + \alpha + 1) \varphi_{n\alpha} . \tag{D.29}$$

The generating function of the functions $\varphi_{n\alpha}$ is

$$\sqrt{2} z^{-\alpha} J_\alpha(2\xi z) \exp\left(z^2 - \frac{1}{2}\xi^2\right) = \sum_{n=0}^{\infty} \frac{z^{2n}}{\sqrt{n!\Gamma(n+\alpha+1)}} \varphi_{n\alpha}(\xi) . \tag{D.30}$$

D.3 Displacement Operators

Consider the Fock space characterized by the annihilation operators a and creation operator $a^\dagger$ for the harmonic oscillator. The number operator is defined by $N \equiv a^\dagger a$. Then

$$[N, a^\dagger] = a^\dagger , \quad [N, a] = -a , \quad [a^\dagger, a] = -1 . \tag{D.31}$$

The basis of the Fock space consists of the number state vectors

$$|n\rangle = \frac{1}{\sqrt{n!}} (a^\dagger)^n |0\rangle , \tag{D.32}$$

where n is a nonnegative integer and $|0\rangle$ is the vacuum state vector defined by

$$a|0\rangle = 0 , \quad \langle 0|0\rangle = 1 . \tag{D.33}$$

By (D.31)–(D.33), we have

$$a|n\rangle = \sqrt{n}|n-1\rangle , \quad a^\dagger|n\rangle = \sqrt{n+1}|n+1\rangle , \quad N|n\rangle = n|n\rangle . \tag{D.34}$$

These number state vectors satisfy the orthogonality and completeness conditions

$$\langle m|n\rangle = \delta_{mn} , \quad \sum_{n=0}^{\infty} |n\rangle\langle n| = 1 . \tag{D.35}$$

The normal form of the displacement operator can be written as

$$D(\alpha) = \exp(-|\alpha|^2/2) \exp(\alpha a^\dagger) \exp(-\alpha^* a) . \tag{D.36}$$

Expanding (D.36) in powers of α and $-\alpha^*$ we have

$$\langle m| D(\alpha) |n\rangle = \exp\left(-|\alpha|^2/2\right) \langle m| \exp(\alpha a^\dagger) \exp(-\alpha^* a) |n\rangle \tag{D.37}$$

$$= \exp\left(-|\alpha|^2/2\right) \sum_{p,q=0}^{\infty} \frac{(-\alpha^*)^p \alpha^q}{p!q!} \langle m| (a^\dagger)^q a^p |n\rangle .$$

Using $a|n'+1\rangle = \sqrt{n'+1}|n'\rangle$ and $a^+|n'\rangle = \sqrt{n'+1}|n'+1\rangle$ for any nonnegative integer n', we get

$$\langle m| (a^\dagger)^q a^p |n\rangle = \sqrt{\frac{m!n!}{(m-q)!(n-p)!}} \delta_{m-q, n-p} , \tag{D.38}$$

for $n \geq p$ and $m \geq q$. Moreover, $\langle m|(a^\dagger)^q a^p|n\rangle = 0$ for $n < p$ and $m < q$.

By (D.37) and (D.38), we obtain

$$\langle m| D(\alpha) |n\rangle = \exp(-|\alpha|^2/2) \alpha^{m-n} \sum_{p=0}^{n} \frac{\sqrt{m!n!}(-\alpha\alpha^*)^p}{p!(n-p)!(m-n+p)!} , \quad m \geq n , \tag{D.39}$$

D Generating Functions for Quantum States

$$\langle m| D(\alpha) |n\rangle = \exp(-|\alpha|^2/2)(-\alpha^*)^{n-m} \sum_{q=0}^{m} \frac{\sqrt{m!n!}\,(-\alpha\alpha^*)^q}{q!(n-p)!(m-n+p)!} \,, \quad n \geq m\,. \tag{D.40}$$

Then the matrix elements of the displacement operator $D(\alpha)$ (see (D.39) and (D.40)) are given by Schwinger's formulae [230, 438, 439]

$$\langle m| D(\alpha) |n\rangle = \sqrt{\frac{n!}{m!}} \alpha^{m-n} L_n^{m-n}(|\alpha|^2) \exp\left(-|\alpha|^2/2\right) \,, \quad m \geq n\,, \tag{D.41}$$

$$\langle m| D(\alpha) |n\rangle = \sqrt{\frac{m!}{n!}} (-\alpha^*)^{n-m} L_m^{n-m}(|\alpha|^2) \exp\left(-|\alpha|^2/2\right) \,, \quad n > m\,,$$

where m and n are nonnegative integers. Here the associated Laguerre polynomial L_n^s is defined by

$$L_n^s(x) = \sum_{k=0}^{n} \frac{(n+s)!(-x)^k}{k!(n-k)!(k+s)!} \,. \tag{D.42}$$

In particular

$$\langle n| D(\alpha) |n\rangle = L_n(|\alpha|^2) \exp\left(-|\alpha|^2/2\right) \,, \tag{D.43}$$

where $L_n = L_n^0$ is the Laguerre polynomial.

D.4 Time Dependent Oscillators

The quantum motion of a charged particle in a quadrupole combined trap can be described in terms of three parametric oscillators.

Consider the parametric oscillator [86] described by the Schrödinger equation

$$\left(i\frac{\partial}{\partial \tau} - H\right)\psi = 0 \,, \tag{D.44}$$

with the Hamiltonian given by

$$H = -\frac{1}{2}\frac{\partial^2}{\partial q^2} + \frac{1}{2}g(\tau)q^2 \,, \tag{D.45}$$

where g is a π-periodic function of τ.

D.4.1 Gaussian Packets

The Gaussian solution of (D.44) is given by

$$\psi_n(q,\tau) = \exp\left[-i\left(n+\frac{1}{2}\right)\gamma\right] \tilde{\psi}_n(q,\tau) \,, \tag{D.46}$$

D.4 Time Dependent Oscillators

$$\tilde{\psi}_n(q,\tau) = \frac{1}{\sqrt{\sqrt{\pi}2^n n!\rho}} H_n\left(\frac{q}{\rho}\right)\exp\left[-\frac{q^2}{2\rho^2}\left(1-\rho\frac{d\rho}{d\tau}\right)\right], \tag{D.47}$$

where H_n is the Hermite polynomial of degree n. Here $\rho = |w|$ and $w = \rho\exp(i\gamma)$ is a stable solution of the Hill equation

$$\frac{d^2 w}{d\tau^2} + g(\tau)w = 0, \tag{D.48}$$

with the Wronskian

$$w^*\frac{dw}{d\tau} - w\frac{dw^*}{d\tau} = 2i. \tag{D.49}$$

Then $d\gamma/d\tau = \rho^{-2}$. According to Floquet's theorem, we can write $w = v\exp(i\beta\tau)$, where v is a π-periodic function of τ and the characteristic exponent β is given by

$$\beta = \frac{1}{\pi}\int_0^\pi \frac{1}{\rho^2}d\tau. \tag{D.50}$$

The quasienergy solutions ψ_n of (D.44) have the scaled quasienergies $(n+1/2)\beta$:

$$\psi_n(q,\tau+\pi) = \exp\left[-i\pi\beta\left(n+\frac{1}{2}\right)\right]\psi_n(q,\tau). \tag{D.51}$$

A Gaussian packet for the parametric oscillator is represented by the generating function

$$\psi^{(\alpha)}(q,\tau) = \exp\left[-\frac{1}{2}|\alpha|^2\right]\sum_{n=0}^\infty \frac{\alpha^n}{\sqrt{n!}}\psi_n(q,\tau), \tag{D.52}$$

where α is a complex parameter. Using (D.46), we have

$$\psi^{(\alpha)}(q,\tau) = \frac{1}{\sqrt{2\pi\sigma_{qq}}}\exp\left[-\frac{1-i\sigma_{pq}}{4\sigma_{qq}}(q-q_c)^2 + ip_c\left(q-\frac{q_c}{2}\right) - \frac{i}{2}\gamma\right], \tag{D.53}$$

where

$$q_c = \langle q \rangle = \frac{1}{\sqrt{2}}(\alpha w^* + \alpha^* w),$$

$$p_c = \langle p \rangle = \frac{1}{\sqrt{2}}\left(\alpha\frac{dw^*}{d\tau} + \alpha^*\frac{dw}{d\tau}\right), \tag{D.54}$$

$$\sigma_{qq} = \langle q^2 \rangle - \langle q \rangle^2 = \frac{\rho^2}{2},$$

$$\sigma_{qp} = \frac{1}{2}\langle qp + pq \rangle - \langle q \rangle\langle p \rangle = \frac{\rho}{2\delta}\frac{d\rho}{d\tau}. \tag{D.55}$$

Here $p = -i\partial/\partial q$ and $\langle\,\rangle$ means the expectation value with respect to $\psi^{(\alpha)}$.

D.4.2 Linear Invariants

In [440,441] suitable time dependent invariants, linear in position and momentum operators, have been used to rederive the Gaussian solution of (D.44). Consider the time-invariant operator

$$A = \frac{1}{\sqrt{2}} \left(wp - \frac{dw}{d\tau} q \right) , \tag{D.56}$$

for which the boson commutation relation $[A, A^\dagger] = 1$ is fulfilled, and $\dot{A} = 0$. Then A is a constant of the motion, which can be used to write the solutions of (D.44) in coherent states, as

$$|\alpha, \tau\rangle = \exp\left(\alpha A - \alpha^* A^\dagger\right) |0, \tau\rangle . \tag{D.57}$$

Here $A|0, \tau\rangle = 0$ and $\langle 0, \tau|0, \tau\rangle = 1$.

In the coordinate representation, the coherent state given by (D.57) is exactly the Gaussian packet (D.52).

D.4.3 Quadratic Invariants

The general solution of the Schrödinger equation (D.44) can be expressed in terms of the eigenstates of the quadratic invariants up to suitable time-dependent phase factors. We consider the operator [88]

$$I = \frac{1}{2} \left[\left(\rho p - \frac{d\rho}{dt} q \right)^2 + \frac{1}{\rho^2} q^2 \right] , \tag{D.58}$$

where ρ is a solution of the Ermakov equation [238]

$$\frac{d^2 \rho}{dt^2} + g(\tau)\rho = \frac{1}{\rho^3} . \tag{D.59}$$

According to (D.58) and (D.59), the operator I is a quantum constant of motion for the Hamiltonian (D.45). The quadratic invariant (D.58) of Lewis and Riesenfeld is the quantum counterpart of the classical Ermakov–Lewis invariant [238].

We define the lowering and raising operators

$$b = \frac{1}{\sqrt{2}} \left[\frac{1}{\rho} q + i(\rho p - \frac{d\rho}{dt} q) \right] , \quad b^\dagger = \frac{1}{\sqrt{2}} \left[\frac{1}{\rho} q - i(\rho p - \frac{d\rho}{dt} q) \right] , \tag{D.60}$$

which satisfy the boson commutation relations $[b, b^\dagger] = 1$. Then (D.58) can be rewritten

$$I = \frac{1}{2} \left(2 b^\dagger b + 1 \right) . \tag{D.61}$$

We now introduce the number state vectors

$$|n,t\rangle = \frac{1}{\sqrt{n!}} \left(b^\dagger\right)^n |0,t\rangle , \qquad (D.62)$$

where $b|0,t\rangle = 0$ and $|0,t\rangle$ is the normalized state vector. These vectors are eigenvectors of I:

$$I|n,t\rangle = \left(n + \frac{1}{2}\right)|n,t\rangle . \qquad (D.63)$$

The quasienergy solutions of (D.44) can be written as

$$\psi_n(q,t) = \exp(i\alpha_n)\langle q|n,t\rangle , \qquad (D.64)$$

where $\alpha_n = (2n+1)\varphi$, $\varphi = \varphi_\mathrm{d} + \varphi_\mathrm{g}$ and

$$\frac{\mathrm{d}\varphi}{\mathrm{d}\tau} = -\frac{1}{2\rho^2} . \qquad (D.65)$$

The dynamical phase φ_d and the geometrical phase factor φ_g are given by

$$\frac{\mathrm{d}\varphi_\mathrm{d}}{\mathrm{d}\tau} = -\langle n,t|H|n,t\rangle = -\frac{1}{4}\left[\left(\frac{\mathrm{d}\rho}{\mathrm{d}\tau}\right)^2 + g(\tau)\rho^2 + \frac{1}{\rho^2}\right] , \qquad (D.66)$$

$$\frac{\mathrm{d}\varphi_\mathrm{g}}{\mathrm{d}\tau} = \mathrm{i}\left\langle n,t\left|\frac{\partial}{\partial\tau}\right|n,t\right\rangle = \frac{1}{4}\left[\left(\frac{\mathrm{d}\rho}{\mathrm{d}\tau}\right)^2 - \rho\frac{\mathrm{d}^2\rho}{\mathrm{d}\tau^2}\right] . \qquad (D.67)$$

D.5 Coherent States for Symplectic Groups

D.5.1 $Sp(2,R)$ Coherent States

The basis of the $Sp(2,R)$ Lie algebra consists of three generators K_0, K_1, and K_2 such that

$$[K_0, K_1] = \mathrm{i}K_2 , \quad [K_2, K_0] = \mathrm{i}K_1 , \quad [K_2, K_1] = \mathrm{i}K_0 . \qquad (D.68)$$

The raising and lowering operators $K_\pm = K_1 \pm \mathrm{i}K_2$ satisfy the commutation relations

$$[K_0, K_\pm] = \pm K_\pm , \quad [K_-, K_+] = 2K_0 . \qquad (D.69)$$

The Casimir operator $C = K_0^2 - K_1^2 - K_2^2$ has eigenvalues $k(k-1)$, where k is the Bargmann index for the unitary irreducible representations of $Sp(2,R)$. For the positive discrete series $2k$ is integer and $k \geq 1$. For the universal covering group representations k is a nonnegative integer [439].

We now recall the construction of the canonical basis for the lowest weight representation space of $Sp(2,R)$ with a fixed positive Bargmann index k [439]. The orthonormal canonical basis consisting of the vectors

$$\phi_{km} = \left[\frac{\Gamma(2k)}{m!\Gamma(2k+m)}\right]^{1/2} (K_+)^m \phi_{k0} , \qquad (D.70)$$

where m is a positive integer and the normalized fundamental vector ϕ_{k0} is characterized by

$$K_0 \phi_{k0} = k\phi_{k0} , \qquad K_- \phi_{k0} = 0 . \qquad (D.71)$$

The action of the generators in the canonical basis is given by

$$\begin{aligned} K_0 \phi_{km} &= (k+m)\phi_{km} , \\ K_+ \phi_{km} &= [(m+1)(m+2k)]^{1/2} \phi_{k\,m+1} , \\ K_- \phi_{km} &= [m(m+2k-1)]^{1/2} \phi_{k\,m-1} , \quad m > 0 . \end{aligned} \qquad (D.72)$$

We now introduce the unitary operators

$$U(z) = \exp(zK_+) \exp(\eta K_0) \exp(-z^* K_-) , \qquad (D.73)$$

where $\eta = \ln(1-zz^*)$ and z is the complex coordinate in the unit disc $|z| \leq 1$. We introduce the following symplectic coherent states for $Sp(2, \mathbf{R})$:

$$\phi_{km}(z) = U(z)|\phi_{km}\rangle . \qquad (D.74)$$

In the case $m = 0$, we obtain the standard geometrical construction of the $Sp(2, \mathbf{R})$ coherent states with the control parameter space given by the unit disc $|z| < 1$ [439].

D.5.2 Linear Dynamical Systems

We consider the $Sp(2, \mathbf{R})$ linear system associated with the dynamical symplectic group $Sp(2, \mathbf{R})$ described by the Hamiltonian

$$H = \hbar (aK_0 + bK_1 + cK_2) , \qquad (D.75)$$

where a, b and c are time-dependent functions. The vector

$$\psi_{km}(z) = \exp(-i\varphi_{km}) \phi_{km} , \qquad (D.76)$$

evolves according to the Schrödinger equation

$$\left(i\hbar \frac{d}{dt} - H\right) \psi_{km}(z) = 0 , \qquad (D.77)$$

where the complex coordinate z and the phase $\varphi_{km} = (k+m)\varphi$ are time-dependent functions satisfying the differential equations

$$i\frac{dz}{dt} = az + \frac{b}{2}(z^2+1) + \frac{ic}{2}(z^2-1) . \qquad (D.78)$$

D.5 Coherent States for Symplectic Groups

$$\frac{d\varphi}{dt} = a + \frac{b}{2}(z + z^*) + \frac{ic}{2}(z - z^*) . \tag{D.79}$$

The unit disc $|z| < 1$ can be mapped onto the Poincaré half plane $\operatorname{Im} \tilde{z} > 0$ by the Cayley transformation $\tilde{z} = (i - z)/(i + z)$. Then (D.78) reduces to the equation

$$\hbar \frac{d\tilde{z}}{dt} = \tilde{z}^2(b - a) + 2c\tilde{z} - a - b . \tag{D.80}$$

The equation (D.80) is linear for $a = b$. If $a \neq b$, then the Riccatti equation (D.80) can be linearized by substituting

$$\tilde{z} = \frac{2}{(b-a)u} \frac{du}{dt} , \tag{D.81}$$

where u satisfies the linear differential equation

$$\frac{d^2 u}{dt^2} + 2c \frac{du}{dt} + \frac{1}{4}\left(b^2 - a^2\right) u = 0 . \tag{D.82}$$

Then the solutions of the Schrödinger equation (D.77) can be written as

$$\psi_{k\,m}(z) = \exp\left[-i\left(k + m\right)\varphi\right] \phi_{k\,m} , \tag{D.83}$$

where φ is obtained from (D.79) and

$$w = \left[(a - b)u - 2i\frac{du}{dt}\right]\left[(a - b)u + 2i\frac{du}{dt}\right]^{-1} , \tag{D.84}$$

with u given by (D.82).

If a and b are time-periodic functions and $c = 0$, then (D.82) is a Hill equation:

$$\frac{d^2 u}{dt^2} + \frac{1}{4}\left(b^2 - a^2\right) u = 0 . \tag{D.85}$$

The quasienergy corresponding to the quasienergy state vector (D.83) is given by

$$E_{km} = 2\hbar\mu(k + m) , \quad m = 0, 1, \ldots , \tag{D.86}$$

where μ is the Floquet exponent for the solution u of (D.85) in a stability region.

E Trap Design and Electronics

One of the early designs of a quadrupole ion trap with ion source still in use is shown in Fig. E.1. The ions of most metallic elements, having a relatively high evaporation point are created inside the trap by the ionization of atoms in a beam intersecting at the trap center an electron or other ionizing beam. Oven sources for atomic beams of the alkaline earth elements for example, are loaded with Mg, Ca, Ba, etc., in metallic form, commonly with natural isotopic abundances, and temperature controlled to operate at the appropriate temperatures, typically a few hundred degrees Celsius. At the operating temperatures, evaporated atoms effuse out through beam-defining apertures and are further collimated by sets of skimmers and apertures, mounted between the oven and the trap. Ions are commonly produced by electron impact ionization using a 1 keV electron beam from an electron gun, or by a photoionizing laser beam. In both cases ionization leads to subsequent trapping of those ions produced at sufficiently low energy. The experimental arrangement of the oven and the electron gun can be seen in Fig. E.2.

Elements that have a relatively high vapor pressure at room temperature (in the context of ultrahigh vacuum) such as Hg and Cs, clearly require a "cold trap" or some type of "getter" to remove atoms after they have passed through the trap. Gaseous elements such as He may simply fill the vacuum system at the desired low pressure.

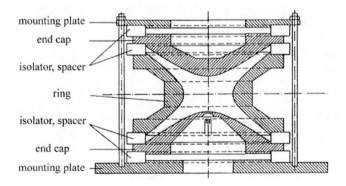

Fig. E.1. Section through the trap electrodes [56]

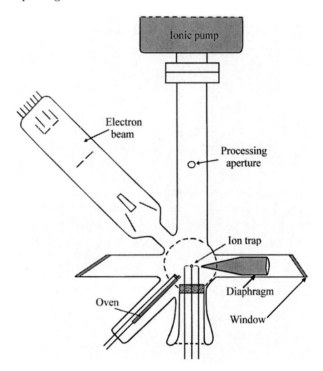

Fig. E.2. Miniaturized ion trap, with an oven and electronic gun as ion source [32]

In order to allow access to the interior of the trap for the purposes of its loading with ions and their detection, holes must be bored in the trap electrodes. For example, the ring electrode often has two opposing holes bored into it, in order to enable (laser) radiation to enter and emerge from the trap for different spectroscopic experiments. Also, ion detection by means of resonance fluorescence requires a wide angle aperture detector, requiring many closely spaced small holes, perhaps several hundred, to be drilled in one end cap electrode (see for example Fig. 1.4). Moreover, in some cases the ion source takes the form of two heated Pt strips mounted parallel to each other, and covered with the desired solution; in this case one end cap has two slits through which the strips pass (Fig. 5.4).

For ion/electron trapping, electromagnetic traps are placed in an ultra high vacuum chamber equipped with an ion getter pump, specifically for each experiment. Two examples of such a set-up are given in Figs. E.2 and E.3. An ion gauge and a residual gas analyzer are usually included for pressure monitoring; normal operating pressure is about 10^{-8} Pa.

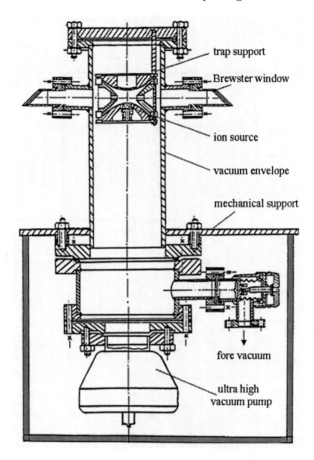

Fig. E.3. Ultrahigh vacuum set-up for ion trap. As ion source, two paralel connected filaments are mounted in the lower end cap electrode of the trap are presented. Brewster windows allow laser experiments with trapped ions

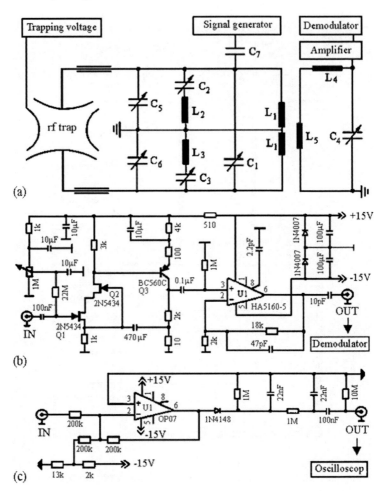

Fig. E.4. (a) Experimental arrangement for trapped ion electronic detection in a trap of 1 cm end cap distance. $C_1 = 40\text{--}70\,\text{pF}$, $C_2 = 1.5\text{--}18\,\text{pF}$, $C_3 = 1.5\text{--}18\,\text{pF}$, $C_4 = 4.5\text{--}70\,\text{pF}$, $C_5 = C_6 = 60\,\text{pF}$, $C_7 = 1\,\text{pF}$, $L_1 = 57\,\text{mH}$, $L_2 = L_3 = 2.5\,\text{mH}$, $L_4 = 100\,\text{mH}$, $L_5 = 0.6\,\text{mH}$; (b) amplifier; (c) demodulator

F Charged Microparticle Trapping

The method of trapping charged particles with time-varying electric fields, as in Paul traps, is not restricted to atomic or molecular ions, but applies equally to charged microparticles. Since these particles can be viewed by the naked eye and the set-up can be realized by simple means and at low cost, this technique is well suited for teaching purposes. The first demonstration of trapped microparticles was performed as early as 1955 by Straubel [163], who confined charged oil drops in a simple ac quadrupole trap operated in air. Wuerker and Langmuir [39] used charged aluminium particles trapped in a low frequency ac electric field and took photographs of the Lissajous-like trajectories of a single particle in vacuum. A few Al-particles formed crystal-like structures, very much like those observed under strong confinement conditions on atomic ions (see Sect. 18.3). Winter and Ortjohann have described a simple device, similar to a Paul trap, in which it is reported that a single dust particle was confined for over two months, and that ordered structures are formed when working with many particles in air [59].

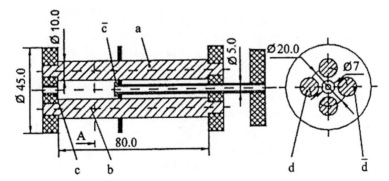

Fig. F.1. Sketch of the linear Paul trap geometry from Fig. F.3 (*left*). The trap consists of four brass rods $a, b, d, \bar{d}$ equidistantly spaced on an approximately 1 cm radius, and two end caps $c, \bar{c}$. The trap length is variable, because one of the end caps $\bar{c}$ is a piston cylinder sliding on the brass bars. To illuminate trapped microparticles with a low power He-Ne laser or with a diode laser travelling along the trap axis, both trap end caps are pierced, thus allowing laser beam incidence and emergence [60]

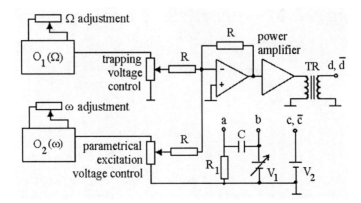

Fig. F.2. Electrical supply scheme for linear microparticle trap from Fig. F.1. The O_1 oscillator delivers an V_{ac} variable voltage of Ω drive frequency, which is applied between opposed pairs of rods in order to achieve trapping. To avoid microparticle escape along the axis, two brass disks $c, \bar{c}$ with a V_2 dc bias (0–500 V) are provided. The O_2 oscillator delivers an ac voltage of ω frequency, for parametrical excitation of microparticle motion, allowing thus their diagnosis. Both voltage amplitudes are variable. A dc variable voltage V_1 (0–500 V) vertically shifts the microparticle position inside the trap [61]

Almost any geometry of trap electrodes with ac voltages applied between them, producing a saddle point in the potential, will provide a pseudo-potential minimum in which charged particles can be trapped. Expansion of the potential in spherical harmonics (see Sect. 4.4) shows that in first order of approximation a quadrupole potential is created, as in an ideal Paul trap. For small amplitudes of the trapped particle motion the higher order components can be neglected.

A simple arrangement most similar to a linear Paul trap would be four cylindrical rods as shown in Fig. F.1. The oppositely placed rods are electrically connected and an ac potential of 1–3 kV amplitude and 20–800 Hz frequency applied to them. This serves for radial confinement of particles having the proper Q/M ratio. Axial confinement is achieved by a dc potential applied to end electrodes on the trap axis (Fig. F.2).

Figure F.3 shows an array of nine particles suspended in this trap in air.

A device similar to the three dimensional Paul trap would consist of a ring electrode and end caps formed of spheres (Fig. F.4), or flat electrodes (Fig. F.3, right).

As shown in the right hand part of Fig. F.3 crystal-like arrangements of dust particles in air are observed. By variation of the air pressure, the damping of the particle motion can easily be varied. The damping constant b is related to the viscosity η at the density ρ of the air by

$$b = \frac{9\eta}{\rho R^2 \Omega}, \tag{F.1}$$

Fig. F.3. Set up for trapping microparticles in air in standard conditions of pressure and temperature. Visible to the naked eye are very stable microparticle ordered structures in two Paul trap geometries: (**a**) linear Paul trap (*left*); (**b**) cylindrical trap (*right*) [442]

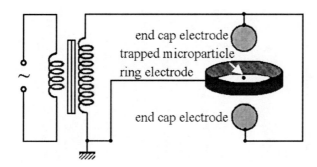

Fig. F.4. Experimental setup for microparticle trapping in air. The ac-voltage is generated with a high-voltage transformer, directly connected to the *line* ($V_{ac} = 1500$ V, $\Omega = 2\pi \times 50$ Hz)

where R is the radius of the dust particle. Solution of the Mathieu differential equation of motion including a damping term by a Runge–Kutta method [443] shows that the damping increases the stability range of the trap. For example, the maximum stability parameter q_{max}, where $q = QV/(Mr_0^2\Omega^2)$, changes from $q_{max} = 0.908$ at $b = 0$ to $q_{max} = 2.13$ at $b = 1.8$, as was experimentally confirmed [59].

References

1. W. Paul: Nobel Lecture Rev. Mod. Phys. **62**, 531 (1990)
2. H.G. Dehmelt: Rev. Mod. Phys. **62**, 525 (1990)
3. P.H. Dawson: Quadrupole Mass Spectrometry and its applications (Elsevier, Amsterdam 1976)
4. *Practical Aspects of Ion Trap Mass Spectrometry*, Vols. 1–3, 2nd edn, Ed. by R.E. March and J.F. Todd (CRC Press, Boca Raton, Florida, 1995)
5. T.M. O'Neil: *Plasma with a Single Sign of Charge*. In *Non-neutral Plasma Physics*, ed. by C.W. Robertson and C.F. Driscoll. (AIP, New York 1988)
6. D. Leibfried, R. Blatt, C. Monroe, D. Wineland: Rev. Mod. Phys. **75**, 281 (2003)
7. L.S. Brown and G. Gabrielse: Rev. Mod. Phys. **58**, 233 (1986)
8. H. Walther: Adv. At. Mol. Opt. Phys. **31**, 137 (1993)
9. *Trapped Charged Particles and Fundamental Physics*, ed. by I. Bergstroem, C. Carlberg, and R. Schuch (World Scientific, **where?** 1995)
10. D.H. Dubin and D. Schneider: (AIP 1999)
11. M. Gross, D. Habs, W. Lange, and H. Walther: (IOP 2003)
12. F.M. Penning: Physica **3**, 873 (1936)
13. J.R. Pierce: *Theory and Design of Electron Beams*, 2nd edn. (van Nostrand, Princeton 1954) p. 40
14. H. Friedburg, W. Paul: Naturwissenschaft **38**, 159 (1951)
15. H.G. Benewitz, W. Paul: Z. Physik **139**, 489 (1954)
16. E.D. Courant, M.S. Livingston, H.S. Snyder: Phys. Rev. **88**, 1190 (1952)
17. J.R. Pierce: *Theory and Design of Electron Beams*, 2nd edn. (van Nostrand, Princeton 1954) p. 194
18. W. Paul, H. Steinwedel: Z. Naturforschung **8a**, 448 (1953)
19. W. Paul, M. Raether: Z. Physik **140**, 262 (1955)
20. W. Paul, O. Osberghaus, E. Fischer: Forsch. Berichte des Wirstschaftsministeriums Nordrhein-Westfalen Nr. **415**, (1958)
21. H.G. Dehmelt, F.G. Major: Phys. Rev. Lett. **8**, 213 (1962)
22. H.G. Dehmelt, K.B. Jefferts: Phys. Rev. **125**, 1318 (1962)
23. E.N. Fortson, F.G. Major, H.G. Dehmelt: Phys. Rev. Lett. **16**, 221 (1966)
24. G. Gräff, F.G. Major, R.W.H. Roeder, G. Werth: Phys. Rev. Lett. **21**, 340 (1968)
25. D. Church, H.G. Dehmelt: J. Appl. Phys. **40**, 3421 (1969)
26. F.G. Major, G. Werth: Phys. Rev. Lett. **30**, 1155 (1973)
27. D. Wineland, P. Ekstrom, H.G. Dehmelt: Phys. Rev. Lett. **31**, 1279 (1973)
28. P. van Dyck, Jr., P. Ekstrom, H.G. Dehmelt: Nature **262**, 776 (1976)
29. R. Ifflaender, G. Werth: Metrologia **13**, 167 (1977)

30. T.W. Hänsch, A.L. Schawlow: Opt. Commun **13**, 68 (1975)
31. D.J. Wineland, H.G. Dehmelt: Bull. APS **20**, 637 (1975)
32. W. Neuhauser, M. Hohenstatt, P.E. Toschek, H.G. Dehmelt: Phys. Rev. Lett. **41**, 233 (1978); Appl. Phys. **17**, 123 (1978)
33. W. Nagourney, J. Sandberg, H. G. Dehmelt: Phys. Rev. Lett. **56**, 2797 (1986)
34. H.G. Dehmelt: "Invariant frequency ratios in electron and positron geonium spectra yield refined data on electron structure". In *Proc. of the Int. Conf. on Atomic Physics, Atomic Physics 7, Cambridge, Massachusetts, 1980*, ed. by D. Kleppner, F.M. Pipkin (Plenum Press, New York London 1981) pp. 337–340
35. H. Häffner et al.: Physica Scripta **46**, 581 (1992)
36. T. Otto et al.: Nucl. Phys. A **567**, 281 (1994)
37. F. Diedrich, E. Peik, J.M. Chen, W. Quint, H. Walther: Phys. Rev. Lett. **59**, 2931 (1987)
38. D.J. Wineland, J.C. Bergquist, W.M. Itano, J.J. Bollinger, C.H. Manney: Phys. Rev. Lett. **59**, 2935 (1987)
39. R.F. Wuerker, H. Shelton, R.V. Langmuir: J. Appl. Phys. **30**, 342 (1959)
40. G. Gabrielse et al.: Phys. Letters B **507**, 1 (2001)
41. F. Diedrich, J.C. Berquist, W.M. Itano, D.J. Wineland: Phys. Rev. Lett. **62**, 403 (1989)
42. D.J. Berkeland, J.D. Miller, J.C. Bergquist, W.M. Itano, D.J. Wineland: Phys. Rev. Lett. **80**, 2089 (1998)
43. U. Tanaka et al.: J. Phys. B **36**, 345 (2003)
44. S.A. Diddams et al.: Science **293**, 825 (2001)
45. E. Schrödinger: Naturwissenschaft **23**, 807 (1935)
46. B.E. King et al.: Phys. Rev. Lett. **81**, 1525 (1998)
47. L.D. Landau, E.M. Lifshitz: *Mechanics*. 2nd edn. (Pergamon Press, 1969) p. 93
48. J.D. Jackson: *Classical Electrodynamics* 2nd edn. (Wiley, New York 1975) p. 132
49. W. Paul, H. Steinwedel: Z. Naturforsch. **A8**, 448 (1953)
50. W. Paul, H. Steinwedel: German Patent No. 944900; U.S. Patent No. 2939952 (1953)
51. R.D. Knight: Int. J. Mass Spectr. Ion Proc **51**, 127 (1983)
52. H. Dehmelt: Advan. At. Mol. Phys. **3**, 53 (1967), ed. by D.R. Bates, I. Estermann (Academic, New York 1967)
53. W. Paul: Rev. Mod. Phys. **62**, 531 (1990)
54. J. Meixner, F.W. Schäffke: *Mathieu'sche Funktionen and Sphäroidfunctionen* (Springer, Berlin Heidelberg 1954)
55. N.W. McLachlan: *Theory and Application of Mathieu Functions* (Clarendon, Oxford 1947)
56. E. Fischer: Z. Phys. **156**, 1 (1959)
57. X.Z. Chu, M. Holzki, R. Alheit, G. Werth: Int. J. Mass Spectrom. Ion Processes **173**, 107 (1998)
58. M. Baril, A. Septier: Rev. Phys. Appl. **9**, 525 (1974)
59. H. Winter, H.W. Ortjohann: Am. J. Phys. **59**, 807 (1991)
60. V.N. Gheorghe, L.C. Giurgiu, O.S. Stoican, B.M. Mihalcea, D.M. Cacicovschi, S. Comanescu: "Linear microparticle trap operating in air". In: *CPEM'96 Digest, Conference on Precision Electromagnetic Measurements, Braunschweig, Germany, 17–20 June 1996.* Ed. by A. Braun, (Institute of Electrical and Electronics Engineers, 1996) pp. 304–305

61. V.N. Gheorghe, L.C. Giurgiu, O.S. Stoican, B.M. Mihalcea, D.M. Cacicovschi, R.V. Molnar: Acta Phys. Polonica A **93**, 1105 (1998)
62. M. Kretzschmar: Z. Naturf. **45a**, 965 (1990)
63. R.D. Knight, M.H. Prior: J. Appl. Phys.**50**, 3044 (1979)
64. H. Schaaf, U. Schmeling, G. Werth: Appl. Phys. **25**, 249 (1981)
65. R. Blatt, P. Zoller, G. Holzmueller, I. Siemers: Z. Phys. D **4**, 121 (1986)
66. I. Siemers, R. Blatt, T. Sauter, W. Neuhauser: Phys. Rev. A **38**, 5121 (1988)
67. Y. Moriwaki, M. Tachikawa, T. Shimizu: Jpn. J. Appl. Phys. **35**, 757 (1996)
68. J. Yu, M. Desaintfuscien, F. Plumelle: Appl. Phys. B **48**, 51 (1989)
69. F. Vedel, J. André: Phys. Rev. A **29**, 2098 (1984)
70. C. Meis, M. Desaintfuscien, M. Jardino: Appl. Phys. B **45**, 59 (1988)
71. R. Alheit, X.Z. Chu, M. Hoefer, M. Holzki, G. Werth: Phys. Rev. A **56**, 4923 (1997)
72. G. Kotowski: Z. Angew. Math., Mech. **23**, 213 (1943)
73. Y. Wang, F. Franzen, K.P. Wanczek: Int. Journ. Mass Spectr. Ion Proc. **124**, 125 (1993)
74. F. Guidugli, P. Traldi: Rapid Comm. Mass Spectrom. **5**, 343 (1991)
75. D.M. Eades, J.V. Johnsen, R.A. Yost: J. Am. Soc. Mass Spectrom. **4**, 917 (1993)
76. F.G. Major, H.G. Dehmelt: Phys. Rev. **170**, 91 (1968)
77. D.C. Burnham: Spin Exchange with Trapped Ions. PhD Thesis, Harvard University, 1967
78. J. André: J. de Physique **37**, 719 (1976)
79. M.N. Gaboriaud, M. Desaintfuscien, F.G. Major: Int. J. Mass Spectr. Ion Physics **41**, 109 (1981)
80. Y. Moriwaki, T. Shimizu: Jpn. J. Appl. Phys. **37**, 344 (1998)
81. T. Hasegawa, K. Uehara: Appl. Phys. B **61**, 159 (1995)
82. M. Nasse, C. Foot: Eur. J. Phys. **22**, 563 (2001)
83. G. Tommaseo et al.: Eur. Phys. J. D (in print)
84. D. Leibfried, R. Blatt, C. Monroe, D. Wineland: Rev. Mod. Phys.: **75**, 281, (2003)
85. P.K. Ghosh: *Ion traps* (Clarendon Press, Oxford 1995)
86. K. Husimi: Prog. Theor. Phys. **9**, 238 (1953)
87. A.M. Perelomov, V.S. Popov: Teor. Mat. Fiz. **1**, 360 (1969)
88. H.R. Lewis, W.B. Riesenfeld: J. Math. Phys. **10**, 1458 (1969)
89. M.I. Malkin, D.A. Trifonov: Phys. Lett. **30**, 127 (1974)
90. M. Combescure: Ann. Inst. Henry Poincaré **44**, 293 (1986)
91. L.S. Brown: Phys. Rev. Lett. **66**, 527 (1991)
92. R.J. Glauber: "Recent Developments in Quantum Optics". In *Proceedings of the International Conference on Quantum Optics, Hyderabad, India, January 1991*, ed. by R. Inguva (Plenum Press, New York 1993) pp. 1
93. S. Stenholm: J. Mod. Optics, **39**, 279 (1992)
94. V.N. Gheorghe, F. Vedel: Phys. Rev. A **45**, 4828 (1992)
95. G. Schrade, V.I. Man'ko, W. Schleich, R.J. Glauber: Quantum Semiclass. Opt. **7**, 307 (1995)
96. M.M. Nieto, D.R. Truax: New J. Phys. **2**, 181 (2000)
97. V.N. Gheorghe, G. Werth: Eur. Phys. Journal D **10**, 197 (2000)
98. E. Schrödinger: Sitzungb. Preuss. Akad. Wiss., 221 (1930)
99. R.J. Cook, D.G. Shankland, A.L. Wells: Phys. Rev A **31**, 564 (1985)

100. M. Levi: Physica D **132**, 150 (1999)
101. M.C. Abbati, R. Cirelli, P. Lanzavecchia, A. Manià: Nuovo Cimento B **83**, 43 (1984)
102. P. Paasche et al.: Eur. Phys. J. D **18**, 295 (2002)
103. G. Gräff, E. Klempt: Z. Naturforschg. **22a**, 1960 (1967)
104. L. Brown, G. Gabrielse: Rev. Mod. Phys. **58**, 233 (1986)
105. G. Bollen, R.B. Moore, G. Savard, H. Stolzenberg: J. Appl. Phys. **68**, 4355 (1990)
106. C. Gerz, D. Wilsdorf, G. Werth: Nucl. Instr. Meth. B **47**, 453 (1990)
107. V.I. Arnold: *Mathematical methods of classical mechanics*. 2nd edn. (Springer, Berlin Heidelberg New York 1983)
108. V.N. Gheorghe, A. Gheorghe, G. Werth: to be published
109. L.S. Brown, G. Gabrielse: Phys. Rev. A **25**, 2423 (1982)
110. V.N. Gheorghe, A. Gheorghe, G. Werth: to be published
111. R.S. Van Dyck, Jr., F.L. Moore, D.L. Farnham, P.B. Schwinberg: Phys. Rev. A **40**, 6308 (1989)
112. J.V. Porto: Phys. Rev. A **64**, 023403 (2001)
113. P. Paasche, C. Angelescu, S. Ananthamurthy, D. Biswas, T. Valenzuela, G. Werth: Eur. Phys. J. D **22**, 183 (2003)
114. L. Schweikhard, J. Ziegler, H. Bopp, K. Lützenkirchen: Int. J. Mass Spectrom. and Ion Processes **141**, 77 (1995)
115. E.A. Beaty: Phys. Rev. A **33**, 3645 (1986)
116. G. Gabrielse: Phys. Rev. A **27**, 2277 (1983)
117. W.M. Itano, D.J. Wineland: Phys. Rev. A **25**, 35 (1982)
118. V. Fock: Z. Phys. **47**, 446 (1928)
119. C.G. Darwin: Proc. Cambridge Phil. Soc. **27**, 86 (1930)
120. A.A. Sokolov, Y.G. Pavlenko: Opt. Spectosc. (USSR) **22**, 1 (1967)
121. A. Feldman, A. Kahn: Phys. Rev. B **1**, 4584 (1970)
122. G. Gräff, E. Klempt, G. Werth: Z. Phys. **222**, 201 (1969)
123. E. Schrödinger: Naturwiss. **14**, 664 (1926)
124. B. Thaller: *The Dirac Equation* (Springer, Berlin Heidelberg New York 1992)
125. L.L. Foldy, S.A. Wouthuysen: Phys. Rev. **78**, 29 (1950)
126. A. Messiah: *Quantum Mechanics*, Vol. 1 (North-Holland, Amsterdam 1961)
127. J.D. Bjorken, S.D. Drell: *Relativistic Quantum Mechanics* (McGraw-Hill, New York San Francisco 1964)
128. S.A. Khan, M. Pusterla: arXiv physics/ 9910034 (1999)
129. G.Z. Li, G. Werth: Physica Scripta **46**, 587 (1992)
130. J. Walz, S.B. Ross, C. Zimmermann, L. Ricci, M. Prevedelli, T.W. Hänsch: Phys. Rev. Lett. **75**, 3257 (1995)
131. K. Dholakia, G. Horvath, D.M. Segal, R.C. Thompson: J. Mod. Opt. **39**, 2179 (1992)
132. M. Yan, X. Luo, X. Zhu: Appl. Phys. **67**, 235 (1998)
133. F.G. Major, private communication
134. M.-N. Benilan, C. Audoin: Int. J. Mass Spectr. Ion Phys. **11**, 421 (1973)
135. G. Gabrielse, F.C. Mackintosh: Int. J. Mass Spectr. Ion Phys. **57**, 1 (1984)
136. J. Byrne, P.S. Farago: Proc. Phys. Soc. London **86**, 808 (1965)
137. G. Gabrielse et al.: Phys. Rev. Lett. **57**, 2504 (1986)
138. H. Häffner et al.: Phys. Rev. Lett. **85**, 5308 (2000)
139. G. Gabrielse, L. Haarsma, S.L. Rolston: Int. J. Mass Spectr. Ion Proc. **88**, 319 (1989)

140. G. Gabrielse, S. Rolston, L. Haarmsma, W. Kells: Phys. Lett. A **129**, 38 (1988)
141. D.S. Hall, G. Gabrielse: Phys. Rev. Lett. **77**, 1962 (1996)
142. G. Gabrielse et al.: Phys. Rev. Lett. **89**, 233401 (2002)
143. M. Amoretti et al.: Nature **419**, 456 (2002)
144. J. Walz, I. Siemers, M. Schubert, W. Neuhauser, R. Blatt, E. Teloy: Phys. Rev. A **50**, 5 (1994)
145. J.D. Prestage, G.J. Dick, L. Malecki: J. Appl. Phys. **66**, 1013 (1989)
146. M.G. Raizen, J.M. Gilligan, J.C. Bergquist, W.M. Itano, D.J. Wineland: J. Mod. Optics **39**, 233 (1992)
147. U. Tanaka, H. Imajo, K. Hayasaka, R. Ohmukai, M. Watanabe, S. Urabe: Opt. Lett. **22**, 1353 (1997)
148. W. Alt, M. Block, P. Seibert, G. Werth: Phys. Rev. A **58**, R23 (1998)
149. M. Drewsen, C. Brodersen, L. Hornekaer, J.S. Hangst: Phys. Rev. Lett. **81**, 2878 (1998)
150. C.F. Roos, D. Leibfried, A. Mundt, F. Schmidt-Kaler, J. Eschner, R. Blatt: Phys. Rev. Lett. **85**, 5547 (2000)
151. H.C. Nägerl et al.: "Linear Ion Traps for Quantum Computation". In: *The Physics of Quantum Information*, ed. by D. Bouwmcester, A. Ekert, A. Zeilinger (Springer, Berlin Heidelberg New York 2000), pp. 163–176
152. K. Matsubara, U. Tanaka, H. Imajo, S. Urabe, M. Watanabe: Appl. Phys. B **76**, 209 (2003)
153. M.G. Raizen, J.M. Gilligan, J.C. Bergquist, W.M. Itano, D.J. Wineland: Phys. Rev. A **45**, 6493 (1992)
154. D. Gerlich: Phys. Scripta T **59**, 256 (1995)
155. D. Gerlich: Adv. Chem. Phys. **LXXXII**, 1 (1992)
156. T. Drees, W. Paul: Z. Phys. **180**, 340 (1964)
157. D.A. Church: J. Appl. Phys. **40**, 3127 (1969)
158. N. Yu, W. Nagourney, H. Dehmelt: J. Appl. Phys. **69**, 3779 (1991)
159. I. Waki et al.: Phys. Rev. Lett. **68**, 310 (1992)
160. G. Birkl, S. Kassner, H. Walther: Nature **357**, 2007 (1992)
161. T. Schätz, U. Schramm, D. Habs: Nature **412**, 717 (2001)
162. R.G. Brewer, R.G. DeVoe, R. Kallenbach: Phys. Rev. A **46**, 11 (1992)
163. H. Straubel: Naturwissenschaften **18**, 506 (1955)
164. D.W.O. Heddle: "Electrostatic Lens Systems" (IOP Publ., 2000)
165. J.D. Jackson: *Classical Electrodynamics*, 3rd edn. (Wiley, New York 1999) p. 140
166. F.G. Major: J. de Phys. Lettres **38**, L-221 (1977)
167. D. Kielpinski: Entanglement and Decoherence in a Trapped-Ion Quantum Register. MA Thesis, Colorado University, Boulder 2001
168. M.A. Rowe et al.: Quantum Information and Computation. Vol. 1 (Rinton Press 2001) p. 1
169. S.R. Jefferts, C. Monroe, E.W. Bell, D.J. Wineland: Phys. Rev. A **51**, 3112 (1995)
170. D. Zajfman et al.: Phys. Rev. A **55**, 3 (1997)
171. K.H. Kingdon: Phys. Rev. **21**, 408 (1923)
172. R.R. Lewis: J. Appl. Phys. **53**, 3975 (1982)
173. R.H. Hooverman: J. Appl. Phys. **34**, 3505 (1963)
174. D.P. Moehs et al.: Rev. Sci. Instr. **69**, 1991 (1998)

175. P.J. Kabarle: Mass Spectrom. **35**, 804 (2000)
176. S.J. Gaskell: J. Mass Spectrom. **32**, 677 (1997)
177. J-T. Watson: *Introduction to Mass Spectrometry*. 3rd edn. (Lippincott-Raven, New York 1997)
178. N. Kjaergaard, L. Hornekaer, A.M. Thommesen, Z. Videsen, M. Drewsen: Appl. Phys. B **71**, 207 (2000)
179. S. Gulde et al.: Appl. Phys. B **73**, 861 (2001)
180. H. Knab, M. Schupp, G. Werth: Europhys. Lett. **4**, 1361 (1987)
181. G. Bollen et al.: Nucl. Instr. Meth. A **368**, 675 (1989)
182. H. Schnatz et al.: Nucl. Instr. Meth. A **251**, 17 (1986)
183. H.A. Schuessler, Chun-sing O: Nucl. Instr. Meth. **186**, 219 (1981)
184. R.B. Moore, M.D.N. Lunney, G. Rouleau: Physica Scripta **46**, 569 (1992)
185. J. Coutandin, G. Werth: Appl. Phys. B **29**, 89 (1982)
186. G. Savard et al.: Phys. Lett. A **158**, 247 (1991)
187. R.S. Van Dyck, P.B. Schwinberg, H.G. Dehmelt: Phys. Rev. Lett. **59**, 26 (1987)
188. G. Gabrielse et al.: Phys. Rev. Lett. **89**, 233401 (2002)
189. R. Ley: Hyperfine Interactions **109**, 167 (1997)
190. H.G. Dehmelt, P. Schwinberg, R.S. Van Dyck: J. Mass Spectr. Ion Phys. **26**, 107 (1978)
191. B. Ghaffari, R. Conti: Phys. Rev. Lett. **75**, 3118 (1995)
192. G. Gabrielse, and the ATRAP Collaboration: Phys. Lett. B **505**, 1 (2001)
193. L. Haarsma, K. Abdullah, G. Gabrielse: Phys. Rev. Lett. **75**, 806 (1995)
194. J. Estrada et al.: Phys. Rev. Lett. **84**, 859 (2000)
195. B.M. Jelenković, A.S. Newbury, J.J. Bollinger, W.M. Itano, T.B. Mitchell: Phys. Rev. A **67**, 063406 (2003)
196. C.M. Surko, R.G. Greaves: Hyperfine Interactions **109**, 181 (1997)
197. R.G. Greaves, C.M. Surko: Phys. Rev. Lett. **85**, 1883 (2000)
198. M. Vedel, I. Rebatel, D. Lunnay, M. Knoop, F. Vedel: Phys. Rev. A **51**, 2294 (1995)
199. M. Vedel, J. Rocher, M. Knoop, F. Vedel: Int. J. Mass Spectrom. Ion Processes **190/191**, 37 (1999)
200. R. Alheit, S. Kleineidam, F. Vedel, M. Vedel, G. Werth: Int. J. Mass Spectr. Ion Proc. **154**, 155 (1996)
201. M. Holzki: Untersuchung von gekoppelten nichtlinearen Oszillationen einer Ionenwolke in Ionenfallen. Diplomarbeit, Johannes Gutenberg University, Mainz 1997
202. F. Vedel, M. Vedel: Phys. Rev. A **41**, 2348 (1990)
203. J. Rocher, M. Vedel, F. Vedel: Int. J. Mass Spectrom. Ion Processes **181**, 173 (1998)
204. H.G. Dehmelt: Bul. Am. Phys. Soc. **7**, 470 (1962)
205. H.G. Dehmelt, F.L. Walls, Phys. Rev. Lett. **21**, 127 (1968)
206. D.A. Church, H.G. Dehmelt, J. Appl. Phys. **40**, 3421 (1969)
207. M. Diederich et al.: Hyperfine interactions **115**, 185 (1998)
208. G. Werth, H. Häffner, W. Quint: Adv. At. Mol. and Opt. Phys. **48**, 191 (2002)
209. A.G. Marshall, P.B. Grosshans: Anal. Chem. **63**, 215 (1991)
210. W.M. Senko, F.M. McLafferty: Annu. Rev. Biophys. Biomol. Struct. **23**, 763 (1994)
211. A.G. Marshall, L. Schweikhard: Int. J. Mass Spectrom. Ion Proc. **118**, 378 (1992)

212. W. Neuhauser, M. Hohenstatt, P.E. Toschek, H. Dehmelt: Phys. Rev. A **22**, 1137 (1980)
213. D.J. Heinzen, D.J. Wineland: Phys. Rev. A **42**, 2977 (1990)
214. J.I. Cirac, L.J. Garay, R. Blatt, A.S. Parkins, P. Zoller: Phys. Rev. A **49**, 421 (1994)
215. J.I. Cirac, A.S. Parkins, R.Blatt, P. Zoller: Adv. Atom. Mol. Opt. Phys. **37**, 237 (1996)
216. A.M. Perelomov: Commun. Math. Phys. **26**, 222 (1972)
217. K. Wódkiewicz, J.H. Eberly: J. Opt. Soc. Am. B **2**, 458 (1985)
218. B. Yurke, S.L. McCall, J.R. Klauder: Phys. Rev. A **33**, 4033 (1986)
219. R.F. Bishop, A. Voudras: J. Phys. A: Math. Gen. **19**, 2525 (1986)
220. J.I. Cirac, A.S. Parkins, R.Blatt, P. Zoller: Phys. Rev. Lett. **70**, 556 (1993)
221. R.L. Matos Filho, W. Vogel: Phys. Rev. A **54**, 4560 (1996)
222. E. Arimondo: Coherent population trapping in laser spectroscopy, *Progress in Optics*, Vol. 35. Ed. E. Wolf (North-Holland, Amsterdam 1996) pp. 257–354
223. R. Wynands, A. Nagel: Appl. Phys. B **68**, 1 (1999)
224. D.M. Meekhof, C. Monroe, B.E. King, W.M. Itano, D.J. Wineland: Phys. Rev. Lett. **76**, 1796 (1996)
225. C. Monroe, D.M. Meekhof, B.E. King, D.J. Wineland: Science **272**, 1131 (1996)
226. J.I. Cirac, A.S. Parkins, R. Blatt, P. Zoller: Phys. Rev. Lett. **70**, 556 (1996)
227. J.I. Cirac, P. Zoller: Phys. Rev. Lett. **74**, 4091 (1995)
228. D.J. Wineland, W.M. Itano, R.S. van Dyck: Adv. At. Mol. Phys. **19**, 135 (1983)
229. R.J. Glauber: Phys. Rev. **131**, 2766 (1963)
230. R.P. Feynman: Phys. Rev. **84**, 108 (1951)
231. R.J. Glauber: Phys. Rev. **84**, 395 (1951)
232. P. Carruthers, M. Nieto: Phys. Rev. Lett **14**, 387 (1965)
233. H.Weyl: *Gruppentheorie und Quantenmechanick* (Hirzel, Leipzig 1928)
234. L.I. Schiff: *Quantum Mechanics*, 3rd edn. (Mc Graw-Hill, New York 1955) pp. 62–65
235. M.M. Nieto, L.M. Simmons: Phys. Rev. D **20**, 1321 (1979)
236. I.A. Malkin, V.I. Man'ko: JETP **28**, 527 (1969)
237. I.A. Malkin, V.I. Man'ko, D.A. Tifonov: Phys. Rev. D **2**, 1371 (1970)
238. V. Ermakov: Univ. Izv. Kiev, Series III **9**, 1 (1880)
239. V.N. Gheorghe, A.C. Gheorghe, G. Werth: "Displaced squeezed states for a single ion confined in a nonlinear electromagnetic trap". In *Europhysics Conference Abstracts 24D, Proc. of the 32nd EGAS Conference*, Vilnius, Lithuania, 4–7 Iuly 2000, Ed.: Z. Rudzikas (Vilnius), Series Editor: R.M. Pick (Paris), Managing Editor: C. Bastian (Mulhouse) p. 237–238
240. D.F. Walls: Nature (London) **324**, 210 (1986)
241. R. Alheit, X.Z. Chu, M. Hoefer, M. Holzki, G. Werth, R. Blümel: Phys. Rev. A **56**, 4023 (1997)
242. D.F. Walls, P. Zoller: Phys. Rev. Lett. **47**, 709 (1981)
243. M. Hillery: Phys. Rev. A **36**, 3796 (1987)
244. N.B. An: Phys. Lett. A **284**, 72 (2001)
245. J.J. Cirac, R. Blatt, A.S. Parkins, P. Zoller: Phys. Rev. Lett. **70**, 556 (1993)
246. M. Combescure: Ann. Phys. **173**, 210 (1987)
247. M. Combescure: Ann. Inst. Henry Poincare **44**, 293 (1987)

248. M. Berry: Proc. Roy. Soc. London A **392**, 45 (1984)
249. F. Calogero: J. Math. Phys. **12**, 419 (1971)
250. R. Blatt, J.I. Cirac, A.S. Parkins, P. Zoller: Phys. Scripta T **59**, 294 (1995)
251. D.J. Wineland et al.: Ann. Phys. **9**, 851 (2000)
252. R.H. Dicke: Phys. Rev. **93**, 99 (1954)
253. M. Tavis and F.W. Cummings: Phys. Rev. **170**, 379 (1968)
254. G. Scharf: Helv. Phys. Acta **43**, 806 (1970)
255. R. Bonifacio, G. Preparta: Phys. Rev. A **2**, 336 (1970)
256. I.R. Senitzky: Phys. Rev. A **3**, 421 (1971)
257. F. Arecchi, E. Courtens, R. Gilmore, H. Thomas: Phys. Rev. A **6**, 2211 (1972)
258. M.E. Smithers, E.C. Lu: Phys. Rev. A **9**, 790 (1974)
259. F. Persico, G. Vetri: Phys. Rev. A **12**, 2083 (1975)
260. S. Kumar, C.L. Mehta: Phys. Rev. A **21**, 1573 (1980)
261. V.N. Gheorghe, C.B. Collins: Phys. Rev. A **24**, 927 (1981)
262. V.N. Gheorghe: Appl. Phys. B **38**, 205 (1985)
263. R.H. Dicke: Phys. Rev. **89**, 472 (1953)
264. W.E. Lamb: Phys. Rev. **51**, 187 (1937)
265. E.T. Jaynes, F.W. Cummings: Proc. IEEE **51**, 89 (1963)
266. B.W. Shore, P.L. Knight: J. Mod. Opt. **40**, 1195 (1993)
267. W.P. Schleich: *Quantum Optics in Phase Space* (Wiley-VCH, Berlin 2001)
268. B.-G. Englert, M. Löffler, O. Benson, B. Varcoe, M. Weidinger, H. Walther: Fortschr. Phys. **46**, 897 (1998)
269. B.T.H. Varcoe, S. Brattke, M. Weidinger, H. Walther: Nature (London) **403**, 743 (2000)
270. J.M. Raimond, M. Brune, S. Haroche: Rev. Mod. Phys. **73**, 565 (2001)
271. J.J. Cirac, R. Blatt, P. Zoller, W.D. Phillips: Phys. Rev. A **46**, 2668 (1992)
272. S. Stenholm: Rev. Mod. Phys. **58**, 699 (1986)
273. Q.A. Turchette et al.: Phys. Rev. A **62**, 053807 (2000)
274. C. Roos et al.: Phys. Rev. Lett. **83**, 4713 (1999)
275. D.J. Wineland et al.: in *Laser Manipulation of Atoms and Ions*, Proceedings of the International School of Physics "Enrico Fermi" Course 118. Eds. E. Arimondo, W.D. Phillips, F. Strumia (North-Holland, Amsterdam 1992) p. 553
276. J. Janszky, Y.Y. Yushin: Opt. Commun. **59**, 151 (1986)
277. F.H. Yi, H.R. Zaidi: Phys. Rev. A **37**, 2985 (1988)
278. D.J. Wineland, J. Dalibard, C. Cohen-Tannoudji: J. Opt. Soc. Am. B **9**, 32 (1992)
279. S. Weigert: Phys. Lett. A **214**, 215 (1996)
280. I. Bialynicki-Birula: Acta Phys. Pol. B **29**, 3569 (1998)
281. D.B. Arvind, N. Mukunda, R. Simon: Phys. Rev. A **52**, 1609 (1995)
282. E.C.G. Sudarshan, C.B. Chiu, G. Bhamathi: Phys. Rev. A **52**, 43 (1995)
283. J. Oz-Vogt; A. Mann, M. Revzen: J. Mod. Opt. **38**, 2339 (1991)
284. C.T. Lee: Phys. Rev. A **52**, 1594 (1995)
285. S. Mancini, D. Vitali, P. Tombesi: J. Opt. B: Quantum Semiclass. Opt. **2**, 190 (2000)
286. M. Combescure: J. Math. Phys. **33**, 3870 (1992)
287. C.K. Law, J.H. Eberly: Phys. Rev. Lett. **76**, 1055 (1996)
288. S.A. Gardiner, J.I. Cirac, P. Zoller: Phys. Rev. A **55**, 1683 (1997)
289. B. Kneer, C.K. Law: Phys. Rev. A **57**, 2096 (1998)

290. A. Ben-Kish et al.: arXiv:quant–ph/0208181 v1, 28 Aug. 2002
291. D.J. Wineland, C. Monroe, W.M. Itano, D. Leibfried, B.E. King, D.M. Meekhof: Jou. Res. Natl. Inst. Stand. Technol. **103**, 259 (1998)
292. W.M. Itano, C. Monroe, D.M. Meekhof, D. Leibfried, B.E. King, D.J. Wineland: "Quantum harmonic oscillator state synthesis and analysis". In *Proc. Conf. on Atom. Optics, SPIE 2995, 43–45 (1997)*
293. C. Monroe, D.M. Meekhof, B.E. King, D.J. Wineland: Science **272**, 1131 (1996)
294. K. Vogel, H. Risken: Phys. Rev. A **40**, R2847 (1989)
295. D.T. Smithey, M.Beck; M.G. Raymer, A. Faridani: Phys. Rev. Lett. **70**, 1244 (1993)
296. P.J. Bardroff; E. Mayr, W.P. Schleich: Phys. Rev. A **51**, 4963 (1995)
297. L.G. Lutterbach, L. Davidovich: Phys. Rev. Lett. **78**, 2547 (1997)
298. G. Schrade, V.I. Man'ko, W.P. Schleich, R.J. Glauber: Quantum Semiclass. Opt. **7**, 307 (1995)
299. E. Wigner: Phys. Rev. **40**, 749 (1932)
300. M. Hillery, R.F. O'Connell, M.O. Scully, E.P. Wigner: Phys. Rev. **106**, 121 (1984)
301. C.C. Garry, P.L. Knight: Am. J. Phys. **65**, 964 (1997)
302. M. Massini, M. Fortunato, S. Mancini, P. Tombesi, D. Vitali: New J. Phys. **2**, 20.1 (2000)
303. D. Leibfried, D.M. Meekhof, B.E. King, C. Monroe, W.M. Itano, D.J. Wineland: Phys. Rev. Lett. **77**, 4281 (1996)
304. S. Wu: Rev. Mod. Phys. **70**, 685 (1998)
305. C. Cohen-Tannoudji: Rev Mod. Phys. **70**, 707 (1998)
306. D. Leibfried, R. Blatt, C. Monroe, D. Wineland: Rev. Mod. Phys. **74**, (2002)
307. R. Blatt: Laser cooling of trapped ions. In: *Fundamental Systems in Quantum Optics, Les Houches LIII, France 1990*. Ed. by I. Dalibard, J.M. Raymond, J. Zinn-Justin (Elsevier, 1992) p. 253
308. J. Javanainen, M. Lindberg, S. Stenholm: J. Opt. Soc. Am. B **1**, 111 (1984)
309. K. Hunger: Z. Astrophysik **39**, 36 (1956)
310. A.G. Mitchell, M.W. Zemansky: Resonance Radiation and Excited Atoms (Cambridge University Press, 1961) p. 319
311. P.A.M. Dirac: Quantum Mechanics. 4th edn. (Oxford 1958) p. 245
312. E. Kienow, E. Klempt, F. Lange, K. Neubecker: Phys. Letters A **46**, 441 (1974)
313. S. Peil, G. Gabrielse: Phys. Rev. Lett. **83**, 1287 (1999)
314. D. Kleppner: Phys. Rev. Lett. **47**, 233 (1981)
315. G. Gabrielse, H.G. Dehmelt: Phys. Rev. Lett. **55**, 67 (1985)
316. R. Blümel, C. Kappler, W. Quint, H. Walther: Phys. Rev. A **40**, 808 (1989)
317. L.S. Cutler, R.P. Giffard, M. D. McGuire: Appl. Phys. B **36**, 137 (1985)
318. F. Arbes, M. Benzing, T. Gudjons, F. Kurth, G. Werth: Z. Physik D **25**, 295 (1993)
319. H.-J. Kluge: Hyperfine Interactions **108**, 207 (1987)
320. C. Lichtenberg et al.: EPJ D **2**, 29 (1998)
321. G. Bollen et al.: Phys. Rev. C **46**, R2140 (1992)
322. D.J. Wineland, H.G. Dehmelt: Bull. Am. Phys. Soc. **20**, 637 (1975)
323. D.J. Wineland, H.G. Dehmelt: J. Appl. Phys. **46**, 919 (1975)
324. D.J. Wineland, H.G. Dehmelt: Int. J. Mass Spectr. Ion Proc. **16**, 338 (1974)

325. H.G. Dehmelt, W. Nagourney, J. Sandberg: Proc. Nat. Acad. Sci. U.S.A. **83**, 5761 (1986)
326. B. D'Urso, B. Odom, G. Gabrielse: Phys. Rev. Lett. **90**, 043001 (2003)
327. D. Möhl, G. Petrucci, L. Thorndahl, S. van der Meer: Phys. Reports **58**, 73 (1980)
328. N. Beverini, V. Lagomarsino, G. Manuzio, F. Scuri, G. Testera, G. Torelli: Hyperfine Interactions **44**, 247 (1988)
329. N. Beverini, V. Lagomarsino, G. Manuzio, F. Scuri, G. Testera, G. Torelli: Phys. Scripta T **22**, 238 (1988)
330. A. Kastler: J. Phys. Radium **11**, 255 (1950)
331. D.J. Wineland, R.E. Drullinger, F.L. Walls: Phys. Rev. Lett. **40**, 1639 (1978)
332. W. Nagourney, G. Janik, H. Dehmelt: Proc. Nat. Acad. Sciences, U.S.A., **80**, 643 (1983)
333. W.M. Itano, D.J. Wineland: Phys. Rev. A **25**, 35 (1982)
334. C. Monroe, D.M. Meekof, B.E. King, S.R. Jefferts, W.M. Itano, D.J. Wineland: Phys. Rev. Lett. **75**, 4011 (1995)
335. D.J. Wineland, J.C. Bergquist, J.J. Bollinger, W.M. Itano: Physica Scripta **T59**, 286 (1995)
336. I.S. Gradshteyn, I.W. Ryzhik: Table of Integrals, Series and Products (Academic Press 1965) p. 710
337. D.J. Wineland, W.M. Itano: Phys. Rev. A **20**, 1521 (1979)
338. J.C. Berquist, W.M. Itano, D.J. Wineland: Phys. Rev. A **36**, 428 (1987)
339. I. Marzoli, J.I. Cirac, R.Blatt, P. Zoller: Phys. Rev. **49**, 2771 (1994)
340. C.A. Sackett et al.: Nature **404**, 256 (2000)
341. H. Rohde et al.: J. Phys. B **3**, 534 (2001)
342. S.E. Harris: Physics Today, **50**, 36 July (1997)
343. G. Morigi, J. Eschner, C. Keitel: Phys. Rev. Lett. **85**, 4458 (2000)
344. G. Alzetta et al.: Nuovo Cimento B **36**, 5 (1976)
345. G. Janik, W. Nagourney, H. Dehmelt: J. Opt. Soc. Am. **B2**, 1251 (1985)
346. M. Schubert, I. Siemers, R. Blatt, W. Neuhauser, P.E. Toschek: Phys. Rev. A **52**, 2994 (1995)
347. P.D. Lett, R.N. Watts, C.I. Westbrook, W.D. Phyllips, P.L. Gould, H.J. Metcalf: Phys. Rev. Lett. **61**, 169 (1988)
348. J. Dalibard et al.: "New Schemes in Laser Cooling". In: *Atomic Physics 11, Proc. of the 11th Int. Conf. on Atomic Physics, Paris, France, 4-8 July 1988.* Ed. by S. Haroche, J.C. Gay, G. Grynberg (World Scientific, Singapore 1989) pp. 199–214
349. S. Chu, D.S. Weiss, Y. Shevy, P.J. Ungar: "Laser Cooling Due to Atomic Dipole Orientation". In: *Atomic Physics 11, Proc. of the 11th Int. Conf. on Atomic Physics, Paris, France, 4-8 July 1988.* Ed. by S. Haroche, J.C. Gay, G. Grynberg (World Scientific, Singapore 1989) pp. 636–638
350. J.P. Barrat, C. Cohen-Tannoudji: J. de Phys. et Radium **22**, 443 (1961)
351. J. Dalibard, C. Cohen-Tanoudji: J. Opt. Soc. Am. B **6**, 2023 (1989)
352. G. Birkl, J.A. Yeazell, R. Rückerl, H. Walther: Europhys. Lett. **27**, 197 (1994)
353. H. Dehmelt, G. Janik, W. Nagourney: Bull. Am Phys Soc **30**, 111 (1988)
354. P.E. Toschek, W. Neuhauser: J. Opt. Soc. Am. B **6**, 2220 (1989)
355. M. Lindberg, J. Javanainen: J. Opt. Soc. Am. B **3**, 1008 (1986)
356. R.E. Drullinger, D.J. Wineland, J.C. Bergquist: Appl. Phys. **22**, 365 (1980)

357. D.J. Wineland, W.M. Itano, J.C. Bergquist, J.J. Bollinger, J.D. Prestage: "Spectroscopy of Stored Atomic Ions". In *Atomic Physics 9, Proc. of the Ninth Intern. Conf. on Atomic Physics, Seatlle, July 23–27, 1984*. Ed. by R.S. Van Dyck, Jr., E.N. Fortson (World Scientific, Singapore 1985) pp. 3–27
358. D.J. Larson, J.C. Bergquist, J.J. Bollinger, W.M. Itano, D.J. Wineland: Phys. Rev. Lett. **57**, 70 (1986)
359. J.J. Bollinger et al.: Ion Trapping Techniques: Laser Cooling and Sympathetic Cooling. In *Intense Position Beams*. Eds. E.H. Ottewitte, W. Kells (World Scientific, Singapore 1988) p. 63
360. T.M. O'Neil: Phys. Fluids **24**, 1447 (1981)
361. L. Spitzer: Physics of Fully Ionized Gases, 2nd edn. (John Wiley and Sons, Hoboken, N.J. 1962)
362. H. Imajo, K. Hayasaka, R. Ohmukai, U. Tanaka, M. Watanabe, S. Urabe: Phys. Rev. A **53**, 122 (1996)
363. L. Gruber et al.: Phys. Rev. Lett. **86**, 636 (2001)
364. M.A. van Eijkelenborg, M.E.M. Storkey, D.M. Segal, R.C. Thompson: Phys. Rev. A **60**, 3903 (1999)
365. T. Baba, I. Waki: Jpn. J. Appl. Phys. **35**, L1134 (1996)
366. K. Mϕlhave, M. Drewsen: Phys. Rev. A **62**, 011401(R) (2000)
367. P. Bowe, L. Hoernekaer, C. Brodersen, M. Drewsen, J.S. Hangst, J.P. Schiffer: Phys. Rev. Lett. **82**, 2071 (1999)
368. J. Steiger, B.R. Beck, L. Gruber, D.A. Church, J.P. Holder, D. Schneider: "Coulomb Cluster in RETRAP". In: *AIP Conference Proceedings No. 457, Trapped Charged Particles and Fundamental Physics, Asilomar, California, 31 Aug–4 Sept 1998*. Ed. by D.H.E. Dubin, D. Schneider (AIP, Woodbury, New York 1999) pp. 284–289
369. G.-Z. Li, R. Poggiani, G. Testera, G. Werth: Z. Phys. D **22**, 375 (1991)
370. F.G. Major: The Quantum Beat: the Physical Principles of Atomic Clocks (Springer, Berlin Heidelberg New York 1998) p. 373
371. R.C. Davidson: Physics of Nonneutral Plasmas (Addison-Wesley, Redwood City, California 1990)
372. D.H.E. Dubin, T.M. O'Neil: Rev. Mod. Phys. **71**, 87 (1999)
373. T.M. O'Neil: Physica Scripta T **59**, 341 (1995)
374. J.H. Malmberg, T.M. O'Neil: Phys. Rev. Lett. **39**, 1333 (1977)
375. T.B. Mitchell, J.J. Bollinger, X.-P. Huang, W.M. Itano: "Mode and Transport Studies of Laser-Cooled Ion Plasmas in a Penning trap". In *Proc. AIP Conf. 457, Trapped Charged Particles and Fundamental Physics, Asilomar, August 31–September 4, 1998*. Ed. by D.H.E. Dubin, D. Schneider (AIP, Woodbury New York 1999) pp. 309–318
376. L.R. Brewer, J.D. Prestage, J.J. Bollinger, M.W. Itano, D.J. Larson, D.J. Wineland: Phys. Rev. A **38**, 859 (1988)
377. L. Brillouin: Phys. Rev. **67**, 260 (1945)
378. J.P. Schiffer, M. Drewsen, J.S. Hangst, L. Hornekaer: Proc. Nat. Acad. Sc. **97**, 10697 (2000)
379. J.J. Bollinger, D.J. Heinzen, F.L. Moore, W.M. Itano, D.J. Wineland: Phys. Rev. A **48**, 525 (1993)
380. S.A. Prasad, T.M. O'Neill: Phys. Fluids **26**, 665 (1983)
381. K.S. Fine: Phys. Rev. Lett. **63**, 2232 (1989)
382. R.G. Greaves, M.D. Tinkle, C.M. Surko: Phys. Rev. Lett. **74**, 90 (1995)

383. A.W. Trivelpiece, R.W. Gould: J. Appl. Phys. **30**, 1784 (1959)
384. D.H.E. Dubin: Phys. Rev. Lett. **66**, 2076 (1991)
385. D.H.E. Dubin, J.P. Schiffer: Phys. Rev. E **53**, 5249 (1996)
386. F. Anderegg et al.: Phys. Rev. Lett. **90**, 115001 (2003)
387. J.J. Bollinger, D.J. Wineland: Phys. Rev. Lett. **53**, 348 (1984)
388. D.J. Wineland, J.J. Bollinger, W.M. Itano, J.D. Prestage: J. Opt. Soc. Am. B **2**, 1721 (1985)
389. D.J. Heinzen et al.: Phys. Rev. Lett. **66**, 2980 (1991)
390. X.-P. Huang, F. Anderegg, E.M. Hollmann, C.F. Driscoll, T.M. O'Neil: Phys. Rev. Lett. **78**, 875 (1997)
391. X.-P. Huang, J.J. Bollinger, T.B. Mitchell, W.M. Itano: Phys. Rev. Lett. **80**, 73 (1998)
392. E.M. Hollmann, F. Anderegg, C.F. Driscoll: Physics of Plasmas **7**, 2776 (2000)
393. P. Hansen: Phys. Rev. A **8**, 3096 (1973)
394. E.L. Pollock, J.P. Hansen: Phys. Rev. A **8**, 3110 (1973)
395. W.L. Slattery, G.D. Doolen, H.E. DeWitt: Phys. Rev. A **21**, 2087 (1980)
396. D.J. Wineland, J.C. Bergquist, W.M. Itano, J.J. Bollinger, C.H. Manney: Phys. Rev. Lett. **59**, 2935 (1987)
397. H. Walther: Adv. At. Mol. and Opt. Phys. **31**, 137 (1993)
398. R. Blümel et al.: Nature, **334**, 309 (1988)
399. R.G. Brewer, J. Hoffnagle, R.G. De Voe: Phys. Rev. Lett. **65**, 2619 (1990)
400. J. Hoffnagle, R.G. Devoe, L. Reyna, R.G. Brewer: Phys. Rev. Lett. **61**, 255 (1988)
401. R. Blümel et al.: "Phase Transition of Stored Laser-Cooled Ions". In: *Proc. 11th Int. Conf. on Atomic Physics (ELICAP), Paris, July 4-8, 1988*. Ed. by S. Haroche, J.C. Gay, G. Grynberg (World Scientific, Singapore 1989) pp. 243–259
402. J.-P. Eckmann, D. Ruelle: Rev. Mod. Phys. **57**, 617 (1985)
403. B.A. Huberman, J.P. Crutchfield: Phys. Rev. Lett. **43**, 1743 (1979)
404. H. Totsuji, T. Kishimoto, C. Totsuji, K. Tsuruta: Phys. Rev. Lett. **88**, 125002 (2002)
405. D.H.E. Dubin: Phys. Rev. A **40**, 1140 (1989)
406. A. Rahman, J.P. Schiffer: Phys. Rev. Lett. **57**, 1133 (1987)
407. D.H.E. Dubin, T.M. O'Neil: Phys. Rev. Lett. **60**, 511 (1988)
408. R.W. Hase, V.V. Avilov: Phys. Rev. A **44**, 4506 (1991)
409. E. Wigner: Trans. Faraday Soc. **34**, 678 (1938)
410. S. Ichimaru: Rev. Mod. Phys. **54**, 1017 (1982)
411. D.H.E. Dubin: Phys. Rev. E **55**, 4017 (1997)
412. M. Block, A. Drakoudis, H. Leuthner, P. Seibert, G. Werth: J. Phys. B (Mol. Opt. Phys.) **33**, L375 (2000)
413. J.P. Schiffer: "Order in Cold Ionic Systems: Dynamics Effects". In *Proc. Workshop on Crystalline Ion Beams, 1988*. Ed. by K.W. Hasse, I. Hoffmann, D. Liesen (GSI Report 89–10, 1989) pp. 2–32
414. J.P. Schiffer, M. Drewsen, J.S. Hangst, L. Hornekaer: Proc. Nat. Ac. Sc. **97**, 10697 (2000)
415. D.H.E. Dubin: Phys. Rev. Lett. **71**, 2753 (1993)
416. J.P. Schiffer: Phys. Rev. Lett. **70**, 818 (1993)
417. L. Hornekaer: Single- and Multi-Species Coulomb ion Crystals: Structure, Dynamics and Sympathetic Cooling. MA Thesis, Aarhus University, Denmark, 2000

418. W.M. Itano, J.C. Bergquist, D.J. Wineland: "Coulomb Clusters of Ions in a Paul Trap". In *Proc. Workshop on Crystalline ion Beams, 1988*. Ed. by R.W. Hasse, I. Hoffmann, D. Liesen (GSI Report 89–10, 1989) pp. 241–254
419. V.M. Bedanov, F.M. Peeters: Phys. Rev. B **49**, 2667 (1994)
420. Y.E. Lozovik, V.A. Mandelshtam: Phys Lett. A **165**, 469 (1992)
421. A.C. Nägerl, R. Blatt, J. Eschner, F. Schmidt-Kahler, D. Leibfried: Optics Express **3**, 89 (1998)
422. M. Block, V.N. Gheorghe, O. Rehm, P. Seibert, G. Werth: "On ion crystals in a Paul trap". In *Europhysics Conference Abstracts (23D), Proc. of the 31st European Group for Atomic Spectroscopy, Marseille 6–9 July 1999*. Ed. by F. Vedel, R.M. Pick, C. Bastian (European Physical Society, 1999) pp. 338–339
423. S.L. Gilbert, I.J. Bollinger, D.J. Wineland: Phys. Rev. Lett. **60**, 2022 (1988)
424. J.N. Tann, J.J. Bollinger, B. Jelenkovik, D.J. Wineland: Phys. Rev. Lett. **75**, 4198 (1995)
425. X.P. Huang, J.J. Bollinger, T.B. Mitchell, W.M. Itano: Phys. Plasmas **5**, 1658 (1998)
426. W.M. Itano, J.J. Bollinger, J.N. Tan, B. Jelenkovic, X.-P. Huang, D.J. Wineland: Science **279**, 686 (1998)
427. T.B. Mitchell, J.J. Bollinger, D.H.E. Dubin, X.P. Huang, W.M. Itano, R.H. Baughman: Science **282**, 1290 (1998)
428. G. Goldoni, F.M. Peeters: Phys. Rev. B **53**, 4591 (1996)
429. U. Schramm, T. Schätz, D. Habs: Phys. Rev. Lett. **87**, 184801 (2001); ibid., Phys. Rev. E **66**, 036501 (2002)
430. U. Schramm, T. Schätz, D. Habs: "Stability of crystalline ion beams". In *AIP Conf. Proc. vol. 606, Int. Workshop on Non-Neutral Plasma Physics IV, San Diego, California, 30 July–2 August 2001*. Ed. by F. Anderegg, L. Schweikhard, C.F. Driscoll, (AIT, Melville, New York 2002) pp. 235–244
431. T. Schätz, U. Schramm, D. Habs: Nature **412**, 717 (2001)
432. E. Mathieu: Course de physique mathématique, Paris, 1873
433. E.H. Bateman, E.A. Erdélyi: Higher transcendental functions, Vol. 3. Automorphic functions (McGraw-Hill 1955)
434. Y. Wang, J. Franzen, K.P. Wanczek: Int. J. Mass Spectrom. Ion Processes **124**, 125 (1993)
435. Y. Wang, J. Franzen: Int. J. Mass Spectrom. Ion Processes **132**, 155 (1994)
436. K. Hübner, H. Klein, Ch. Lichtenberg, G. Marx, G. Werth: Europhys. Lett. **37**, 459 (1997)
437. V.N. Gheorghe, A. Gheorghe, G. Werth, L. Windholz: "Nonlinear Resonances in Ion Traps". In: *Europhysics Conference Abstracts 34th EGAS, Sofia 9–12 July 2002*. Ed. by K. Blagoev, R.M. Pick, P. Helfenstein (European Physical Society, 2002) pp. 272–273
438. J. Schwinger: Phys. Rev. **91**, 728 (1953)
439. A. Perelomov: *Generalized Coherent States and Their Applications* (Springer, Berlin Heidelberg New York 1986) pp. 67–71, 73–76, 208–210, 213–216
440. I.A. Malkin, V.I. Man'ko: JETP **28**, 527 (1969)
441. I.A. Malkin, V.I. Man'ko, D.A. Tifonov: Phys. Rev. D **2**, 1371 (1970)
442. V.N. Gheorghe, L.C. Giurgiu, O.S. Stoican, B.M. Mihalcea, D.M. Cacicovschi, R.V. Molnar: *Miniaturized set-up for the charged microparticle trapping* Romanian Patent Nr. 109684 (1995)
443. M.K. Jain: in *Numerical Solution of Differential Equations* (Wiley, New Delhi, 1978), pp. 80-84

Index

adiabatic approximation 24, 30, 37, 279
anharmonicities 96, 98
annihilation / creation operator 170
anomalous magnetic moment 79
antihydrogen 135
antimatter 266
antiprotons 217
attractor 279, 280
average kinetic energy 30
axial motion 25, 32
    collective resonance 33
    individual resonance 33

background gas 36
Bessel functions 96, 114, 227, 317
    modified Bessel function 229
bolometric technique 211
Boltzmann distribution function 193, 265
Bragg diffraction 290
Brewster window 329
Brillouin
    density / limit 266, 270
buffer gas 38

canonical transformations 311
carrier frequency 194, 221
chaos-order transitions 191
chaotic regime 204
classical equations of motion in
    ideal combined trap 87
    ideal cylindrical trap 97
    ideal Paul trap 18
    ideal Penning trap 52
    Kingdon trap 125
    linear trap 111
    octupole trap 109

real cylindrical trap 98
real Paul trap 30
real Penning trap 59, 62, 64
classical harmonic lattice vibration theory 278
coherent population trapping 234
cold fluid model 270
collisions
    cross sections 195
    damping constant 37
    ion–ion collisions 278
    viscous damping force 37
combined traps 87
    magnetron-free operation 90
confined ion states
    gaseous / liquid / crystalline state 275
confinement region 13
constants of the motion 157
    linear invariants 158
    quadratic invariants 158
Coulomb coupling parameter 275
Coulomb crystals
    (concentric) shells 281, 284, 286
    anisotropy parameter 283
    bcc / fcc / hcp lattice 278, 280
    breathing mode 286
    center-of-mass mode 286
    helicale crystal 283, 287
    linear string 284, 287
    microparticle crystals 277, 331
    modes of vibrations 285
    oblate crystal 283, 284
    prolate crystal 283
    spherical crystal 283
    zig-zag crystal 283, 287
Coulomb scattering 251, 253
crystalline regime 278

Index

cylindrical traps
  cavity 14, 95
  guard rings 100
  miniature traps 99

dark resonance 234
Debye–Scherrer "powder technique" 290
density matrix 178, 183, 185, 235
deterministic chaos 267
Dicke effect 191
Dirac Hamiltonian 79
distribution function 29
Doppler effect 194
  the first order effect 226
  the second-order (relativistic) effect 226
Doppler shift 191
Doppler width 275
dust particle 332

effective potential 11
eigenfrequencies
  in Paul trap 23
eigenfrequencies in Penning trap
  axial frequency 52
  cyclotron frequency 51
  free cyclotron frequency 266
  magnetron frequency 52
  modified cyclotron frequency 52
eigenfrequency shifts
  in Paul trap 33
  in Penning trap 59, 60, 62, 67
elastic scattering 203
electromagnetic-induced transparency (EIT) 233
electrostatic traps 123
error function 49
excitation spectrum 195, 276
expectation value 315, 316

Fermi's Golden Rules 199
Floquet's theorem 299
fluorescence spectrum 226, 275
Fock space 153
Fokker–Planck functions 223
Foldy–Wouthuysen transformation 80
Fourier spectrum 221

generating functions
  Hermite functions 316
  Laguerre functions 317
geonium 99
group representation theory
  dynamical group 161, 162
gyromagnetic factor 78

harmonic oscillator 20
harmonic polynomials 309
Heisenberg equation 92
Heisenberg's uncertainty relation 154
Heisenberg-picture 153
Hermite polynomials 41
Hilbert space 177
Hill equation 41, 44, 157

instabilities 39
  in an imperfect Paul trap 33
ion cloud
  average space charge field 30
  Brownian-motion model 30
  center-of-charge 30
  center-of-mass oscillations 32
  charged plasma model 30
  energy loss mechanisms 30
  equilibrium temperature 30
  Gaussian density distribution 29, 30
  heating effects 30
  individual ion oscillations 32
  Maxwellian velocity distribution 30
  plasma frequency 30
  radius 31
  space charge potential 31
  temperature 30
  thermodynamic equilibrium 28
ion getter pump 328
ion metastable state 297
ion motion 21
  axial motion 52
  center of mass motion 278
  classical trajectories 37
  cyclotron motion 52
  damping 37
  kinetic energy 37
  Lissajous pattern 97
  macromotion (secular motion) 22, 25, 37

Index   351

magnetron motion 52
micromotion 25, 37
phase space trajectories 21
radial motion 52
stable orbits 36
unstable orbits 68
velocity 37
ionic interaction potential
  Coulombian potential 29
ionisation potential 133

Kingdon trap 124

Lagrangian 43
Lamb–Dicke
  parameter 222, 227
  regime 194, 229
Laplace transform method 172
Larmor's theorem 266
laser
  beam incidence / emergence 331
  beam profile / scattering 29, 30
  counter-propagating beams 291
  detuning 280
  diode laser 331
  He-Ne laser 331
  induced fluorescence 225
Legendre polynomials 309
line shape
  broadening 195
  Doppler (Gaussian) profile 195
  Lorentz collision broadening 195
  Lorentz profile 195, 228
  Voigt profile 195, 229
linear traps
  ideal trap 109
  multipole trap 115
lines of instabilities
  theoretical / experimental lines 35, 36

magnetic moment 78
magnetic moment anomaly 79
many-body problem 263
mass separator 133
mass spectrometry 209
Mathieu equation
  characteristic curves 20
  characteristic exponent 19, 95, 299

convergent series 300
  emittance ellipse 301
  general solution 299, 300
  homogeneous type 18
  parametric oscillator 299
  periodic solutions 19
  phase space 301
  recurrence relationship 300
  stability diagram 19, 89, 111
  stable / unstable regions 19, 300
  Wronskian 300
Mathieu functions
  even / odd functions 48
Mathieu regime 268
Mendeleev periodic table 288
metal work function 133
method of normal forms 310
microplasmas 275
microwave cavity 96
molecular dynamics (MD) simulation 270, 278
Monte Carlo method 277, 288
motion general solutions in
  ideal combined trap 90
  ideal Paul trap 23
  ideal Penning trap 56
multipolar traps 108

nested traps 107
neutral (electron–ion) plasmas
  Coulomb correlation parameter 263
  Debye length / shielding 261
  highly correlated plasmas 263
  Langmuir oscillations 261
  long range mean Coulomb field 261
  weakly correlated plasmas 263
nonclassical (quantum) states 151
  arbitrary states 176
  coherent states 174, 187
  creation / preparation 151, 173
  dark states 151
  detection / recognition / reconstruction 151
  Fock (number) states 151, 153, 154, 173, 186, 233
  nonlinear coherent states 151
    even / odd states 181
  oscillator coherent states 154
  quasienergy states 162

Schrödinger-cat states 151, 178, 179, 182
squeezed states 151, 159, 175
state superposition 155
SU(1, 1) squeezed states 151
symplectic coherent states
  multiparticle states 165
  thermal distribution 177
noninteracting particles 29
nonlinear Coulomb interaction 278
nonlinear resonances 312
nonrelativistic Hamiltonian 80
nonrelativistic limit 80

observables 315
octupole trap 108
  trapping potential 109
one component plasma (OCP) 264
  high density OCP 277
one-particle Hamiltonian 265
optical absorption spectrum 221
optical molasses 237
optical pumping 226, 237
optical resonance profile 222
optical spectrometry 209
ordered (crystalline) state 277
oscillating electric potential 17

Paul trap
  effective potential approach 46
  ideal trap 17
  Lissajous trajectories 22
  rapidly oscillating fields 46, 266
  real trap 27
  rf-heating 204, 267
  saddle potential surface 18
Pauli Hamiltonian 79
Pauli matrices 79, 170
Penning trap 51
  (modified) cyclotron motion 207
  axial motion 206
  cyclotron / magnetron energy 208
  cyclotron / magnetron motion
    coupling 207
  ideal trap 51
  mode energy transfer 208
  motional spectrum 56
  periodic orbits 54
  quasiperiodic orbits 54
  real traps 57

perturbations 310
  (un)perturbed Hamiltonian 81
  anharmonic perturbation 82
  dodecapole 82
  electric / magnetic perturbations 81, 83
  ellipticity / misalignment 84
  expectation value 81
  first-order perturbation theory 81
  octupole 82
  oscillator basis 81
  perturbing potentials 33
  small anharmonic perturbations 49
  theory of stationary perturbations 81
phase space 161
phase transition
  bistability 280
  chaos-order (cloud-crystal) 275, 276, 291
  hysteresis 280
photon count rate 279
photon counting imaging system 275
photon recoil 221, 279
photon statistics 176
planar traps 118
plasma diffusion 265
plasma oscillations 269
  breathing mode 269
  center-of-mass mode 286
  diocotron waves 269
  dipole oscillation 270
  eigenmodes 270
  high frequency waves 269
  low order modes 270
  mode eigenfrequency 272
  normal modes 269
  quadrupole oscillation 270
Poincaré section 279
Poisson's equation 270
potential energy minimum 12
potential interaction
  Calogero potential 168
  Coulomb potential 168
precooling 234

quadrupole potential 13
quadrupole transitions 207
quantum (parametric) oscillator 40

quantum equations of motion
   in ideal combined trap   91
   in ideal Paul trap   43
   in ideal Penning trap   72, 79
   in real Penning trap   81, 83
quantum Hamiltonian
   atom-field Hamiltonian   169
      asymptotic solutions   169
      coupling parameters   170
      interaction Hamiltonian   170
      two-level Hamiltonian   170
   center-of-mass Hamiltonian   162
quantum mechanical system   315
quantum nondemolishing technique   201
quantum state engineering   151
quasienergy spectrum
   discrete / continuous spectrum   162

Rabi-frequency   172, 173, 176, 232
radiation pressure   221
relativistic corrections   80
residual gas   36
resonance condition   310, 311
resonance fluorescence   195
resonance spectrum   13, 95
resonant charge exchange   37
ring traps   117
   storage ring   118, 291
rotating wall technique   272, 290

scattering rate   236
Schrödinger equation   156, 172
   continuous spectra   41
   discrete spectra   41
   Gaussian solutions   157
   orthonormal solutions   41
   quasienergy functions   41
   stationary   155
Schrödinger / Heisenberg inequality   316
Schwarz inequality   315
semiconductor
   quantum dots   288
sideband frequencies   194, 222
sideband Raman transitions   232
spherical harmonics   33, 332
spin motion / operator   78
spinor   79

spontaneous emission noise   279
spontaneous Raman process   240
stability domain
   for different charges   21, 23
   in ideal combined trap   89
   in ideal Paul trap   20
   in linear trap   112
   in real Paul trap   27
stable / unstable trapping   41, 264
standard temperature and pressure   277
statistical energy distribution   37
storage time   38
strong interparticle correlation   270
synchrotron radiation   251

thermodynamic equilibrium   143, 263, 275
transient chaotic regime   278
trap
   (ultra-high) vacuum system   36, 193
   (un)compensated trap   99, 101
   confinement space   15
   correcting electrodes   99, 100
   driving frequency   25
   electrostatic boundary-value problem   96, 101, 105, 114
   end caps   14
   equipotential surfaces   95
   hyperbolic trap   13, 98
   image charges   68
   imperfect hyperbolic trap   98
   multipole   309
   operating point   34, 39
   potential depth   15, 25, 26, 31
   potential perturbations   35
   potential well   29
   pseudo-potential (minimum)   26, 332
   quality factor   96
   ring   14
   saddle point   332
   stability parameter   21
   stable confinement   18
   time averaged restoring force   17
   trapping conditions   30
   trapping field   17
   trapping volume   14, 95
   tunability   71

354   Index

trap design   327
   atomic beam source (oven)   131, 327
   electron gun   131, 327
   ion source (filament)   133
   miniaturized trap   328
   trap electrodes   327
trap electronics   327
   ion electronic detection   330
trapped ion crystallization
   local density approximation   281
trapped ion detection
   bolometric detection   142
   extraction pulse   140
   Fourier transform detection   145
   nondestructive detection   141
   rf (weak) field   23, 140
trapped ion heating   275
trapped ion–laser interaction   169
trapped nonneutral plasmas   263
   Coulomb correlation parameter   263
   strongly coupled plasmas   263, 291
   thermodynamic equilibrium   261
   weakly coupled plasmas   263
   Wigner crystal   264
trapped particle cooling
   absorption probability   222
   absorption-emission cycles   222
   adiabatic cooling   257
   average recoil momentum   222
   collisional (buffer gas) cooling   37, 191, 203
   Doppler cooling regime   223
   Doppler sidebands   226
   EIT cooling   233
   electronic transition   195
   excited state radiative lifetime   195
   laser cooling   192, 221
   negative feedback   191, 215, 216
   particle recoil   222
   radiative cooling   192, 197
   resistive cooling   191, 211, 215

   resolved sideband cooling   226
   resonance width   222
   Sisyphus cooling   236
   spontaneous photon re-emission   222
   stimulated Raman cooling   246
   stochastic cooling   216
   sympathetic cooling   192, 250
   theoretical temperature limit   221
   weak binding Doppler regime   222
trapped particle motion
   chaotic motion   278
trapped particle orbits   303
   (shortened / elongated) epitrochoid   305
   epicycloid   306
   nonperiodic orbits   312, 314
   periodic orbits   19, 54, 305–307, 312
   phase space trajectories   303
   quasiperiodic orbits   54, 305, 307
trapped positrons   135
Trivelpiece–Gould dispersion relation   270
two-level atomic system   170

ultrahigh vacuum set-up   329
uncertainty relations   315

vapor pressure   327
variance / covariance   315
variational method   281
viscous drag   270

wave packet center   166
Wigner crystals   281
Wigner function   183, 185–188
Wigner–Seitz radius   263
Wronskian   44

X-ray diffraction analysis   290

Zeeman sublevels   237

Springer Series on
ATOMS + PLASMAS

Editors: G. Ecker   P. Lambropoulos   I.I. Sobel'man   H. Walter
Founding Editor: H.K.V. Lotsch

1 **Polarized Electrons**
   2nd Edition
   By J. Kessler

2 **Multiphoton Processes**
   Editors: P. Lambropoulos and S.J. Smith

3 **Atomic Many-Body Theory**
   2nd Edition
   By I. Lindgren and J. Morrison

4 **Elementary Processes
   in Hydrogen-Helium Plasmas**
   Cross Sections
   and Reaction Rate Coefficients
   By R.K. Janev, W.D. Langer, K. Evans Jr.,
   and D.E. Post Jr.

5 **Pulsed Electrical Discharge in Vacuum**
   By G.A. Mesyats and D.I. Proskurovsky

6 **Atomic and Molecular Spectroscopy**
   Basic Aspects and Practical Applications
   3rd Edition
   By S. Svanberg

7 **Interference of Atomic States**
   By E.B. Alexandrov, M.P. Chaika,
   and G.I. Khvostenko

8 **Plasma Physics**
   Basic Theory with Fusion Applications
   3rd Edition
   By K. Nishikawa and M. Wakatani

9 **Plasma Spectroscopy**
   The Influence of Microwave and Laser Fields
   By E. Oks

10 **Film Deposition by Plasma Techniques**
   By M. Konuma

11 **Resonance Phenomena
   in Electron–Atom Collisions**
   By V.I. Lengyel, V.T. Navrotsky,
   and E.P. Sabad

12 **Atomic Spectra and Radiative Transitions**
   2nd Edition
   By I.I. Sobel'man

13 **Multiphoton Processes in Atoms**
   2nd Edition
   By N.B. Delone and V.P. Krainov

14 **Atoms in Plasmas**
   By V.S. Lisitsa

15 **Excitation of Atoms
   and Broadening of Spectral Lines**
   2nd Edition, By I.I. Sobel'man, L. Vainshtein,
   and E. Yukov

16 **Reference Data on Multicharged Ions**
   By V.G. Pal'chikov and V.P. Shevelko

17 **Lectures on Non-linear Plasma Kinetics**
   By V.N. Tsytovich

18 **Atoms and Their Spectroscopic Properties**
   By V.P. Shevelko

19 **X-Ray Radiation of Highly Charged Ions**
   By H.F. Beyer, H.-J. Kluge, and V.P. Shevelko

20 **Electron Emission
   in Heavy Ion–Atom Collision**
   By N. Stolterfoht, R.D. DuBois,
   and R.D. Rivarola

21 **Molecules
   and Their Spectroscopic Properties**
   By S.V. Khristenko, A.I. Maslov,
   and V.P. Shevelko

22 **Physics of Highly Excited Atoms and Ions**
   By V.S. Lebedev and I.L. Beigman

23 **Atomic Multielectron Processes**
   By V.P. Shevelko and H. Tawara

24 **Guided-Wave-Produced Plasmas**
   By Yu.M. Aliev, H. Schlüter, and A. Shivarova

25 **Quantum Statistics
   of Strongly Coupled Plasmas**
   By D. Kremp, W. Kraeft, and M. Schlanges

26 **Atomic Physics with Heavy Ions**
   By H.F. Beyer and V.P. Shevelko

Springer Series on
ATOMIC, OPTICAL, AND PLASMA PHYSICS

Editors-in-Chief:

Professor G.W.F. Drake
Department of Physics, University of Windsor
401 Sunset, Windsor, Ontario N9B 3P4, Canada

Professor Dr. G. Ecker
Ruhr-Universität Bochum, Fakultät für Physik und Astronomie
Lehrstuhl Theoretische Physik I
Universitätsstrasse 150, 44801 Bochum, Germany

Editorial Board:

Professor W.E. Baylis
Department of Physics, University of Windsor
401 Sunset, Windsor, Ontario N9B 3P4, Canada

Professor R.N. Compton
Oak Ridge National Laboratory
Building 4500S MS6125, Oak Ridge, TN 37831, USA

Professor M.R. Flannery
School of Physics, Georgia Institute of Technology
Atlanta, GA 30332-0430, USA

Professor B.R. Judd
Department of Physics, The Johns Hopkins University
Baltimore, MD 21218, USA

Professor K.P. Kirby
Harvard-Smithsonian Center for Astrophysics
60 Garden Street, Cambridge, MA 02138, USA

Professor P. Lambropoulos, Ph.D.
Max-Planck-Institut für Quantenoptik, 85748 Garching, Germany, and
Foundation for Research and Technology – Hellas (F.O.R.T.H.),
Institute of Electronic Structure & Laser (IESL),
University of Crete, PO Box 1527, Heraklion, Crete 71110, Greece

Professor G. Leuchs
Friedrich-Alexander-Universität Erlangen-Nürnberg
Lehrstuhl für Optik, Physikalisches Institut
Staudtstrasse 7/B2, 91058 Erlangen, Germany

Professor P. Meystre
Optical Sciences Center, The University of Arizona
Tucson, AZ 85721, USA

Professor Dr. H. Walther
Sektion Physik der Universität München
Am Coulombwall 1, 85748 Garching/München, Germany